VARIATIONAL PRINCIPLES

B. L. MOISEIWITSCH

The Department of Applied Mathematics
and Theoretical Physics
The Queen's University of Belfast

DOVER PUBLICATIONS, INC.
Mineola, New York

Copyright

Copyright © 1966, 2004 by B. L. Moiseiwitsch
All rights reserved.

Bibliographical Note

This Dover edition, first published in 2004, is an unabridged and slightly corrected republication of the work originally published in the series "Interscience Monographs and Texts in Physics and Astronomy," Volume XX, by Interscience Publishers/John Wiley & Sons, London and New York, in 1966. A new Preface to the Dover Edition, written by the author, has been specially prepared for the present volume. In addition, several corrections have been made within the text.

Library of Congress Cataloging-in-Publication Data

Moiseiwitsch, B. L. (Benjamin Lawrence)
    Variational principles / B.L. Moiseiwitsch.
        p. cm.
    Originally published: London ; New York : Interscience Publishers, 1966, in series: Interscience monographs and texts in physics and astronomy ; 20. With new pref.
    Includes bibliographical references and index.
    ISBN 0-486-43817-1 (pbk.)
    1. Mathematical physics. 2. Variational principles. 3. Calculus of variations. I. Title.

QC20.M584 2004
530.15—dc22

2004056234

Manufactured in the United States of America
Dover Publications, Inc., 31 East 2nd Street, Mineola, N.Y. 11501

Preface to the Dover Edition

This edition is the same as the original edition of my book apart from the correction of a few minor errors and misprints.

The search for unifying concepts and general principles has been one of the most important aims of mathematicians from early times and, as before, I would like to emphasize that some of the most aesthetically satisfying and illuminating general principles are the *variational principles* that are the subject of this book. Variational principles are concerned with the maximum and minimum properties, or more generally the stationary properties, of an extensive range of quantities of mathematical and physical interest spanning a wide field of applications.

Not only are these variational principles characterized by an elegant mathematical structure, but also they often possess the greatest practical utility in the solution of important problems in physics and chemistry and provide the most reliable method of accurately determining the values of many physical quantities of fundamental importance such as energy eigenvalues.

Variational principles have an ancient history and arose in antiquity in connection with the solutions of the isoperimetric and the geodesic problems. The isoperimetric problem, proposed by Greek mathematicians of the second century B.C., was originally concerned with the determination of the shape of a plane closed curve possessing a given perimeter encompassing the greatest possible area, namely the circle, while the geodesic problem was concerned with determining the curve that produced the shortest path between a pair of given points on the surface of the earth, namely an arc of a great circle on a sphere.

These two problems have many generalizations. For example, the triangle with a given perimeter enclosing the greatest area is equilateral and, in general, the polygon having a given perimeter that encloses the greatest area is a regular polygon. Also, the closed surface possessing a given area that surrounds the greatest volume is a sphere. Further, the shortest path connecting two points on a circular cylinder is a helix, a result that is related to the shape of the DNA molecule.

Another ancient variational problem, first solved by the Greek mathematician and scientist Heron of Alexandria in the first century A.D., is the determination of the path of a light ray reflected by a plane mirror. This is an example of the principle of least time that was later named after the seventeenth century mathematician Fermat who examined the refraction of a light ray at an interface separating two different media and showed that it takes the route of the shortest possible time.

However, this variational principle is not necessarily a minimum principle since a light ray that passes from one focal point to the other focal point of an ellipse by reflection at a mirror within the ellipse, which is tangential to the ellipse at its minor axis, takes the trajectory with the maximum possible total distance between the foci connected by two straight-line segments with a vertex at the mirror. Thus we have here an example of a principle of maximum time.

Many elementary examples of variational principles also occur in mechanics. There is the hanging chain problem that is concerned with finding the shape of a uniform chain hanging between two points under gravity and was found to be the catenary by Jakob Bernoulli in the seventeenth century and can be solved by minimizing the potential energy of the chain. Also, there is the brachistochrone problem proposed and solved by Jakob Bernoulli for determining the shape of a smooth wire connecting two points such that a bead takes the shortest possible time to travel from the initial to the final point of the wire. The solution to this problem is a cycloid and was solved by a number of other mathematicians including Leibnitz, de l'Hôpital, and Isaac Newton as well as Jakob's brother Johann.

An attempt to generalize Fermat's principle in optics to mechanics was first made by Maupertuis who introduced in 1744 the concept of *action* that he defined as *mvs*, the product of mass, velocity and distance. This was put on a firm mathematical basis by Euler, also in 1744, and led to the principle of least action where the action was defined to be $\int mv\, ds$, and later to the work of Lagrange and Hamilton and to Hamilton's principle, giving rise to the development of the subject of the *calculus of variations*.

The determination of frequency eigenvalues was discussed by Lord Rayleigh in his book *The Theory of Sound* in 1877 and led to Rayleigh's principle, which states that the fundamental frequency of a vibrating system is necessarily increased if constraints are imposed upon it and this gave rise to variational principles for determining the frequencies of vibration of strings and membranes as well as other mechanical systems, and subsequently to the Ritz variational method that has often been employed for evaluating the energy eigenvalues of quantum mechanical systems such as atoms and molecules. More recently, variational principles have been used with much success in the solution of various atomic scattering problems.

Thus variational principles are of fundamental importance for our understanding of the physical world and are of the greatest value for the accurate determination of its properties and I hope that my book goes some way towards demonstrating this.

B. L. Moiseiwitsch

Belfast
January 2004

Preface

The two main objectives of this book are on the one hand to show how the equations of the various branches of mathematical physics may be expressed in the succinct and elegant form of variational principles and thereby to illuminate the relationship between these equations, and on the other hand to demonstrate how variational principles may be employed to determine the discrete eigenvalues for stationary state problems and to find the values of quantities such as the phase shifts which arise in the theory of scattering.

The first chapter of this book is devoted to the variational formulation of classical as well as relativistic mechanics. It introduces variational principles through Hamilton's principle and the principle of least action, showing their equivalence to the dynamical equations of Lagrange and Hamilton, and also includes an examination of mechanics from the view point of differential geometry. This leads to a variational treatment of geodesic lines in Riemannian space and of the motion of a particle in a gravitational field.

In the second chapter we turn to the subject of optics and consider Fermat's principle of least time. The analogy between dynamics and geometrical optics is then discussed and the wave equations of Schrödinger, Klein–Gordon and Dirac are evolved. This is followed by an examination of the role of Hamilton's principle in quantum mechanics.

Through the use of the variational principle, the next chapter develops the Lagrangian and Hamiltonian formulations of the general field equations of physics and then considers particular applications to the equations of wave motion in classical dynamics, to the electromagnetic field equations, to the diffusion equation and to the various equations of wave mechanics. A brief discussion of the Schwinger dynamical principle in the theory of quantized fields concludes the part of the book dealing with the field equations of mathematical physics.

The remaining part of the book is concerned with discrete and continuous eigenvalue problems. After summarizing the theory of the small oscillations of a dynamical system at the beginning of the fourth chapter, Rayleigh's principle is proved and the Ritz variational method developed for the Sturm–Liouville equation. The more general problem of the eigenenergies of a quantum mechanical system is now discussed, upper bounds to the eigenenergies obtained and lower bounds to the ground state eigenenergy derived. The problem of determining the eigenenergies of atomic systems is then investigated and the special case of the two-electron system treated in considerable detail, the remarkable accuracy with which the energies of such systems have been calculated by using the Ritz variational method being emphasized. We then turn our attention to the energy curves of molecules and consider the cases of the hydrogen molecular ion and the hydrogen molecule as examples. The chapter ends with a discussion of the relationship between perturbation theory and the variational method.

The last chapter deals with the use of variational principles in the theory of scattering, a subject which has received much attention recently. Variational principles for the scattering amplitude and for the phase shift due to Hulthén, Kohn, Schwinger and others are established. The special case of the scattering of particles having vanishing energy is treated in some detail, upper bounds to the scattering length being derived and application being made to the elastic scattering of electrons and positrons by hydrogen atoms and to the elastic scattering of neutrons by deuterons. Finally we examine time-dependent scattering theory and look at variational principles for the collision operator and the transition matrix which arise therein.

Inevitably, in order to remain within the confines of a small volume, it has proved necessary to omit much interesting material from the present work owing to the very large number of different applications of variational principles that have been carried out recently. An effort has been made to keep the treatment of variational principles at a reasonably elementary level, the general aim being to provide a fairly broad view of the way in which variational principles have been applied to various fundamental problems in theoretical physics, although a certain amount of emphasis has been given to their role in the quantum theory of scattering.

The author wishes to express his gratitude to Stanford Research Institute for providing him with the opportunity to complete the writing of this book while he was a visitor there. Thanks are also due to Dr. K. L. Bell for his help in the reading of the proofs.

B. L. Moiseiwitsch

Contents

Preface. v

1. Analytical Dynamics 1
   1.1. Newtonian dynamics 1
   1.2. Generalized coordinates 5
   1.3. D'Alembert's principle 6
   1.4. Lagrange's equations 8
   1.5. Conservation of energy 12
   1.6. Euler–Lagrange equations 14
   1.7. Hamilton's principle 17
   1.8. Virial theorem 20
   1.9. Principle of least action 22
   1.10. Hamilton's equations 24
   1.11. Variational principle for Hamilton's equations . . 25
   1.12. Special theory of relativity 26
   1.13. Relativistic mechanics 28
   1.14. Contact transformations 33
   1.15. Hamilton–Jacobi equation 35
   1.16. Hamilton's integral 36
   1.17. Hamilton's characteristic function 37
   1.18. Geometrical mechanics 41
   1.19. Gravitational field 43
   1.20. Geodesic lines in Riemannian space 45
   1.21. Motion of a particle in a gravitational field . . . 48

2. Optics, Wave Mechanics and Quantum Mechanics . . . 51
   2.1. Huygens' principle 51
   2.2. Fermat's principle of least time 53
   2.3. Wave motion 55
   2.4. Schrödinger equation 57
   2.5. Klein–Gordon equation 60
   2.6. Dirac equation 61
   2.7. Quantum mechanics 64
   2.8. Hamilton's principle in quantum mechanics . . . 73

3. Field Equations 77
   3.1. Lagrangian formalism 77

3.2.	Hamiltonian formalism	81
3.3.	Laplace's equation	83
3.4.	Poisson's equation	86
3.5.	Scalar wave equation	90
3.6.	Subsidiary conditions	100
3.7.	Helmholtz equation	101
3.8.	Sturm–Liouville equation	103
3.9.	Electromagnetic field	105
3.10.	Diffusion equation	112
3.11.	Complex field components	115
3.12.	Schrödinger equation	117
3.13.	Klein–Gordon equation	119
3.14.	Vector meson field	123
3.15.	Dirac equation	126
3.16.	Linear equations	129
3.17.	Quantum field equations	131

4. Eigenvalue Problems 134
4.1.	Small oscillations of a dynamical system	134
4.2.	Stationary property of angular frequencies	140
4.3.	Rayleigh's principle	142
4.4.	Vibrating string	144
4.5.	Sturm–Liouville eigenvalue problem	147
4.6.	Ritz variational method	153
4.7.	Wave motion	156
4.8.	Schrödinger equation for a particle	161
4.9.	Eigenenergies of a quantum mechanical system	163
4.10.	Atomic eigenenergies	170
4.11.	Hartree–Fock equations	183
4.12.	Molecular energy curves	187
4.13.	Virial theorem	195
4.14.	Variational principle for a general eigenvalue equation	197
4.15.	Perturbation theory	198
4.16.	Variational principle for an arbitrary operator	207

5. Scattering Theory 213
5.1.	One-dimensional potential barrier	213
5.2.	Scattering at a surface	216
5.3.	Scattering of particles in wave mechanics	227
5.4.	Variational principles for the scattering amplitude	231
5.5.	Scattering phase shifts	240
5.6.	Variational principles for the phase shifts	245
5.7.	Scattering length and effective range	261
5.8.	Elastic scattering of electrons by hydrogen atoms	274

5.9. Elastic scattering of neutrons by deuterons . . . 283
5.10. Inelastic scattering of electrons by hydrogen atoms . . 287
5.11. Virial theorem 289
5.12. Time-dependent scattering theory 292

Bibliography 303

Index 304

CHAPTER 1

Analytical Dynamics

Although variational principles first made their appearance with Fermat's principle of least time in optics, it will be most convenient to postpone the discussion of this principle until Chapter 2 and commence by investigating the variational principles of classical dynamics. These comprise the principle of least action discovered by Maupertuis and first stated in precise mathematical form by Euler, and the principle due to Hamilton. They form the foundations of analytical dynamics and lead to the elegant Hamilton–Jacobi theory of mechanics. However, before we begin to discuss the variational principles of dynamics, we shall give a brief résumé of those parts of elementary classical mechanics which are needed for the purposes of this book, starting with Newton's laws of motion and proceeding to their generalization to the form of Lagrange's equations.

1.1 Newtonian dynamics

The fundamental equation of particle dynamics is *Newton's second law of motion*

$$\mathbf{F} = \frac{d\mathbf{p}}{dt} \tag{1.1}$$

which defines the *force* **F** acting on a particle as the time rate of change of its *momentum* **p** given by

$$\mathbf{p} = m\mathbf{v}, \tag{1.2}$$

where m is the *inertial mass* and **v** is the *velocity* of the particle.

This velocity is given by
$$\mathbf{v} = \frac{d\mathbf{r}}{dt}, \qquad (1.3)$$
where $\mathbf{r}$ denotes the *position vector* of the particle referred to the origin of an *inertial frame of reference*, relative to which a particle moves in a straight line with constant speed in the absence of forces.

For a particle of constant mass, Newton's second law of motion takes the form
$$\mathbf{F} = m\mathbf{a}, \qquad (1.4)$$
where
$$\mathbf{a} = \frac{d\mathbf{v}}{dt} \qquad (1.5)$$
is the *acceleration* of the particle.

We now note that the mass m has not been defined during the course of the preceding considerations, and so if we are to determine the mass of a given particle it is necessary to introduce further information characterizing the force $\mathbf{F}$. Such information is provided by *Newton's third law of motion*. Thus consider a pair of particles denoted by 1 and 2 and suppose that $\mathbf{F}_{12}$ and $\mathbf{F}_{21}$ are the forces acting on particles 1 and 2 due to the influence of particles 2 and 1 respectively. Then Newton's third law asserts that
$$\mathbf{F}_{12} = -\mathbf{F}_{21}. \qquad (1.6)$$
If the masses m_1 and m_2 of the particles are constant, it follows from (1.4) that
$$m_1 \mathbf{a}_1 = -m_2 \mathbf{a}_2, \qquad (1.7)$$
where $\mathbf{a}_1$ and $\mathbf{a}_2$ are the accelerations of the particles. Measurement of $\mathbf{a}_1$ and $\mathbf{a}_2$ when the two particles are moving solely under their mutual influence gives the ratio m_1/m_2. Hence, by assigning a given value to the mass of an arbitrarily chosen particle, the mass of any other particle may be determined.

Principle of energy

Suppose now that we have under consideration a system of N particles with constant masses m_i ($i = 1, \ldots, N$). According to

Newton's second law, the equation of motion of the ith particle is

$$\mathbf{F}_i = m_i \frac{d\mathbf{v}_i}{dt}, \tag{1.8}$$

where $\mathbf{F}_i$ is the total force acting on the ith particle and $\mathbf{v}_i$ is the velocity of this particle. Then the *work* done by all the forces, both internal and external, in moving the system of particles from a configuration denoted by 1 to a configuration denoted by 2 is given by

$$\sum_{i=1}^{N} \int_{1}^{2} \mathbf{F}_i \cdot d\mathbf{r}_i = \sum_{i=1}^{N} \int_{1}^{2} m_i \frac{d\mathbf{v}_i}{dt} \cdot \mathbf{v}_i \, dt = \sum_{i=1}^{N} \int_{1}^{2} \frac{d}{dt} (\tfrac{1}{2} m_i \mathbf{v}_i \cdot \mathbf{v}_i) \, dt$$
$$= \left[\sum_{i=1}^{N} \tfrac{1}{2} m_i v_i^2 \right]_{1}^{2}.$$

Now the *kinetic energy* of a particle of mass m and speed v is defined to be $\tfrac{1}{2} m v^2$, and so the kinetic energy of the system of particles is the sum

$$T = \sum_{i=1}^{N} \tfrac{1}{2} m_i v_i^2. \tag{1.9}$$

It follows that

$$\sum_{i=1}^{N} \int_{1}^{2} \mathbf{F}_i \cdot d\mathbf{r}_i = T_2 - T_1, \tag{1.10}$$

where T_1 and T_2 are the values of the kinetic energy of the system in configurations 1 and 2 respectively.

If $\mathbf{F}_{ij}$ denotes the force acting on the ith particle due to the influence of the jth particle of the system, and if $\mathbf{F}_i^E$ denotes the force acting on the ith particle of the system due to the influence of all the particles *external* to the system, we have

$$\mathbf{F}_i = \sum_{\substack{j=1 \\ j \neq i}}^{N} \mathbf{F}_{ij} + \mathbf{F}_i^E. \tag{1.11}$$

By Newton's third law of motion

$$\mathbf{F}_{ij} = -\mathbf{F}_{ji} \tag{1.12}$$

for any pair of particles i and j, and so

$$\mathbf{F}_{ij} \cdot d\mathbf{r}_i + \mathbf{F}_{ji} \cdot d\mathbf{r}_j = \mathbf{F}_{ij} \cdot (d\mathbf{r}_i - d\mathbf{r}_j).$$

In the case of a system of particles composing a rigid body, the distances $r_{ij} = |\mathbf{r}_i - \mathbf{r}_j|$ are constant, and so $d\mathbf{r}_i - d\mathbf{r}_j$ is perpendicular to $\mathbf{r}_i - \mathbf{r}_j$. It follows that

$$\mathbf{F}_{ij}.d\mathbf{r}_i + \mathbf{F}_{ji}.d\mathbf{r}_j = 0$$

on making the further assumption that $\mathbf{F}_{ij}$ lies along the line joining the particles i and j. Hence

$$\sum_{\substack{i=1 \\ i \neq j}}^{N} \sum_{j=1}^{N} \int_{1}^{2} \mathbf{F}_{ij}.d\mathbf{r}_i = 0 \tag{1.13}$$

which means that the total work done by the internal forces of the system vanishes so that we get

$$\sum_{i=1}^{N} \int_{1}^{2} \mathbf{F}_i.d\mathbf{r}_i = \sum_{i=1}^{N} \int_{1}^{2} \mathbf{F}_i^E.d\mathbf{r}_i. \tag{1.14}$$

Using (1.10) this yields the *principle of energy*

$$\sum_{i=1}^{N} \int_{1}^{2} \mathbf{F}_i^E.d\mathbf{r}_i = T_2 - T_1, \tag{1.15}$$

which states that the work done by the external forces acting on a rigid body is equal to the change in the kinetic energy of the body.

We now suppose that the external forces can be written as the gradients of *potentials* V_i, i.e. that we may write

$$\mathbf{F}_i^E = -\nabla_i V_i, \tag{1.16}$$

V_i being a scalar function of the position of the ith particle. Then

$$\sum_{i=1}^{N} \int_{1}^{2} F_i^E.d\mathbf{r}_i = - \sum_{i=1}^{N} \int_{1}^{2} \nabla_i V_i.d\mathbf{r}_i = - \left[\sum_{i=1}^{N} V_i \right]_{1}^{2}$$

and hence it follows from the principle of energy that

$$T_1 + \sum_{i=1}^{N} (V_i)_1 = T_2 + \sum_{i=1}^{N} (V_i)_2. \tag{1.17}$$

Introducing the total potential energy V defined by

$$V = \sum_{i=1}^{N} V_i, \tag{1.18}$$

we see that the *total energy*

$$E = T + V \tag{1.19}$$

is *conserved*. Any force which can be expressed in the form (1.16), as the gradient of a potential, is therefore called a *conservative force*.

Now suppose that the system of particles does not form a rigid body but that the internal forces of the system can be written as the gradients of potentials, in which case we may put

$$\mathbf{F}_{ij} = -\nabla_i V_{ji}, \qquad (1.20)$$

where the potential V_{ij} is chosen to be a function of r_{ij} only, thus ensuring that Newton's third law of motion is satisfied. In fact we have

$$\mathbf{F}_{ij} = -\nabla_{\mathbf{r}_{ij}} V_{ij} = -\mathbf{F}_{ji}, \qquad (1.21)$$

where $\nabla_{\mathbf{r}_{ij}}$ denotes the gradient operator with respect to the vector $\mathbf{r}_{ij} = \mathbf{r}_i - \mathbf{r}_j$. Consequently

$$\int_1^2 \mathbf{F}_{ij} \cdot d\mathbf{r}_i + \int_1^2 \mathbf{F}_{ji} \cdot d\mathbf{r}_j = -\int_1^2 \nabla_{\mathbf{r}_{ij}} V_{ij} \cdot d\mathbf{r}_{ij}$$

and so we obtain

$$\sum_{\substack{i=1 \\ i \neq j}}^{N} \sum_{j=1}^{N} \int_1^2 \mathbf{F}_{ij} \cdot d\mathbf{r}_i = -\tfrac{1}{2} \sum_{\substack{i=1 \\ i \neq j}}^{N} \sum_{j=1}^{N} \int_1^2 \nabla_{\mathbf{r}_{ij}} V_{ij} \cdot d\mathbf{r}_{ij} \qquad (1.22)$$

$$= -\left[\tfrac{1}{2} \sum_{\substack{i=1 \\ i \neq j}}^{N} \sum_{j=1}^{N} V_{ij} \right]_1^2$$

from which it follows by the principle of energy that the total energy $E = T + V$ is conserved, as before, with the total potential energy V now being defined according to the formula

$$V = \sum_{i=1}^{N} V_i + \tfrac{1}{2} \sum_{\substack{i=1 \\ i \neq j}}^{N} \sum_{j=1}^{N} V_{ij}. \qquad (1.23)$$

1.2 Generalized coordinates

The particles forming a rigid body are constrained by the internal forces to move so that the distances between them remain

constant. Consequently for any pair of particles i, j of a rigid body we have
$$(\mathbf{r}_i - \mathbf{r}_j)^2 - d_{ij}^2 = 0, \qquad (1.24)$$
where d_{ij} is the constant distance between the two particles.

Now any constraint imposed on a system of particles which can be specified by a set of equations possessing the general form
$$f_s(\mathbf{r}_1, \mathbf{r}_2, \ldots, \mathbf{r}_N, t) = 0 \qquad (s = 1, \ldots, c) \qquad (1.25)$$
involving the coordinates of the particles and the time t is termed a *holonomic constraint*, and accordingly we see from (1.24) that a rigid body presents us with an example of a system subject to holonomic constraints.

A system consisting of N particles without any constraints requires $3N$ coordinates to determine its configuration completely. If the system of particles is subject to holonomic constraints specified by c equations having the form (1.25), then just $n = 3N - c$ of these $3N$ coordinates are independent and the dynamical system is said to have n *degrees of freedom*. Using equations (1.25) we may eliminate c of the $3N$ coordinates and then introduce n independent parameters $q_1, q_2, \ldots, q_n$ such that the position vectors $\mathbf{r}_1, \mathbf{r}_2, \ldots, \mathbf{r}_N$ of the N particles comprising the system may be written in the form
$$\begin{aligned}\mathbf{r}_1 &= \mathbf{r}_1(q_1, q_2, \ldots, q_n, t), \\ \mathbf{r}_2 &= \mathbf{r}_2(q_1, q_2, \ldots, q_n, t), \\ &\vdots \\ \mathbf{r}_N &= \mathbf{r}_N(q_1, q_2, \ldots, q_n, t).\end{aligned} \qquad (1.26)$$

The parameters $q_1, q_2, \ldots, q_n$ which we have used to specify the system of particles are called *generalized coordinates*.

1.3 D'Alembert's principle

We now consider a change in the configuration of the system of particles arising from arbitrary infinitesimal changes $\delta \mathbf{r}_i$ in the position vectors $\mathbf{r}_i$ of the particles at a given time t, which do not violate the constraints imposed on the system. Such a change of configuration is called a *virtual displacement*.

ANALYTICAL DYNAMICS

The *virtual work* done by the force $\mathbf{F}_i$ acting on the ith particle of the system as a result of the displacement $\delta\mathbf{r}_i$ is $\mathbf{F}_i \cdot \delta\mathbf{r}_i$ and so, setting

$$\mathbf{F}_i = \mathbf{F}_i^C + \mathbf{F}_i^A, \tag{1.27}$$

where $\mathbf{F}_i^C$ is the force of *constraint* and F_i^A is the force *applied* to the ith particle, we see that the virtual work done on the entire system is

$$\sum_{i=1}^{N} \mathbf{F}_i \cdot \delta\mathbf{r}_i = \sum_{i=1}^{N} \mathbf{F}_i^C \cdot \delta\mathbf{r}_i + \sum_{i=1}^{N} \mathbf{F}_i^A \cdot \delta\mathbf{r}_i. \tag{1.28}$$

For the case of a rigid body we have shown on p. 4 that the virtual work done by the internal forces of constraint vanishes. The vanishing of the virtual work also occurs for many other types of constraining force, and we shall therefore make the supposition here that

$$\sum_{i=1}^{N} \mathbf{F}_i^C \cdot \delta\mathbf{r}_i = 0. \tag{1.29}$$

Now, for a system in equilibrium, the total force $\mathbf{F}_i$ acting on each individual particle vanishes and so

$$\sum_{i=1}^{N} \mathbf{F}_i \cdot \delta\mathbf{r}_i = 0. \tag{1.30}$$

Hence, using (1.28) and (1.29), we obtain

$$\sum_{i=1}^{N} \mathbf{F}_i^A \cdot \delta\mathbf{r}_i = 0 \tag{1.31}$$

so that for a system of particles in equilibrium the virtual work performed by the applied forces is zero, which is known as the *principle of virtual work*.

If we rewrite Newton's second law of motion in the form

$$\mathbf{F}_i - \frac{d\mathbf{p}_i}{dt} = 0$$

and regard $-(d\mathbf{p}_i/dt)$ as a force which added to $\mathbf{F}_i$ produces equilibrium, we may replace (1.30) by

$$\sum_{i=1}^{N}\left(\mathbf{F}_i - \frac{d\mathbf{p}_i}{dt}\right)\cdot\delta\mathbf{r}_i = 0$$

which yields *D'Alembert's principle*:

$$\sum_{i=1}^{N}\left(\mathbf{F}_i^A - \frac{d\mathbf{p}_i}{dt}\right)\cdot\delta\mathbf{r}_i = 0 \qquad (1.32)$$

on using (1.29) again. The quantity $-(d\mathbf{p}_i/dt)$ is sometimes referred to as the *force of inertia*.

1.4 Lagrange's equations

We shall now employ D'Alembert's principle to derive the generalized equations of motion for a system of particles due to Lagrange. If the system can be completely specified by n generalized coordinates $q_1, q_2, \ldots, q_n$, we have

$$\mathbf{r}_i = \mathbf{r}_i(q_1, q_2, \ldots, q_n, t) \qquad (i = 1, 2, \ldots, N). \qquad (1.33)$$

Hence

$$\delta\mathbf{r}_i = \sum_{r=1}^{n} \frac{\partial \mathbf{r}_i}{\partial q_r} \delta q_r \qquad (1.34)$$

where the δq_r are infinitesimal changes in the q_r, and therefore

$$\sum_{i=1}^{N} \frac{d\mathbf{p}_i}{dt}\cdot\delta\mathbf{r}_i = \sum_{i=1}^{N}\sum_{r=1}^{n} \frac{d}{dt}(m_i\mathbf{v}_i)\cdot\frac{\partial \mathbf{r}_i}{\partial q_r}\delta q_r$$

$$= \sum_{i=1}^{N}\sum_{r=1}^{n}\left\{\frac{d}{dt}\left(m_i\mathbf{v}_i\cdot\frac{\partial \mathbf{r}_i}{\partial q_r}\right) - m_i\mathbf{v}_i\cdot\frac{d}{dt}\left(\frac{\partial \mathbf{r}_i}{\partial q_r}\right)\right\}\delta q_r.$$

But

$$\mathbf{v}_i = \frac{d\mathbf{r}_i}{dt} = \sum_{r=1}^{n} \frac{\partial \mathbf{r}_i}{\partial q_r}\dot{q}_r + \frac{\partial \mathbf{r}_i}{\partial t}$$

and so

$$\frac{\partial \mathbf{v}_i}{\partial \dot{q}_r} = \frac{\partial \mathbf{r}_i}{\partial q_r}, \qquad (1.35)$$

ANALYTICAL DYNAMICS

the time derivatives being denoted by dots. Also

$$\frac{d}{dt}\left(\frac{\partial \mathbf{r}_i}{\partial q_r}\right) = \sum_{s=1}^{n} \frac{\partial^2 \mathbf{r}_i}{\partial q_s \, \partial q_r} \dot{q}_s + \frac{\partial^2 \mathbf{r}_i}{\partial t \, \partial q_r} = \frac{\partial \mathbf{v}_i}{\partial q_r} \quad (1.36)$$

since we may interchange the order of the derivatives. It follows that

$$\sum_{i=1}^{N} \frac{d\mathbf{p}_i}{dt} \cdot \delta \mathbf{r}_i = \sum_{i=1}^{N} \sum_{r=1}^{n} \left\{ \frac{d}{dt}\left(m_i \mathbf{v}_i \cdot \frac{\partial \mathbf{v}_i}{\partial \dot{q}_r}\right) - m_i \mathbf{v}_i \cdot \frac{\partial \mathbf{v}_i}{\partial q_r} \right\} \delta q_r$$

$$= \sum_{r=1}^{n} \left\{ \frac{d}{dt}\left(\frac{\partial T}{\partial \dot{q}_r}\right) - \frac{\partial T}{\partial q_r} \right\} \delta q_r, \quad (1.37)$$

where

$$T = \sum_{i=1}^{N} \tfrac{1}{2} m_i v_i^2$$

is the total kinetic energy of the system of particles.

Now we may write

$$\sum_{i=1}^{N} \mathbf{F}_i^A \cdot \delta \mathbf{r}_i = \sum_{i=1}^{N} \sum_{r=1}^{n} \mathbf{F}_i^A \cdot \frac{\partial \mathbf{r}_i}{\partial q_r} \delta q_r = \sum_{r=1}^{n} Q_r \, \delta q_r, \quad (1.38)$$

where

$$Q_r = \sum_{i=1}^{N} \mathbf{F}_i^A \cdot \frac{\partial \mathbf{r}_i}{\partial q_r} \quad (1.39)$$

is termed the rth component of the *generalized force*. Hence, using (1.37) and (1.38), we see that D'Alembert's principle may be expressed in the form

$$\sum_{r=1}^{n} \left\{ \frac{d}{dt}\left(\frac{\partial T}{\partial \dot{q}_r}\right) - \frac{\partial T}{\partial q_r} - Q_r \right\} \delta q_r = 0. \quad (1.40)$$

Since the generalized coordinates q_r form a set of completely independent parameters, the virtual displacements δq_r are independent of each other and therefore it must follow that

$$\frac{d}{dt}\left(\frac{\partial T}{\partial \dot{q}_r}\right) - \frac{\partial T}{\partial q_r} = Q_r \quad (r = 1, \ldots, n). \quad (1.41)$$

In the case of a conservative system of applied forces we have

$$\mathbf{F}_i^A = -\nabla_i V, \quad (1.42)$$

where $V(\mathbf{r}_1, \mathbf{r}_2, \ldots, \mathbf{r}_N)$ is the total potential of the system of particles, and so

$$Q_r = -\sum_{i=1}^{N} \nabla_i V \cdot \frac{\partial \mathbf{r}_i}{\partial q_r} = -\frac{\partial V}{\partial q_r}. \tag{1.43}$$

Hence, for a conservative system, equations (1.41) yield

$$\frac{d}{dt}\left(\frac{\partial T}{\partial \dot{q}_r}\right) - \frac{\partial T}{\partial q_r} = -\frac{\partial V}{\partial q_r} \quad (r = 1, \ldots, n). \tag{1.44}$$

Introducing the function

$$L = T - V, \tag{1.45}$$

called the *Lagrangian function* of the system, we may rewrite (1.44) in the form

$$\frac{d}{dt}\left(\frac{\partial L}{\partial \dot{q}_r}\right) = \frac{\partial L}{\partial q_r} \quad (r = 1, \ldots, n) \tag{1.46}$$

since V is a function only of the generalized coordinates q_r and not of the generalized velocities $\dot{q}_r$.

The above set of n second-order differential equations are known as *Lagrange's equations of motion*.

Generalized potentials

In certain cases the components of the generalized force can be written in the form

$$Q_r = \frac{d}{dt}\left(\frac{\partial U}{\partial \dot{q}_r}\right) - \frac{\partial U}{\partial q_r}, \tag{1.47}$$

where U depends upon the generalized velocities $\dot{q}_1, \ldots, \dot{q}_n$ as well as the generalized coordinates $q_1, \ldots, q_n$. Then we have

$$\frac{d}{dt}\left(\frac{\partial T}{\partial \dot{q}_r}\right) - \frac{\partial T}{\partial q_r} = \frac{d}{dt}\left(\frac{\partial U}{\partial \dot{q}_r}\right) - \frac{\partial U}{\partial q_r}$$

and so the Lagrange equations (1.46) are still applicable, with the Lagrangian function now being expressed in terms of a generalized potential U by the formula

$$L = T - U. \tag{1.48}$$

Motion of a charged particle in an electromagnetic field

An example of a potential which depends upon the velocities as well as the coordinates is provided by the motion of a charged particle in an electromagnetic field. In order to investigate this, we shall require *Maxwell's electromagnetic field equations* for the *electric field* **E** and the *magnetic field* **H**, which are

$$\text{curl } \mathbf{H} - \frac{1}{c}\frac{\partial \mathbf{E}}{\partial t} = \frac{4\pi}{c}\mathbf{j}, \tag{1.49}$$

$$\text{curl } \mathbf{E} + \frac{1}{c}\frac{\partial \mathbf{H}}{\partial t} = 0, \tag{1.50}$$

$$\text{div } \mathbf{E} = 4\pi\rho, \tag{1.51}$$

and

$$\text{div } \mathbf{H} = 0, \tag{1.52}$$

where ρ is the *electric charge density* and **j** is the *current density*.

By virtue of equation (1.52) we may express the magnetic field vector in the form

$$\mathbf{H} = \text{curl } \mathbf{A}, \tag{1.53}$$

where **A** is a quantity called the *vector potential*. It then follows from (1.50) that

$$\text{curl}\left(\mathbf{E} + \frac{1}{c}\frac{\partial \mathbf{A}}{\partial t}\right) = 0$$

and so

$$\mathbf{E} + \frac{1}{c}\frac{\partial \mathbf{A}}{\partial t} = -\nabla V, \tag{1.54}$$

where V is called the *scalar potential*.

Now the force acting on a particle of *charge e*, moving with velocity **v** under the influence of an electric field **E** and a magnetic field **H**, is given by

$$\mathbf{F} = e\mathbf{E} + \frac{e}{c}\mathbf{v} \times \mathbf{H}, \tag{1.55}$$

which is referred to as the *Lorentz force*. Substituting for **H** using (1.53) and **E** using (1.54) gives

$$\begin{aligned}\mathbf{F} &= e\left[-\nabla V - \frac{1}{c}\frac{\partial \mathbf{A}}{\partial t} + \frac{1}{c}\mathbf{v} \times (\nabla \times \mathbf{A})\right] \\ &= e\left[-\nabla V - \frac{1}{c}\frac{\partial \mathbf{A}}{\partial t} + \frac{1}{c}\{\nabla(\mathbf{v}.\mathbf{A}) - \mathbf{v}.\nabla\mathbf{A}\}\right] \\ &= e\left[-\nabla V - \frac{1}{c}\frac{d\mathbf{A}}{dt} + \frac{1}{c}\nabla(\mathbf{v}.\mathbf{A})\right] \end{aligned} \quad (1.56)$$

since

$$\frac{d\mathbf{A}}{dt} = \frac{\partial \mathbf{A}}{\partial t} + \mathbf{v}.\nabla\mathbf{A}.$$

Hence we may write the x component of the Lorentz force in the form

$$F_x = \frac{d}{dt}\left(\frac{\partial U}{\partial \dot{x}}\right) - \frac{\partial U}{\partial x}, \quad (1.57)$$

where the generalized potential U is given by

$$U = eV - \frac{e}{c}\mathbf{v}.\mathbf{A}, \quad (1.58)$$

with analogous expressions for the other two components of **F**.

We see, therefore, that the Lagrangian function for a charged particle moving in an electromagnetic field with scalar potential V and vector potential **A** has the form

$$L = T - eV + \frac{e}{c}\mathbf{v}.\mathbf{A}, \quad (1.59)$$

where T is the kinetic energy of the particle.

1.5 Conservation of energy

If the dynamical system is subject to invariable constraints, the Lagrangian function L will not depend explicitly on the time t. In this case, L will be a function only of the generalized coordinates q_r and the generalized velocities $\dot{q}_r$, and so

$$\frac{dL}{dt} = \sum_{r=1}^{n}\left(\frac{\partial L}{\partial q_r}\dot{q}_r + \frac{\partial L}{\partial \dot{q}_r}\ddot{q}_r\right). \quad (1.60)$$

Employing Lagrange's equations (1.46) then gives

$$\frac{dL}{dt} = \sum_{r=1}^{n} \left\{ \frac{d}{dt}\left(\frac{\partial L}{\partial \dot{q}_r}\right)\dot{q}_r + \frac{\partial L}{\partial \dot{q}_r}\ddot{q}_r \right\} = \sum_{r=1}^{n} \frac{d}{dt}\left(\dot{q}_r \frac{\partial L}{\partial \dot{q}_r}\right)$$

which may be rewritten in the form

$$\frac{d}{dt}\left(\sum_{r=1}^{n} \dot{q}_r \frac{\partial L}{\partial \dot{q}_r} - L\right) = 0, \qquad (1.61)$$

and hence we may put

$$\sum_{r=1}^{n} \dot{q}_r \frac{\partial L}{\partial \dot{q}_r} - L = H \qquad (1.62)$$

where H is a constant.

We now remember that for a conservative system the Lagrangian function L is given by $T - V$. Further, since we have made the supposition that the dynamical system is subject to invariable constraints, the position vectors $\mathbf{r}_i$ of the particles of the system will not depend explicitly on the time t, so that

$$\mathbf{v}_i = \sum_{r=1}^{n} \frac{\partial \mathbf{r}_i}{\partial q_r}\dot{q}_r$$

and consequently

$$T = \sum_{r=1}^{n} \sum_{s=1}^{n} a_{rs}\dot{q}_r\dot{q}_s, \qquad (1.63)$$

where

$$a_{rs} = \sum_{i=1}^{N} \tfrac{1}{2}m_i \frac{\partial \mathbf{r}_i}{\partial q_r} \cdot \frac{\partial \mathbf{r}_i}{\partial q_s} = a_{sr}. \qquad (1.64)$$

It follows that

$$\sum_{r=1}^{n} \dot{q}_r \frac{\partial L}{\partial \dot{q}_r} = \sum_{r=1}^{n} \dot{q}_r \frac{\partial T}{\partial \dot{q}_r} = 2T \qquad (1.65)$$

and hence

$$H = T + V = E. \qquad (1.66)$$

Thus the total energy of a conservative system subject to invariable constraints is constant.

1.6 Euler–Lagrange equations

We are now ready to turn our attention to the problem of expressing the equations of motion of classical mechanics in the form of a variational principle. To this end we introduce the integral

$$I = \int_{t_1}^{t_2} L \, dt, \tag{1.67}$$

where L is a function of the time t, the generalized coordinates $q_1, \ldots, q_n$ and their derivatives $\dot{q}_1, \ldots, \dot{q}_n$ with respect to t. Next we choose a variation of the coordinates given by

$$q_r(t) \to q_r(t) + \epsilon \eta_r(t) \qquad (r = 1, \ldots, n), \tag{1.68}$$

where the $\eta_r(t)$ are any set of functions satisfying the conditions

$$\eta_r(t_1) = \eta_r(t_2) = 0 \qquad (r = 1, \ldots, n) \tag{1.69}$$

which ensure that the coordinates remain fixed at the end points, and ϵ is an infinitesimal parameter. The resulting change in the integral I is then given by

$$\delta I = \int_{t_1}^{t_2} L(q_r + \epsilon \eta_r, \dot{q}_r + \epsilon \dot{\eta}_r, t) \, dt - \int_{t_1}^{t_2} L(q_r, \dot{q}_r, t) \, dt$$

$$= \epsilon \int_{t_1}^{t_2} \sum_{r=1}^{n} \left(\frac{\partial L}{\partial q_r} \eta_r + \frac{\partial L}{\partial \dot{q}_r} \dot{\eta}_r \right) dt$$

to the first order in ϵ. An integration by parts yields

$$\int_{t_1}^{t_2} \frac{\partial L}{\partial \dot{q}_r} \dot{\eta}_r \, dt = \left[\frac{\partial L}{\partial \dot{q}_r} \eta_r \right]_{t_1}^{t_2} - \int_{t_1}^{t_2} \frac{d}{dt} \left(\frac{\partial L}{\partial \dot{q}_r} \right) \eta_r \, dt$$

and so we obtain

$$\delta I = \epsilon \int_{t_1}^{t_2} \sum_{r=1}^{n} \left\{ \frac{\partial L}{\partial q_r} - \frac{d}{dt} \left(\frac{\partial L}{\partial \dot{q}_r} \right) \right\} \eta_r \, dt \tag{1.70}$$

because of the conditions (1.69) imposed at $t = t_1$ and t_2.

Let us now suppose that I is *stationary* with respect to the arbitrary small changes (1.68) in the coordinates q_r. Then to the first order in ϵ we have

$$\delta I = 0 \tag{1.71}$$

and hence

$$\int_{t_1}^{t_2} \sum_{r=1}^{n} \left\{ \frac{\partial L}{\partial q_r} - \frac{d}{dt}\left(\frac{\partial L}{\partial \dot{q}_r}\right) \right\} \eta_r \, dt = 0. \tag{1.72}$$

Since the functions η_r are completely independent of each other, it follows that

$$\frac{\partial L}{\partial q_r} - \frac{d}{dt}\left(\frac{\partial L}{\partial \dot{q}_r}\right) = 0 \qquad (r = 1, \ldots, n) \tag{1.73}$$

which are known as the *Euler–Lagrange equations*.

If L is the Lagrangian function of a dynamical system, the Euler–Lagrange equations are just the Lagrange equations of motion derived in section 1.4. Thus we see that the equations of classical mechanics are equivalent to the stationary property (1.71) of the time integral (1.67) of the Lagrangian function.

Method of Lagrange undetermined multipliers

So far we have supposed that the coordinates $q_1 \ldots, q_n$ are entirely independent. Let us now consider the case when they are subject to *m auxiliary conditions*

$$\sum_{r=1}^{n} A_{rs} \, dq_r + T_s \, dt = 0 \qquad (s = 1, \ldots, m) \tag{1.74}$$

having the form of relations between the differentials dq_r and dt. If the coefficients A_{rs} can be expressed as derivatives $\partial f_s/\partial q_r$ of functions f_s of the coordinates and time, and further if $T_s = \partial f_s/\partial t$, the left-hand side of (1.74) becomes the perfect differential df_s, in which case the auxiliary conditions correspond to holonomic constraints since they may be immediately integrated to give m relations of the type

$$f_s(q_1, \ldots, q_n, t) = \text{constant}.$$

On the other hand, if the coefficients A_{rs} and T_s cannot be expressed in this manner, the auxiliary conditions (1.74) cannot be integrated and the constraints are then said to be *non-holonomic*.

To deal with the case of non-holonomic constraints we introduce m undetermined quantities λ_s ($s = 1, \ldots, m$) called *Lagrange multipliers*. Since the variations $\epsilon \eta_r$ in the coordinates q_r do not involve any change in the time t, we may put $dt = 0$ and $dq_r = \epsilon \eta_r$. Then, multiplying (1.74) across by λ_s, summing over s and integrating with respect to t, we obtain

$$\int_{t_1}^{t_2} \sum_{s=1}^{m} \sum_{r=1}^{n} \lambda_s A_{rs} \eta_r \, dt = 0$$

which together with (1.72) enables us to write

$$\int_{t_1}^{t_2} \sum_{r=1}^{n} \left\{ \frac{\partial L}{\partial q_r} - \frac{d}{dt}\left(\frac{\partial L}{\partial \dot{q}_r}\right) - \sum_{s=1}^{m} \lambda_s A_{rs} \right\} \eta_r \, dt = 0. \quad (1.75)$$

We cannot immediately put the coefficients of all the functions η_r equal to zero here, since only $n - m$ of these functions are independent as a consequence of the m auxiliary conditions which have to be obeyed. However, without loss of generality, we can suppose that the functions η_r for $r = m + 1, \ldots, n$ are independent. Then, choosing the m Lagrange multipliers λ_s so that

$$\frac{\partial L}{\partial q_r} - \frac{d}{dt}\left(\frac{\partial L}{\partial \dot{q}_r}\right) = \sum_{s=1}^{m} \lambda_s A_{rs} \quad (1.76)$$

for $r = 1, \ldots, m$, it follows that the remaining equations (1.76) for $r = m + 1, \ldots, n$ must also be satisfied owing to the independence of the functions η_r for these values of r.

The Euler–Lagrange equations (1.76), together with the auxiliary conditions (1.74) expressed in the form of the differential equations

$$\sum_{r=1}^{n} A_{rs} \dot{q}_r + T_s = 0 \quad (s = 1, \ldots, m), \quad (1.77)$$

may be used to determine the $n + m$ unknowns q_r ($r = 1, \ldots, n$) and λ_s ($s = 1, \ldots, m$).

1.7 Hamilton's principle

We showed in section 1.6 that the Euler–Lagrange equations arising as a result of the stationary property of a certain time integral could be identified with Lagrange's equations of motion for a dynamical system. In the present section we shall take the converse approach to dynamics. Starting with Lagrange's equations in the form (1.46), we shall show that a certain time integral is necessarily stationary with respect to a suitable arbitrary variation of the coordinates.

Let $L(q_1, \ldots, q_n; \dot{q}_1, \ldots, \dot{q}_n; t)$ be the Lagrangian function for a conservative dynamical system, subject to holonomic constraints, which is completely specified by n generalized coordinates $q_1, \ldots, q_n$. We may represent the motion of this dynamical system by a curve in the n dimensional space, called *configuration space*, which has $q_1, \ldots, q_n$ as rectangular Cartesian coordinates, the time t being regarded as a parameter. This curve is referred to as a *trajectory* of the dynamical system in configuration space.

Suppose that the arc $A_1 A_2$ in configuration space represents a part of a trajectory of the dynamical system, and let $B_1 B_2$ be an adjacent arc which need not necessarily be a trajectory. Suppose further that t_1, t_2 and $t_1 + \Delta t_1, t_2 + \Delta t_2$ are the times corresponding to the end-points A_1, A_2 and B_1, B_2 of the arcs respectively. If δ denotes the variation of a quantity between a point of the arc $A_1 A_2$ and a point of the arc $B_1 B_2$ associated with the same time t, then

$$\int_{B_1 B_2} L \, dt - \int_{A_1 A_2} L \, dt = L_{A_2} \Delta t_2 - L_{A_1} \Delta t_1 + \int_{t_1}^{t_2} \delta L \, dt$$
$$= L_{A_2} \Delta t_2 - L_{A_1} \Delta t_1$$
$$+ \int_{t_1}^{t_2} \sum_{r=1}^{n} \left(\frac{\partial L}{\partial \dot{q}_r} \delta \dot{q}_r + \frac{\partial L}{\partial q_r} \delta q_r \right) dt, \quad (1.78)$$

where the variation δq_r is analogous to the variation $\epsilon \eta_r$ introduced in section 1.6. Now using Lagrange's equations (1.46) we get

$$\int_{B_1B_2} L\,dt - \int_{A_1A_2} L\,dt = L_{A_2}\Delta t_2 - L_{A_1}\Delta t_1$$
$$+ \int_{t_1}^{t_2} \sum_{r=1}^{n} \left\{ \frac{\partial L}{\partial \dot{q}_r} \delta \dot{q}_r + \frac{d}{dt}\left(\frac{\partial L}{\partial \dot{q}_r}\right) \delta q_r \right\} dt$$
$$= L_{A_2}\Delta t_2 - L_{A_1}\Delta t_1$$
$$+ \int_{t_1}^{t_2} \frac{d}{dt}\left(\sum_{r=1}^{n} \frac{\partial L}{\partial \dot{q}_r} \delta q_r \right) dt$$

since

$$\delta \dot{q}_r = \frac{d}{dt}(q_r + \delta q_r) - \frac{d}{dt}q_r = \frac{d}{dt}\delta q_r, \quad (1.79)$$

and so

$$\int_{B_1B_2} L\,dt - \int_{A_1A_2} L\,dt = \left[L\Delta t + \sum_{r=1}^{n} \frac{\partial L}{\partial \dot{q}_r} \delta q_r \right]_{A_1}^{A_2}. \quad (1.80)$$

If we denote the changes in q_r between the points A_1 and B_1 and between the points A_2 and B_2 by $(\Delta q_r)_{A_1}$ and $(\Delta q_r)_{A_2}$ respectively, we have

$$\begin{aligned}(\Delta q_r)_{A_1} &= (\delta q_r)_{A_1} + (\dot{q}_r)_{A_1}\Delta t_1, \\ (\Delta q_r)_{A_2} &= (\delta q_r)_{A_2} + (\dot{q}_r)_{A_2}\Delta t_2\end{aligned} \quad (1.81)$$

and we may then rewrite (1.80) in the form

$$\int_{B_1B_2} L\,dt - \int_{A_1A_2} L\,dt = \left[\sum_{r=1}^{n} \frac{\partial L}{\partial \dot{q}_r} \Delta q_r + \left(L - \sum_{r=1}^{n} \frac{\partial L}{\partial \dot{q}_r} \dot{q}_r \right) \Delta t \right]_{A_1}^{A_2}. \quad (1.82)$$

We now consider the circumstance when the initial points A_1, B_1 of the two arcs are coincident and the final points A_2, B_2 of the two arcs are coincident, in which case we have $\Delta q_r = 0$ ($r = 1, \ldots, n$). Further, we suppose that A_1, B_1 correspond to the same time t_1 and A_2, B_2 correspond to the same time t_2, so that $\Delta t = 0$ at A_1 and A_2. Then we obtain

$$\int_{B_1B_2} L\,dt = \int_{A_1A_2} L\,dt \quad (1.83)$$

which means that the integral $\int L\, dt$ is stationary with respect to any small variation from an actual trajectory of a conservative holonomic system provided the end-points remain unaltered in space and time. This is known as *Hamilton's principle*.

We have shown in section 1.5 that

$$H = \sum_{r=1}^{n} \dot{q}_r \frac{\partial L}{\partial \dot{q}_r} - L$$

is a constant when the dynamical system has invariable constraints, so that the Lagrangian is not an explicit function of the time t. In this case the requirement that the end-points of the two arcs should have the same time values may be replaced by the requirement that the total time of description of the two arcs $A_1 A_2$ and $B_1 B_2$ should be the same, without destroying the stationary property of the integral $\int L\, dt$.

Hamilton's principle can be readily generalized to the case in which the dynamical system is non-conservative. We let an arc $A_1 A_2$ in configuration space represent a part of a trajectory of the system and choose a neighbouring arc $B_1 B_2$ such that the initial points A_1, B_1 of the two arcs are coincident and the final points A_2, B_2 of the two arcs are likewise coincident, and such that A_1, B_1 correspond to the same time t_1 and A_2, B_2 correspond to the same time t_2. Then, keeping the generalized forces Q_r acting on the system unaltered during the variation denoted by δ between a point of the arc $A_1 A_2$ and a synchronous point of the arc $B_1 B_2$, we have

$$\int_{B_1 B_2} \left(T + \sum_{r=1}^{n} Q_r q_r\right) dt - \int_{A_1 A_2} \left(T + \sum_{r=1}^{n} Q_r q_r\right) dt$$

$$= \int_{t_1}^{t_2} \left(\delta T + \sum_{r=1}^{n} Q_r\, \delta q_r\right) dt$$

$$= \int_{t_1}^{t_2} \sum_{r=1}^{n} \left(\frac{\partial T}{\partial \dot{q}_r} \delta \dot{q}_r + \frac{\partial T}{\partial q_r} \delta q_r + Q_r\, \delta q_r\right) dt$$

$$= \int_{t_1}^{t_2} \sum_{r=1}^{n} \left\{\frac{\partial T}{\partial \dot{q}_r} \delta \dot{q}_r + \frac{d}{dt}\left(\frac{\partial T}{\partial \dot{q}_r}\right) \delta q_r\right\} dt, \quad (1.84)$$

using equations (1.41). Hence

$$\int_{t_1}^{t_2} \left(\delta T + \sum_{r=1}^{n} Q_r \, \delta q_r \right) dt = \int_{t_1}^{t_2} \frac{d}{dt} \left(\sum_{r=1}^{n} \frac{\partial T}{\partial \dot{q}_r} \delta q_r \right) dt$$

$$= \left[\sum_{r=1}^{n} \frac{\partial T}{\partial \dot{q}_r} \delta q_r \right]_{t_1}^{t_2} = 0$$

since $\delta q_r(t_1) = \delta q_r(t_2) = 0$ for $r = 1, \ldots, n$.

Thus we see that the general form of Hamilton's principle can be written as

$$\int_{t_1}^{t_2} \left(\delta T + \sum_{r=1}^{n} Q_r \, \delta q_r \right) dt = 0. \quad (1.85)$$

If we put

$$Q_r = -\frac{\partial V}{\partial q_r},$$

(1.85) immediately assumes the form

$$\int_{t_1}^{t_2} (\delta T - \delta V) \, dt = 0$$

which is just Hamilton's principle for a conservative system derived previously.

1.8 Virial theorem

An interesting result known as the virial theorem may be established by introducing a variation of coordinates $\delta q_r = \epsilon q_r$ which corresponds to the change of scale given by

$$q_r(t) \to (1 + \epsilon) q_r(t). \quad (1.86)$$

Such a variation does not satisfy the terminal conditions that δq_r should vanish at $t = t_1$ and t_2 and yields

$$\int_{t_1}^{t_2} \left(\delta T + \sum_{r=1}^{n} Q_r \, \delta q_r \right) dt = \epsilon \left[\sum_{r=1}^{n} \frac{\partial T}{\partial \dot{q}_r} q_r \right]_{t_1}^{t_2}$$

in place of equation (1.85), giving rise to the formula

$$\int_{t_1}^{t_2} \sum_{r=1}^{n} \left(\frac{\partial T}{\partial \dot{q}_r} \dot{q}_r + \frac{\partial T}{\partial q_r} q_r + Q_r q_r \right) dt = \left[\sum_{r=1}^{n} \frac{\partial T}{\partial \dot{q}_r} q_r \right]_{t_1}^{t_2} \quad (1.87)$$

ANALYTICAL DYNAMICS

on using $\delta q_r = \epsilon q_r$ once again. Dividing by the time interval $t_2 - t_1$, allowing $t_2 - t_1$ to tend to infinity and assuming that the generalized coordinates and velocities of the dynamical system remain finite so that the quantity on the right-hand side of (1.87) is bounded, leads to the result

$$\overline{\sum_{r=1}^{n} \left(\frac{\partial T}{\partial \dot{q}_r} \dot{q}_r + \frac{\partial T}{\partial q_r} q_r + Q_r q_r \right)} = 0, \tag{1.88}$$

the bar signifying a time average. For a dynamical system subject to time-independent constraints, we may use equation (1.65), in which case (1.88) becomes

$$2\overline{T} + \overline{\sum_{r=1}^{n} \left(\frac{\partial T}{\partial q_r} + Q_r \right) q_r} = 0. \tag{1.89}$$

Let us now suppose that the system is composed of N masses m_i with position vectors $\mathbf{r}_i$ moving under the action of forces $\mathbf{F}_i$. Then

$$T = \sum_{i=1}^{N} \tfrac{1}{2} m_i \dot{\mathbf{r}}_i^2$$

and we may replace the second term of (1.89) by

$$\overline{\sum_{i=1}^{N} \mathbf{F}_i \cdot \mathbf{r}_i}$$

which reduces (1.89) to

$$2\overline{T} + \overline{\sum_{i=1}^{N} \mathbf{F}_i \cdot \mathbf{r}_i} = 0. \tag{1.90}$$

This result is known as the *virial theorem*, and the quantity

$$-\tfrac{1}{2} \overline{\sum_{i=1}^{N} \mathbf{F}_i \cdot \mathbf{r}_i}$$

is called the *virial of Clausius*.

If the forces acting on the system are conservative forces, so that we may write
$$\mathbf{F}_i = -\nabla_i V,$$
then the virial theorem takes the form
$$2\overline{T} = \overline{\sum_{i=1}^{N} \mathbf{r}_i \cdot \nabla_i V}. \tag{1.91}$$

In addition, if the potential V is a homogeneous function of the mth degree in the radial distances r_i, then
$$\sum_{i=1}^{N} \mathbf{r}_i \cdot \nabla_i V = mV \tag{1.92}$$

and hence the virial theorem now assumes the very simple form
$$2\overline{T} = m\overline{V}. \tag{1.93}$$

In the special case $m = -1$, associated with Newton's law of gravitation and Coulomb's law of force in electrostatics, we have
$$2\overline{T} = -\overline{V}. \tag{1.94}$$

1.9 Principle of least action

We confine our attention once more to the motion of a dynamical system for which the Lagrangian function does not depend explicitly on the time, so that
$$\sum_{r=1}^{n} \dot{q}_r \frac{\partial L}{\partial \dot{q}_r} - L = H$$
is a constant.

Let the arc $A_1 A_2$ be a part of an actual trajectory of the system and let $B_1 B_2$ be a neighbouring arc for which
$$\sum_{r=1}^{n} \dot{q}_r \frac{\partial L}{\partial \dot{q}_r} - L = H + \Delta H,$$

ANALYTICAL DYNAMICS

where ΔH is a small constant. Then, using equation (1.82), we see that

$$\int_{B_1 B_2} \left(\sum_{r=1}^{n} \dot{q}_r \frac{\partial L}{\partial \dot{q}_r} \right) dt - \int_{A_1 A_2} \left(\sum_{r=1}^{n} \dot{q}_r \frac{\partial L}{\partial \dot{q}_r} \right) dt$$

$$= \int_{B_1 B_2} (H + \Delta H) \, dt - \int_{A_1 A_2} H \, dt + \int_{B_1 B_2} L \, dt - \int_{A_1 A_2} L \, dt$$

$$= H(\Delta t_2 - \Delta t_1) + \Delta H(t_2 - t_1) + \left[\sum_{r=1}^{n} \frac{\partial L}{\partial \dot{q}_r} \Delta q_r - H \Delta t \right]_{A_1}^{A_2}$$

$$= \left[\sum_{r=1}^{n} \frac{\partial L}{\partial \dot{q}_r} \Delta q_r + t \Delta H \right]_{A_1}^{A_2}. \tag{1.95}$$

If the constant H is the same for both arcs so that $\Delta H = 0$, and if the terminal points of the two arcs coincide so that $\Delta q_r = 0$ ($r = 1, \ldots, n$) at A_1 and A_2, it follows that

$$\int_{B_1 B_2} \left(\sum_{r=1}^{n} \dot{q}_r \frac{\partial L}{\partial \dot{q}_r} \right) dt = \int_{A_1 A_2} \left(\sum_{r=1}^{n} \dot{q}_r \frac{\partial L}{\partial \dot{q}_r} \right) dt. \tag{1.96}$$

Hence for variations satisfying the above conditions the integral

$$A = \int \left(\sum_{r=1}^{n} \dot{q}_r \frac{\partial L}{\partial \dot{q}_r} \right) dt, \tag{1.97}$$

referred to as the *action*, is stationary. This result is called the *principle of least action*.

If the dynamical system is conservative and subject to invariable constraints, it follows from (1.65) that

$$A = 2 \int T \, dt. \tag{1.98}$$

In the absence of external forces the kinetic energy of the system is constant and then the principle of least action becomes

$$(t_2 + \Delta t_2) - (t_1 + \Delta t_1) = t_2 - t_1, \tag{1.99}$$

which means that the total time of description is stationary for an actual trajectory of the system.

For a single particle of mass m, the action integral may be readily expressed in an alternative form by converting to an integration over the path length s. Since the speed of the particle is given by $v = ds/dt$, we see that

$$A = \int mv \, ds = \int \sqrt{(2mT)} \, ds = \int \sqrt{\{2m(E - V)\}} \, ds, \quad (1.100)$$

where E is the total energy of the particle and V is the potential function. The stationary property of (1.100) is Jacobi's form of the principle of least action.

1.10 Hamilton's equations

The introduction of generalized momenta given by

$$p_r = \frac{\partial L}{\partial \dot{q}_r} \quad (r = 1, \ldots, n), \quad (1.101)$$

enables us to write Lagrange's equations in the form

$$\dot{p}_r = \frac{\partial L}{\partial q_r} \quad (r = 1, \ldots, n). \quad (1.102)$$

As a consequence of equations (1.101) any function of $\dot{q}_1, \dot{q}_2, \ldots, \dot{q}_n$ may be expressed equally well in terms of $p_1, p_2, \ldots, p_n$. We shall refer to the p_r as the *canonical momenta* conjugate to the coordinates q_r. Since the generalized coordinates need not necessarily have the dimensions of length, it follows that the canonical momenta may not have the dimensions of linear momentum.

Now consider the change δL in the Lagrangian function due to independent increments in the coordinates q_r and the velocities $\dot{q}_r$ or, alternatively, in the coordinates q_r and the canonical momenta p_r. Using (1.101) and (1.102), we obtain

$$\begin{aligned} \delta L &= \sum_{r=1}^{n} \left(\frac{\partial L}{\partial q_r} \delta q_r + \frac{\partial L}{\partial \dot{q}_r} \delta \dot{q}_r \right) \\ &= \sum_{r=1}^{n} (\dot{p}_r \delta q_r + p_r \delta \dot{q}_r) \\ &= \delta \left(\sum_{r=1}^{n} p_r \dot{q}_r \right) + \sum_{r=1}^{n} (\dot{p}_r \delta q_r - \dot{q}_r \delta p_r) \end{aligned}$$

ANALYTICAL DYNAMICS

and so
$$\delta\left(\sum_{r=1}^{n} p_r \dot{q}_r - L\right) = \sum_{r=1}^{n} (\dot{q}_r \, \delta p_r - \dot{p}_r \, \delta q_r). \quad (1.103)$$

Let
$$H = \sum_{r=1}^{n} p_r \dot{q}_r - L, \quad (1.104)$$

where H is expressed in terms of the q_r, p_r, t rather than the $q_r, \dot{q}_r, t$. H is known as the *Hamiltonian function*. Then we have that the change in the Hamiltonian function due to independent increments in the q_r and p_r is given by

$$\delta H = \sum_{r=1}^{n} (\dot{q}_r \, \delta p_r - \dot{p}_r \, \delta q_r). \quad (1.105)$$

But this may be also expressed in the form
$$\delta H = \sum_{r=1}^{n} \left(\frac{\partial H}{\partial p_r} \delta p_r + \frac{\partial H}{\partial q_r} \delta q_r\right) \quad (1.106)$$

and so, comparing (1.105) with (1.106), we see that
$$\dot{q}_r = \frac{\partial H}{\partial p_r}, \quad \dot{p}_r = -\frac{\partial H}{\partial q_r} \quad (r = 1, \ldots, n). \quad (1.107)$$

These are known as *Hamilton's equations*. They form a set of $2n$ first-order differential equations in contrast to Lagrange's equations, which consist of n second-order equations. This difference arises because in the derivation of Hamilton's equations the $2n$ infinitesimal changes δq_r and δp_r are regarded as independent, whereas only the n variations δq_r are taken to be independent in the derivation of Lagrange's equations.

1.11 Variational principle for Hamilton's equations

Instead of representing the motion of a dynamical system by a trajectory in configuration space, we now describe the motion of a dynamical system by a path in the $2n$-dimensional space called *phase space* having $q_1, \ldots, q_n$ and $p_1, \ldots, p_n$ as Cartesian coordinates, the time t being regarded as a parameter. Let the arc

A_1A_2 in phase space represent a portion of the actual path of the dynamical system and let B_1B_2 be a neighbouring arc whose initial point B_1 coincides with A_1 and whose final point B_2 coincides with A_2. Further, taking A_1 and B_1 to be associated with the same time t_1, and taking A_2 and B_2 to be associated with the same time t_2, we have

$$\int_{B_1B_2} \left(\sum_{r=1}^{n} \dot{q}_r p_r - H\right) dt - \int_{A_1A_2} \left(\sum_{r=1}^{n} \dot{q}_r p_r - H\right) dt$$

$$= \int_{t_1}^{t_2} \delta\left(\sum_{r=1}^{n} \dot{q}_r p_r - H\right) dt$$

$$= \int_{t_1}^{t_2} \sum_{r=1}^{n} \left\{ p_r \delta\dot{q}_r + \dot{q}_r \delta p_r - \left(\frac{\partial H}{\partial p_r} \delta p_r + \frac{\partial H}{\partial q_r} \delta q_r\right)\right\} dt.$$

But

$$\int_{t_1}^{t_2} p_r \delta\dot{q}_r \, dt = \int_{t_1}^{t_2} p_r \frac{d}{dt}(\delta q_r) \, dt$$

$$= [p_r \delta q_r]_{t_1}^{t_2} - \int_{t_1}^{t_2} \dot{p}_r \delta q_r \, dt = -\int_{t_1}^{t_2} \dot{p}_r \delta q_r \, dt$$

and so

$$\int_{t_1}^{t_2} \delta\left(\sum_{r=1}^{n} \dot{q}_r p_r - H\right) dt$$

$$= \int_{t_1}^{t_2} \sum_{r=1}^{n} \left\{\left(\dot{q}_r - \frac{\partial H}{\partial p_r}\right) \delta p_r - \left(\dot{p}_r + \frac{\partial H}{\partial q_r}\right) \delta q_r\right\} dt. \quad (1.108)$$

Since the variations δp_r and δq_r are independent, the assumption of a stationary property for the integral

$$\int_{t_1}^{t_2} \left(\sum_{r=1}^{n} \dot{q}_r p_r - H\right) dt$$

leads to Hamilton's equations (1.107).

1.12 Special theory of relativity

At this stage of the present chapter it is convenient to consider the modifications produced in classical dynamics by the special theory of relativity, and to this end we begin by briefly discussing the Lorentz transformation.

The basic assertion of the special theory of relativity is that the velocity of light *in vacuo* has the same value c for all inertial frames of reference. Consider now two such inertial frames of reference S and S' moving with uniform relative velocity **v** and let any given event be characterized by Cartesian coordinates x, y, z referred to S and a time t determined by means of a clock fixed in S, or by Cartesian coordinates x', y', z' referred to S' and a time t' as measured by a clock fixed in S'. Suppose further that the origins O and O' of the two frames of reference coincide at the instant $t = t' = 0$ and that at this instant a light signal is emitted from their common origin. Then the resulting spherical light wave has the equation

$$s^2 \equiv c^2 t^2 - x^2 - y^2 - z^2 = 0 \qquad (1.109)$$

referred to S, and the equation

$$s'^2 \equiv c^2 t'^2 - x'^2 - y'^2 - z'^2 = 0 \qquad (1.110)$$

referred to S', the velocity of light c being the same for both reference frames. Now a particle moving in a straight line relative to S must also be moving in a straight line relative to S', since both S and S' are inertial frames of reference. Hence the relations connecting the coordinates x, y, z, t and x', y', z', t' must be linear, in which case, since $s'^2 = 0$ when $s^2 = 0$, we can write

$$s^2 = s'^2. \qquad (1.111)$$

The formulae relating the coordinates x, y, z, t and x', y', z', t' are known as the *Lorentz transformation equations*. If the corresponding Cartesian axes of S and S' are chosen parallel to each other, and if S' is moving relative to S with velocity **v** in the direction of the positive x-axis, the Lorentz transformation takes the form

$$x' = \kappa(x - vt), \quad y' = y, \quad z' = z, \quad t' = \kappa(t - vx/c^2) \qquad (1.112)$$

where

$$\kappa = \frac{1}{\sqrt{\{1 - (v/c)^2\}}}. \qquad (1.113)$$

Alternatively, the Lorentz transformation equations can be expressed in the form

$$x = \kappa(x' + vt'), \qquad y = y', \qquad z = z', \qquad t = \kappa(t' + vx'/c^2) \tag{1.114}$$

which can be readily seen to follow from equations (1.112) by replacing v by $-v$ and interchanging the primed and unprimed coordinates, which has the effect of exchanging the roles of the two frames of reference.

The formula (1.111) only applies to the case of the homogeneous Lorentz transformation for which the origins O and O' of the two frames of reference coincide at the instant $t = t' = 0$. If this is not the case, equation (1.111) becomes replaced by the relation

$$ds^2 = ds'^2, \tag{1.115}$$

where

$$ds^2 = c^2\, dt^2 - dx^2 - dy^2 - dz^2, \tag{1.116}$$

so that now ds^2 is invariant under the Lorentz transformation instead of s^2.

1.13 Relativistic mechanics

In the special theory of relativity, the momentum **p** of a particle moving with velocity **v** relative to an observer situated at the origin O of an inertial frame of reference S and measuring time t by means of a clock at rest in S is defined, as in non-relativistic mechanics, by the formula

$$\mathbf{p} = m\mathbf{v}, \tag{1.117}$$

where m is the mass of the particle. Now, in Newtonian mechanics, the momentum of a system of interacting particles is necessarily conserved in the absence of external forces. However, in relativistic mechanics, in order to conserve the momentum of interacting particles, the mass of a given particle must depend upon its speed relative to an observer, and is given by

$$m = \frac{m_0}{\sqrt{\{1 - (v/c)^2\}}}, \tag{1.118}$$

where m_0 is the mass of the particle when it is at rest in S and is therefore called the *rest mass* of the particle.

According to the special theory of relativity, the total energy of a particle of mass m relative to the frame of reference S takes the form

$$E = mc^2. \qquad (1.119)$$

Since

$$\mathbf{p}^2 = \frac{m_0^2 v^2}{1 - (v/c)^2},$$

we see that the total energy may be expressed in terms of the momentum by the formula

$$E^2 = \frac{m_0^2 c^4}{1 - (v/c)^2} = c^2(m_0^2 c^2 + \mathbf{p}^2). \qquad (1.120)$$

The energy of the particle when it is at rest relative to S is called the *rest energy* of the particle and is given by

$$E_0 = m_0 c^2. \qquad (1.121)$$

It follows that the energy of the particle due to its motion relative to S is

$$T = E - E_0 = m_0 c^2 \left(\frac{1}{\sqrt{\{1 - (v/c)^2\}}} - 1 \right) \qquad (1.122)$$

which is the relativistic expression for the kinetic energy of a free particle. If $v/c \ll 1$, (1.122) reduces to

$$T = \tfrac{1}{2} m_0 v^2$$

and so we regain the non-relativistic formula for the kinetic energy.

As in Newtonian dynamics, the force acting on a particle is given by

$$\mathbf{F} = \frac{d\mathbf{p}}{dt} \qquad (1.123)$$

which we may rewrite in the form

$$\mathbf{F} = \frac{d}{dt}\left(\frac{m_0 \mathbf{v}}{\sqrt{\{1-(v/c)^2\}}}\right) \quad (1.124)$$

using the relativistic expression for the momentum **p**.

Lagrangian function for a particle

We are now in a position to derive an expression for the Lagrangian function L appropriate to the relativistic motion of a particle under the action of a conservative force given by

$$\mathbf{F} = -\nabla V. \quad (1.125)$$

As in the non-relativistic case, Hamilton's principle

$$\delta \int_{t_1}^{t_2} L \, dt = 0 \quad (1.126)$$

results in the Euler–Lagrange equations

$$\frac{d}{dt}\left(\frac{\partial L}{\partial \dot{q}_r}\right) = \frac{\partial L}{\partial q_r} \quad (r = 1, 2, 3), \quad (1.127)$$

where q_1, q_2, q_3 form a set of three generalized coordinates for the particle.

Now employing Cartesian coordinates x, y, z and putting

$$L = K - V, \quad (1.128)$$

where K is assumed to be a function of the velocities $\dot{x}, \dot{y}, \dot{z}$ only, we see that

$$\frac{d}{dt}\left(\frac{\partial K}{\partial \dot{x}}\right) = -\frac{\partial V}{\partial x}, \quad \frac{d}{dt}\left(\frac{\partial K}{\partial \dot{y}}\right) = -\frac{\partial V}{\partial y}, \quad \frac{d}{dt}\left(\frac{\partial K}{\partial \dot{z}}\right) = -\frac{\partial V}{\partial z}.$$

These equations are equivalent to the equations of motion (1.123) if

$$p_x = \frac{\partial K}{\partial \dot{x}}, \quad p_y = \frac{\partial K}{\partial \dot{y}}, \quad p_z = \frac{\partial K}{\partial \dot{z}}. \quad (1.129)$$

ANALYTICAL DYNAMICS

It then follows that

$$\frac{dK}{dt} = \frac{\partial K}{\partial \dot{x}}\ddot{x} + \frac{\partial K}{\partial \dot{y}}\ddot{y} + \frac{\partial K}{\partial \dot{z}}\ddot{z} = p_x\ddot{x} + p_y\ddot{y} + p_z\ddot{z}$$

$$= \frac{m_0(\dot{x}\ddot{x} + \dot{y}\ddot{y} + \dot{z}\ddot{z})}{\sqrt{\{1 - (v/c)^2\}}} = \frac{m_0 v}{\sqrt{\{1 - (v/c)^2\}}} \frac{dv}{dt}$$

and hence

$$K = -m_0 c^2 \sqrt{\{1 - (v/c)^2\}} \tag{1.130}$$

choosing the arbitrary constant of integration to vanish. Hence the Lagrangian function for a relativistic particle has the form

$$L = -m_0 c^2 \sqrt{\{1 - (v/c)^2\}} - V. \tag{1.131}$$

In the case of a particle having charge e moving under the action of an electromagnetic field with scalar potential V and vector potential $\mathbf{A}$, it can be readily shown, by following an analogous procedure to that described on pp. 11 and 12, that the Lagrangian function is

$$L = -m_0 c^2 \sqrt{\{1 - (v/c)^2\}} - eV + \frac{e}{c}\mathbf{v}\cdot\mathbf{A}. \tag{1.132}$$

Since the Cartesian components of momentum are defined by the formulae

$$p_x = \frac{\partial L}{\partial \dot{x}}, \quad p_y = \frac{\partial L}{\partial \dot{y}}, \quad p_z = \frac{\partial L}{\partial \dot{z}}$$

we obtain

$$\mathbf{p} = \frac{m_0 \mathbf{v}}{\sqrt{\{1 - (v/c)^2\}}} + \frac{e}{c}\mathbf{A}, \tag{1.133}$$

from which it follows that the total energy

$$E = m_0 c^2 + T + eV \tag{1.134}$$

of the particle is given by

$$(E - eV)^2 = c^2\left\{m_0^2 c^2 + \left(\mathbf{p} - \frac{e}{c}\mathbf{A}\right)^2\right\}. \tag{1.135}$$

Hamiltonian function for a particle

By performing a similar analysis to that described in section 1.10 for the non-relativistic case, Hamilton's equations can be derived in the form

$$\dot{q}_r = \frac{\partial H}{\partial p_r}, \quad \dot{p}_r = -\frac{\partial H}{\partial q_r} \quad (r = 1, 2, 3), \quad (1.136)$$

where

$$p_r = \frac{\partial L}{\partial \dot{q}_r} \quad (1.137)$$

are the generalized components of momentum, and

$$H = \sum_{r=1}^{3} \dot{q}_r \frac{\partial L}{\partial \dot{q}_r} - L \quad (1.138)$$

is the Hamiltonian function.

Now we have

$$H = \sum_{r=1}^{3} \dot{q}_r \frac{\partial K}{\partial \dot{q}_r} - L$$

$$= \sum_{r=1}^{3} \dot{q}_r \left[\frac{1}{2} \frac{\partial (v^2)}{\partial \dot{q}_r} \frac{m_0}{\sqrt{\{1 - (v/c)^2\}}} \right] - K + V,$$

and since, by Euler's theorem,

$$\sum_{r=1}^{3} \dot{q}_r \frac{\partial (v^2)}{\partial \dot{q}_r} = 2v^2$$

owing to the fact that v^2 is a homogeneous function of the second degree in $\dot{q}_1, \dot{q}_2, \dot{q}_3$, we see that

$$H = \frac{m_0 v^2}{\sqrt{\{1 - (v/c)^2\}}} + m_0 c^2 \sqrt{\{1 - (v/c)^2\}} + V$$

$$= \frac{m_0 c^2}{\sqrt{\{1 - (v/c)^2\}}} + V$$

$$= m_0 c^2 + T + V, \quad (1.139)$$

which is just the total energy E. Apart from the fact that E includes the rest energy $m_0 c^2$, this is the same result as we obtained in classical non-relativistic mechanics.

Principle of least action

We conclude our discussion of relativistic mechanics by considering the action integral for a particle

$$A = \int \sum_{r=1}^{3} \dot{q}_r \frac{\partial L}{\partial \dot{q}_r} dt. \qquad (1.140)$$

Now

$$\sum_{r=1}^{3} \dot{q}_r \frac{\partial L}{\partial \dot{q}_r} = H + L = m_0 c^2 + T + K$$

and hence the action integral takes the form

$$A = \int (m_0 c^2 + T + K) \, dt \qquad (1.141)$$

so that the principle of least action in relativistic mechanics becomes

$$\delta \int (T + K) \, dt = 0. \qquad (1.142)$$

Another way of expressing the action integral can be obtained by noting that

$$m_0 c^2 + T + K = \frac{m_0 v^2}{\sqrt{\{1 - (v/c)^2\}}},$$

for then we have

$$A = \int \frac{m_0 v}{\sqrt{\{1 - (v/c)^2\}}} \, ds = \int mv \, ds \qquad (1.143)$$

since $v = ds/dt$, where s is the path length, and

$$m = m_0/\sqrt{\{1 - (v/c)^2\}}.$$

This expression is formally the same as one of the non-relativistic expressions for the action integral derived in section 1.9.

1.14 Contact transformations

Let us consider a transformation which converts the variables q_r, p_r ($r = 1, \ldots, n$) of a dynamical system whose motion is determined by Hamilton's equations

$$\dot{q}_r = \frac{\partial H}{\partial p_r}, \qquad \dot{p}_r = -\frac{\partial H}{\partial q_r}, \qquad (1.144)$$

into the variables Q_r, P_r ($r = 1, \ldots, n$) of another dynamical system with Hamilton's equations

$$\dot{Q}_r = \frac{\partial K}{\partial P_r}, \qquad \dot{P}_r = -\frac{\partial K}{\partial Q_r}, \tag{1.145}$$

where $H(q_r, p_r, t)$ and $K(Q_r, P_r, t)$ are the respective Hamiltonian functions of the two systems. The transformation defined by the equations

$$Q_r = Q_r(q_1, \ldots, q_n, p_1, \ldots, p_n, t),$$

$$P_r = P_r(q_1, \ldots, q_n, p_1, \ldots, p_n, t) \tag{1.146}$$

satisfying (1.144) and (1.145) is called a *contact transformation*.

By virtue of the variational principle derived in section 1.11, we have that

$$\delta \int_{t_1}^{t_2} \left(\sum_{r=1}^{n} \dot{q}_r p_r - H \right) dt = 0 \tag{1.147}$$

and

$$\delta \int_{t_1}^{t_2} \left(\sum_{r=1}^{n} \dot{Q}_r P_r - K \right) dt = 0. \tag{1.148}$$

Hence we may write

$$\sum_{r=1}^{n} \dot{q}_r p_r - H = \sum_{r=1}^{n} \dot{Q}_r P_r - K + \frac{dW}{dt}, \tag{1.149}$$

where W is an arbitrary function of any $2n$ of the variables q_r, p_r, Q_r, P_r ($r = 1, \ldots, n$) as well as the time t, because

$$\delta \int_{t_1}^{t_2} \frac{dW}{dt} dt = [\delta W]_{t_1}^{t_2} = 0 \tag{1.150}$$

as a consequence of the requirement that the variation vanishes at the terminal points.

Let us now choose W to be an arbitrary function of the q_r, Q_r, t. Then we have

$$\frac{dW}{dt} = \sum_{r=1}^{n} \left(\frac{\partial W}{\partial q_r} \dot{q}_r + \frac{\partial W}{\partial Q_r} \dot{Q}_r \right) + \frac{\partial W}{\partial t} \tag{1.151}$$

and so, using (1.149), we get

$$\sum_{r=1}^{n} \left(p_r - \frac{\partial W}{\partial q_r}\right)\dot{q}_r - \sum_{r=1}^{n} \left(P_r + \frac{\partial W}{\partial Q_r}\right)\dot{Q}_r + \left(K - H - \frac{\partial W}{\partial t}\right) = 0. \tag{1.152}$$

Remembering that the q_r and Q_r are independent variables in the present case, we see that the coefficients of $\dot{q}_r$ and $\dot{Q}_r$ must vanish, yielding

$$p_r = \frac{\partial W}{\partial q_r}, \tag{1.153}$$

$$P_r = -\frac{\partial W}{\partial Q_r}, \tag{1.154}$$

and

$$K = H + \frac{\partial W}{\partial t}. \tag{1.155}$$

It follows that for a contact transformation from the set of variables q_r, p_r ($r = 1, \ldots, n$) to the set of variables Q_r, P_r ($r = 1, \ldots, n$), any variation of W in which the time t remains unchanged can be expressed in the form

$$\begin{aligned}\delta W &= \sum_{r=1}^{n} \left(\frac{\partial W}{\partial q_r}\delta q_r + \frac{\partial W}{\partial Q_r}\delta Q_r\right) \\ &= \sum_{r=1}^{n} (p_r\,\delta q_r - P_r\,\delta Q_r).\end{aligned} \tag{1.156}$$

The function W characterizing the contact transformation is called the *generating function* of the transformation.

1.15 Hamilton–Jacobi equation

A contact transformation producing considerable simplification is that which converts the original dynamical system into a system whose Hamiltonian function K vanishes. In this case, we have

$$\frac{\partial}{\partial t} W(q_r, Q_r, t) + H(q_r, p_r, t) = 0 \tag{1.157}$$

and so W, considered as a function of the $n + 1$ variables q_r and t, must satisfy the first-order partial differential equation

$$\frac{\partial W}{\partial t} + H\left(q_r, \frac{\partial W}{\partial q_r}, t\right) = 0 \qquad (1.158)$$

which is known as the *Hamilton–Jacobi equation*.

Suppose that the complete solution $W(q_r, \alpha_r, t)$, containing n arbitrary constants $\alpha_1, \ldots, \alpha_n$ as well as an additive constant, has been obtained for the Hamilton–Jacobi equation. This solution is called *Hamilton's principal function*. Now owing to the vanishing of K, Hamilton's equations for the new system take the form

$$\dot{Q}_r = \frac{\partial K}{\partial P_r} = 0, \qquad \dot{P}_r = -\frac{\partial K}{\partial Q_r} = 0 \qquad (1.159)$$

and so the P_r, Q_r ($r = 1, \ldots, n$) are all constant. Hence identifying the constants α_r with the coordinates Q_r of the new dynamical system and putting $P_r = \beta_r$ ($r = 1, \ldots, n$) where the β_r are another set of constants, we may carry out a contact transformation from the q_r, p_r to the α_r, β_r specified by

$$p_r = \frac{\partial W}{\partial q_r} \qquad (r = 1, \ldots, n) \qquad (1.160)$$

and

$$\beta_r = -\frac{\partial W}{\partial \alpha_r} \qquad (r = 1, \ldots, n). \qquad (1.161)$$

Since these equations, together with the initial values of the coordinates q_r and the momenta p_r, determine the constants α_r and β_r, we see that the solution of the original dynamical problem has been found, since the equations (1.161) enable us to express the coordinates q_r as functions of the time.

1.16 Hamilton's integral

A solution of the Hamilton–Jacobi equation in the form of an integral can be obtained by choosing the constants α_r, β_r as the values of q_r, p_r respectively at the initial time $t = t_1$.

ANALYTICAL DYNAMICS

The change in *Hamilton's integral*

$$\int_{t_1}^{t} L \, dt \qquad (1.162)$$

arising from infinitesimal variations $\delta\alpha_r$, $\delta\beta_r$ in the initial values of q_r, p_r is given by (1.80) to be

$$\delta \int_{t_1}^{t} L \, dt = \sum_{r=1}^{n} (p_r \, \delta q_r - \beta_r \, \delta\alpha_r). \qquad (1.163)$$

Denoting Hamilton's integral (1.162) when expressed as a function of q_r, α_r, t by $W(q_r, \alpha_r, t)$, we regain equations (1.160) and (1.161), and so the transformation from q_r, p_r to α_r, β_r must be a contact transformation. Since the inverse transformation from α_r, β_r to q_r, p_r must also be a contact transformation, we see that the evolution with time of a dynamical system can be regarded as the unfolding of a contact transformation.

Now

$$\frac{dW}{dt} = \frac{\partial W}{\partial t} + \sum_{r=1}^{n} \frac{\partial W}{\partial q_r} \dot{q}_r$$

so that

$$L = \frac{\partial W}{\partial t} + \sum_{r=1}^{n} p_r \dot{q}_r.$$

Hence

$$\frac{\partial W}{\partial t} + H = 0 \qquad (1.164)$$

from which it follows that Hamilton's integral (1.162) is a solution of the Hamilton–Jacobi equation.

1.17 Hamilton's characteristic function

Suppose next that the Hamiltonian function H of the dynamical system is a function of the coordinates q_r and the momenta p_r but is independent of the time. We choose a contact transformation with generating function $S(q_r, Q_r)$ for which the new Hamiltonian function K is a function of the new coordinates Q_r only, and not of the new momenta P_r. Since

$$\dot{Q}_r = \frac{\partial K}{\partial P_r} = 0$$

we have $Q_r = \alpha_r$, where the α_r are arbitrary constants, and hence K is a constant.

Now $\partial S/\partial t = 0$ for the chosen transformation and so equation (1.155) gives

$$H(q_r, p_r) = K(\alpha_r)$$

which, on substituting

$$p_r = \frac{\partial S}{\partial q_r}, \qquad (1.165)$$

yields the first-order partial differential equation

$$H\left(q_r, \frac{\partial S}{\partial q_r}\right) = K(\alpha_r). \qquad (1.166)$$

We also have that

$$\dot{P}_r = -\frac{\partial K}{\partial Q_r}$$

which, on rewriting it in the form

$$\dot{P}_r = -\frac{\partial K}{\partial \alpha_r} = -\omega_r,$$

gives

$$P_r = -\omega_r t - \beta_r$$

where β_r is an arbitrary constant. But

$$P_r = -\frac{\partial S}{\partial Q_r} \qquad (1.167)$$

and so

$$\frac{\partial S}{\partial \alpha_r} = \omega_r t + \beta_r. \qquad (1.168)$$

Let us choose the generating function S so that $K = Q_1$. Then we get the equation

$$H\left(q_r, \frac{\partial S}{\partial q_r}\right) = \alpha_1 \qquad (1.169)$$

whose solution has the form $S(q_r, \alpha_r) +$ constant. In the present case we have $\omega_1 = 1$, $\omega_r = 0$ $(r \neq 1)$ and so

$$\frac{\partial S}{\partial \alpha_1} = t + \beta_1, \tag{1.170}$$

$$\frac{\partial S}{\partial \alpha_r} = \beta_r \quad (r \neq 1). \tag{1.171}$$

$S(q_r, \alpha_r)$ is called *Hamilton's characteristic function*. A comparison between equations (1.158) and (1.169) shows that it is related to Hamilton's principal function $W(q_r, \alpha_r, t)$ by the formula

$$W(q_r, \alpha_r, t) = S(q_r, \alpha_r) - \alpha_1 t. \tag{1.172}$$

For a conservative system subject to invariable constraints, the constant value of the Hamiltonian function H is the total energy E, and then $\alpha_1 = E$.

As a particular example of some considerable interest, we now investigate the motion of a particle having mass m under the influence of a potential V. The Hamiltonian function for such a particle is given by

$$H = \frac{1}{2m}\mathbf{p}^2 + V, \tag{1.173}$$

where $\mathbf{p}$ is the momentum of the particle. If E is the total energy, Hamilton's characteristic function S satisfies the equation

$$\frac{1}{2m}(\nabla S)^2 + V = E \tag{1.174}$$

since

$$\mathbf{p} = \nabla S. \tag{1.175}$$

Now consider two adjacent surfaces over which the characteristic function S has the constant values S_0 and $S_0 + dS$ respectively. Then the distance between these surfaces in the direction of the normal at any point of the first surface is given by

$$ds = \frac{dS}{|\nabla S|} = \frac{dS}{\sqrt{\{2m(E - V)\}}}. \tag{1.176}$$

Hence, if the surface over which the characteristic function has the constant value S_0 is known, we can construct the neighbouring surface over which it has the value $S_0 + dS$. In this way we can form a particular solution of equation (1.174) extending over a given region of space, by starting with an arbitrary surface $S = S_0$ and determining successive neighbouring surfaces over which the characteristic function has constant values differing by infinitesimal amounts dS.

Since the gradient of S is perpendicular to surfaces of constant S, it follows from (1.175) that the trajectory of the particle is perpendicular to surfaces of constant S. Hence a family of possible trajectories of the particle can be obtained by constructing paths which are normal to the surfaces of constant S.

Hamilton's principal function W is given by

$$W(\mathbf{r}, t) = S(\mathbf{r}) - Et \qquad (1.177)$$

and so a surface of constant W travels between the two surfaces over which the characteristic function has the values S_0 and $S_0 + dS$ in a time interval

$$dt = \frac{dS}{E}. \qquad (1.178)$$

Hence the speed of propagation of the W surface is

$$\frac{ds}{dt} = \frac{E}{\sqrt{\{2m(E - V)\}}} = \frac{E}{mv}, \qquad (1.179)$$

where v is the speed of the particle.

We now turn to the special case of the motion of a free particle. Then the potential V vanishes at all points of space and so

$$(\nabla S)^2 = 2mE. \qquad (1.180)$$

Hence

$$S = \sqrt{(2mE)}\hat{\mathbf{n}}.\mathbf{r} \qquad (1.181)$$

and therefore the path of the particle is a straight line in the direction of the unit vector $\hat{\mathbf{n}}$ which is perpendicular to the planes of constant S. It then follows from (1.177) that Hamilton's principal function is given by

$$W = \sqrt{(2mE)}\hat{\mathbf{n}}.\mathbf{r} - Et. \qquad (1.182)$$

1.18 Geometrical mechanics

In this section we shall look at the motion of a dynamical system from the viewpoint of differential geometry. The *Riemannian geometry of* an n-dimensional space having $x_1, \ldots, x_n$ as coordinates is characterized by an infinitesimal line element ds which is given by the quadratic differential form

$$ds^2 = \sum_{i=1}^{n} \sum_{k=1}^{n} g_{ik} \, dx_i \, dx_k \tag{1.183}$$

where g_{ik}, known as the *metric tensor*, is symmetric so that

$$g_{ik} = g_{ki} \tag{1.184}$$

and the elements of the metric tensor are functions of $x_1, \ldots, x_n$.

For the geometrical interpretation of a dynamical system composed of N particles having masses m_i and rectangular Cartesian coordinates x_i, y_i, z_i ($i = 1, \ldots, N$), we choose a Riemannian space of $3N$ dimensions having a line element ds defined by

$$ds^2 = \sum_{i=1}^{N} m_i \{(dx_i)^2 + (dy_i)^2 + (dz_i)^2\}. \tag{1.185}$$

Since the speed v_i of the ith particle is given by

$$v_i^2 = \left(\frac{dx_i}{dt}\right)^2 + \left(\frac{dy_i}{dt}\right)^2 + \left(\frac{dz_i}{dt}\right)^2, \tag{1.186}$$

it follows that the total kinetic energy of the dynamical system can be expressed in the form

$$T = \tfrac{1}{2} \sum_{i=1}^{N} m_i v_i^2 = \tfrac{1}{2} \left(\frac{ds}{dt}\right)^2 \tag{1.187}$$

and so we may interpret T as the kinetic energy of a particle of unit mass moving in the $3N$-dimensional space.

If the generalized coordinates $q_1, \ldots, q_n$ completely specify the dynamical system, so that we may write

$$\begin{aligned} x_i &= x_i(q_1, \ldots, q_n), \\ y_i &= y_i(q_1, \ldots, q_n), \quad (i = 1, \ldots, N) \\ z_i &= z_i(q_1, \ldots, q_n), \end{aligned} \tag{1.188}$$

(1.185) can be put into the general Riemannian form for a line element

$$ds^2 = \sum_{r=1}^{n} \sum_{s=1}^{n} a_{rs}\, dq_r\, dq_s, \qquad (1.189)$$

where

$$a_{rs} = a_{sr} = \sum_{i=1}^{N} m_i\!\left(\frac{\partial x_i}{\partial q_r}\frac{\partial x_i}{\partial q_s} + \frac{\partial y_i}{\partial q_r}\frac{\partial y_i}{\partial q_s} + \frac{\partial z_i}{\partial q_r}\frac{\partial z_i}{\partial q_s}\right). \qquad (1.190)$$

For a conservative system with potential function V and total energy E, the action integral

$$A = 2\int T\, dt$$

may be expressed as

$$A = \int \sqrt{\{2(E-V)\}}\, ds \qquad (1.191)$$

using the result (1.187). The stationary property of this version of the action integral is known as the *Jacobi form* of the principle of least action and has already been derived for the case of a single particle in section 1.9. Introducing another line element $d\sigma$ defined by

$$d\sigma^2 = 2(E-V)\, ds^2 \qquad (1.192)$$

and substituting into (1.191) we get

$$A = \int d\sigma. \qquad (1.193)$$

It follows that the principle of least action requires that the dynamical system moves along the shortest path between any two points of the Riemannian space having line element $d\sigma$ given by (1.192). Such a shortest path is called a *geodesic line*.

In the absence of external forces, the potential function V vanishes at all points of space, in which case we have

$$A = \sqrt{(2E)} \int ds \qquad (1.194)$$

ANALYTICAL DYNAMICS 43

so that the least action principle is now equivalent to the statement that the dynamical system must move along a geodesic line in configuration space. Furthermore

$$\left(\frac{ds}{dt}\right)^2 = 2E, \tag{1.195}$$

and so the representative particle of unit mass moves with constant speed in configuration space.

1.19 Gravitational field

We begin our discussion of the gravitational field by considering a particle P of *gravitational mass* m moving under the gravitational attraction of a system of N particles P_i having gravitational masses m_i ($i = 1, \ldots, N$). According to *Newton's law of gravitation*, the force of attraction between two particles is proportional to the gravitational masses of the particles and is inversely proportional to the square of the distance between them. Hence the force acting on the particle P takes the form

$$-\gamma m \sum_{i=1}^{N} \frac{m_i}{r_i^3} \mathbf{r}_i, \tag{1.196}$$

where $\mathbf{r}_i$ is the position vector of particle P referred to particle P_i and γ is the constant of gravitation. By Newton's second law of motion, the acceleration $\mathbf{a}$ of the particle P relative to an inertial frame of reference is given by

$$m\mathbf{a} = -\gamma m \sum_{i=1}^{N} \frac{m_i}{r_i^3} \mathbf{r}_i \tag{1.197}$$

from which it follows at once that $\mathbf{a}$ is independent of the mass m, the assumption being made that the gravitational and inertial masses of P are identical.

Since the acceleration due to a gravitational field is independent of the mass of the particle which, as we have seen above, arises from the equivalence of the gravitational and inertial masses, it is impossible to distinguish between an acceleration resulting from a gravitational field and an acceleration arising from the choice of a non-inertial frame of reference, i.e. a reference frame relative to which a free particle appears accelerated. This assumption is known as the *principle of equivalence*.

We have already noted in section 1.12 that if we characterize an event by the spacial Cartesian coordinates x, y, z and the time t relative to an inertial frame of reference, the infinitesimal four-dimensional line element ds, given by

$$ds^2 = c^2 dt^2 - dx^2 - dy^2 - dz^2,$$

is invariant under a Lorentz transformation to the space-time coordinates of any other inertial frame of reference. Putting $ct = X^0$ and $x = X^1, y = X^2, z = X^3$, we may write

$$ds^2 = \sum_{\mu=0}^{3} \sum_{\nu=0}^{3} G_{\mu\nu} dX^\mu dX^\nu, \qquad (1.198)$$

where the elements of the tensor $G_{\mu\nu}$ are given by $G_{00} = 1$, $G_{11} = G_{22} = G_{33} = -1$ and $G_{\mu\nu} = 0$ for $\mu \neq \nu$. We now introduce the set of curvilinear coordinates x^μ ($\mu = 0, 1, 2, 3$), associated with a non-inertial frame of reference, defined by

$$x^\mu = x^\mu(X^0, X^1, X^2, X^3) \qquad (\mu = 0, 1, 2, 3) \qquad (1.199)$$

where the x^μ are continuous and differentiable functions of X^0, X^1, X^2, X^3. Inverting the set of relations (1.199) so that they take the form

$$X^\mu = X^\mu(x^0, x^1, x^2, x^3) \qquad (\mu = 0, 1, 2, 3), \qquad (1.200)$$

we see that

$$dX^\mu = \sum_{\nu=0}^{3} \frac{\partial X^\mu}{\partial x^\nu} dx^\nu$$

and hence, substituting into (1.198), we get

$$ds^2 = \sum_{\mu=0}^{3} \sum_{\nu=0}^{3} g_{\mu\nu} dx^\mu dx^\nu, \qquad (1.201)$$

where

$$g_{\mu\nu} = g_{\nu\mu} = \sum_{\mu'=0}^{3} \sum_{\nu'=0}^{3} G_{\mu'\nu'} \frac{\partial X^{\mu'}}{\partial x^\mu} \frac{\partial X^{\nu'}}{\partial x^\nu}. \qquad (1.202)$$

ANALYTICAL DYNAMICS

The tensor $g_{\mu\nu}$ is called the metric tensor and its elements are functions of the curvilinear coordinates x^0, x^1, x^2, x^3.

The metric tensor (1.202) associated with the above non-inertial frame of reference was obtained by making a coordinate transformation (1.199) from an inertial frame of reference and does not correspond to any actual gravitational field. An actual gravitational field arising from the presence of matter is associated with a metric tensor $g_{\mu\nu}$ such that no coordinate transformation of the form (1.199) is able to convert the general quadratic expression for ds^2 given by (1.201) into the form (1.198) corresponding to a *Euclidean* or flat four-dimensional space, except within an infinitesimal region. The space determined by the metric tensor is then called a *non-Euclidean* or curved space and is described by a general Riemannian geometry.

1.20 Geodesic lines in Riemannian space

A path in a Riemannian space of four dimensions with curvilinear coordinates x^μ ($\mu = 0, 1, 2, 3$) can be represented parametrically by the set of equations

$$x^\mu = x^\mu(\theta) \qquad (\mu = 0, 1, 2, 3), \tag{1.203}$$

where the x^μ are functions of the parameter θ. Choosing a Lagrangian function $L(x^\mu, \dot{x}^\mu)$ which depends upon the coordinates x^μ and their derivatives $\dot{x}^\mu$ with respect to θ, we investigate the conditions for the integral

$$I = \int_{\theta_1}^{\theta_2} L(x^\mu, \dot{x}^\mu)\, d\theta \tag{1.204}$$

to be stationary with respect to an arbitrary infinitesimal variation of the coordinates

$$x^\mu(\theta) \to x^\mu(\theta) + \delta x^\mu(\theta) \qquad (\mu = 0, 1, 2, 3) \tag{1.205}$$

satisfying the terminal conditions

$$\delta x^\mu(\theta_1) = \delta x^\mu(\theta_2) = 0 \qquad (\mu = 0, 1, 2, 3) \tag{1.206}$$

at the end-points $\theta = \theta_1$ and θ_2. The change in I due to this variation in the coordinates $x^\mu(\theta)$ is given by

$$\delta I = \int_{\theta_1}^{\theta_2} \sum_{\mu=0}^{3} \left(\frac{\partial L}{\partial x^\mu} \delta x^\mu + \frac{\partial L}{\partial \dot{x}^\mu} \delta \dot{x}^\mu \right) d\theta$$

which may be written in the form

$$\delta I = \int_{\theta_1}^{\theta_2} \sum_{\mu=0}^{3} \left\{ \frac{\partial L}{\partial x^\mu} \delta x^\mu + \frac{\partial L}{\partial \dot{x}^\mu} \frac{d}{d\theta}(\delta x^\mu) \right\} d\theta$$

since

$$\delta \dot{x}^\mu = \frac{d}{d\theta}(\delta x^\mu).$$

Integrating by parts and using the terminal conditions (1.206), we arrive at the result

$$\delta I = \int_{\theta_1}^{\theta_2} \sum_{\mu=0}^{3} \left\{ \frac{\partial L}{\partial x^\mu} - \frac{d}{d\theta}\left(\frac{\partial L}{\partial \dot{x}^\mu}\right) \right\} \delta x^\mu \, d\theta \qquad (1.207)$$

and so the integral I is stationary if

$$\frac{\partial L}{\partial x^\mu} - \frac{d}{d\theta}\left(\frac{\partial L}{\partial \dot{x}^\mu}\right) = 0 \qquad (\mu = 0, 1, 2, 3). \qquad (1.208)$$

These are the Euler–Lagrange equations and the method of their derivation is closely analogous to that employed in section 1.6, where it was shown that Lagrange's dynamical equations follow as the result of the stationary property of an integral over time.

The distance along a path connecting two points of the Riemannian space is given by

$$\int_{\theta_1}^{\theta_2} \frac{ds}{d\theta} d\theta. \qquad (1.209)$$

This path is called a geodesic line if the integral (1.209) is stationary. Hence, taking

$$L = \frac{ds}{d\theta} = \left(\sum_{\mu=0}^{3} \sum_{\nu=0}^{3} g_{\mu\nu} \dot{x}^\mu \dot{x}^\nu \right)^{1/2} \qquad (1.210)$$

as the Lagrangian function, it follows from the Euler–Lagrange equations (1.208) that the geodesic line is determined by the equations

$$\frac{\frac{1}{2}\sum_{\mu=0}^{3}\sum_{\nu=0}^{3}(\partial g_{\mu\nu}/\partial x^{\lambda})\dot{x}^{\mu}\dot{x}^{\nu}}{\left(\sum_{\mu=0}^{3}\sum_{\nu=0}^{3}g_{\mu\nu}\dot{x}^{\mu}\dot{x}^{\nu}\right)^{1/2}} - \frac{d}{d\theta}\left\{\frac{\sum_{\mu=0}^{3}g_{\mu\lambda}\dot{x}^{\mu}}{\left(\sum_{\mu=0}^{3}\sum_{\nu=0}^{3}g_{\mu\nu}\dot{x}^{\mu}\dot{x}^{\nu}\right)^{1/2}}\right\} = 0. \tag{1.211}$$

By putting the parameter θ equal to the path distance s, (1.210) yields

$$\sum_{\mu=0}^{3}\sum_{\nu=0}^{3} g_{\mu\nu} \frac{dx^{\mu}}{ds}\frac{dx^{\nu}}{ds} = 1 \tag{1.212}$$

and the equations determining the geodesic line become

$$\frac{d}{ds}\left(\sum_{\mu=0}^{3} g_{\mu\lambda}\frac{dx^{\mu}}{ds}\right) = \frac{1}{2}\sum_{\mu=0}^{3}\sum_{\nu=0}^{3}\frac{\partial g_{\mu\nu}}{\partial x^{\lambda}}\frac{dx^{\mu}}{ds}\frac{dx^{\nu}}{ds} \tag{1.213}$$

which may be rewritten in the form

$$\sum_{\mu=0}^{3} g_{\mu\lambda}\frac{d^{2}x^{\mu}}{ds^{2}} + \sum_{\mu=0}^{3}\sum_{\nu=0}^{3} \Gamma_{\lambda,\mu\nu}\frac{dx^{\mu}}{ds}\frac{dx^{\nu}}{ds} = 0, \tag{1.214}$$

where

$$\Gamma_{\lambda,\mu\nu} = \frac{1}{2}\left(\frac{\partial g_{\mu\lambda}}{\partial x^{\nu}} + \frac{\partial g_{\nu\lambda}}{\partial x^{\mu}} - \frac{\partial g_{\mu\nu}}{\partial x^{\lambda}}\right). \tag{1.215}$$

We now introduce the tensor $g^{\mu\nu}$ whose elements satisfy the relation

$$\sum_{\lambda=0}^{3} g_{\mu\lambda}g^{\lambda\nu} = \delta_{\mu}^{\nu}, \tag{1.216}$$

where δ_{μ}^{ν} is the Kronecker delta symbol which vanishes when $\mu \neq \nu$ and is equal to unity when $\mu = \nu$. The tensor $g^{\mu\nu}$ is therefore the reciprocal of the tensor $g_{\mu\nu}$. Multiplying equations (1.214) by $g^{\lambda\rho}$ and summing over λ, we get

$$\frac{d^{2}x^{\rho}}{ds^{2}} + \sum_{\lambda=0}^{3}\sum_{\mu=0}^{3}\sum_{\nu=0}^{3}\frac{1}{2}\left(\frac{\partial g_{\mu\lambda}}{\partial x^{\nu}} + \frac{\partial g_{\nu\lambda}}{\partial x^{\mu}} - \frac{\partial g_{\mu\nu}}{\partial x^{\lambda}}\right)g^{\lambda\rho}\frac{dx^{\mu}}{ds}\frac{dx^{\nu}}{ds} = 0 \tag{1.217}$$

for the equations of the geodesic line, which may be expressed in the form

$$\frac{d^2x^\rho}{ds^2} + \sum_{\mu=0}^{3}\sum_{\nu=0}^{3} \Gamma^\rho_{\mu\nu} \frac{dx^\mu}{ds}\frac{dx^\nu}{ds} = 0, \qquad (1.218)$$

where

$$\Gamma^\rho_{\mu\nu} = \tfrac{1}{2} \sum_{\lambda=0}^{3} \left(\frac{\partial g_{\mu\lambda}}{\partial x^\nu} + \frac{\partial g_{\nu\lambda}}{\partial x^\mu} - \frac{\partial g_{\mu\nu}}{\partial x^\lambda}\right) g^{\lambda\rho} = \sum_{\lambda=0}^{3} \Gamma_{\lambda,\mu\nu} g^{\lambda\rho}. \qquad (1.219)$$

$\Gamma_{\lambda,\mu\nu}$ and $\Gamma^\rho_{\mu\nu}$ are known as *Christoffel three-index symbols*.

1.21 Motion of a particle in a gravitational field

We have shown on p. 31 that the Lagrangian function for a free particle of rest mass m_0 moving with uniform speed v relative to an inertial frame of reference is given by

$$L = -m_0 c^2 \sqrt{\{1 - (v/c)^2\}}$$

which we may rewrite in the form

$$L = -m_0 c \frac{ds}{dt},$$

where ds is an infinitesimal four-dimensional line element.

In view of the discussion in the preceding section we must take, therefore,

$$L = -m_0 c \frac{ds}{d\theta} = -m_0 c \left(\sum_{\mu=0}^{3}\sum_{\nu=0}^{3} g_{\mu\nu} \dot{x}^\mu \dot{x}^\nu\right)^{1/2} \qquad (1.220)$$

as the Lagrangian function for a particle of rest mass m_0 moving under the influence of a gravitational field characterized by a metric tensor $g_{\mu\nu}$. The path of the particle is then described by the equations (1.218). Since the components of the acceleration of the particle referred to the four-dimensional space are d^2x^ρ/ds^2 ($\rho = 0, 1, 2, 3$), we see that the four components of the gravitational force acting on the particle must be

$$-m_0 \sum_{\mu=0}^{3}\sum_{\nu=0}^{3} \Gamma^\rho_{\mu\nu} \frac{dx^\mu}{ds}\frac{dx^\nu}{ds}.$$

Introducing the four components of momentum defined by

$$p_\mu = \frac{\partial L}{\partial \dot{x}^\mu} = \frac{m_0^2 c^2 \sum_{\nu=0}^{3} g_{\mu\nu} \dot{x}^\nu}{L}, \quad (1.221)$$

we see that

$$\sum_{\mu=0}^{3} p_\mu g^{\mu\nu} = \frac{m_0^2 c^2 \dot{x}^\nu}{L}$$

and so

$$\sum_{\mu=0}^{3} \sum_{\nu=0}^{3} p_\mu g^{\mu\nu} p_\nu = m_0^2 c^2. \quad (1.222)$$

Now putting

$$p_\mu = \frac{\partial W}{\partial x^\mu} \quad (\mu = 0, 1, 2, 3), \quad (1.223)$$

where W is Hamilton's principal function, we arrive at the Hamilton–Jacobi equation

$$\sum_{\mu=0}^{3} \sum_{\nu=0}^{3} \frac{\partial W}{\partial x^\mu} g^{\mu\nu} \frac{\partial W}{\partial x^\nu} = m_0^2 c^2. \quad (1.224)$$

For a free particle moving relative to an inertial frame of reference we have $g^{\mu\nu} = G^{\mu\nu}$, $x^\mu = X^\mu$, and so (1.224) takes the form

$$(\nabla W)^2 - \frac{1}{c^2}\left(\frac{\partial W}{\partial t}\right)^2 + m_0^2 c^2 = 0. \quad (1.225)$$

Since the Christoffel symbols vanish for this case, the path of the particle is a straight line given by

$$\frac{d^2 X^\rho}{ds^2} = 0 \quad (\rho = 0, 1, 2, 3) \quad (1.226)$$

which is, of course, the equation of a geodesic in a four-dimensional Euclidean space.

To conclude, we note that in the case of a light ray *in vacuo* we have along its path

$$ds^2 = c^2 dt^2 - dx^2 - dy^2 - dz^2 = \sum_{\mu=0}^{3} \sum_{\nu=0}^{3} G_{\mu\nu} dX^\mu dX^\nu = 0$$

when referred to an inertial frame of reference, and so we can no longer use s as a parameter. Taking

$$L = \left(\frac{ds}{d\theta}\right)^2 = \sum_{\mu=0}^{3}\sum_{\nu=0}^{3} g_{\mu\nu} \frac{dx^\mu}{d\theta}\frac{dx^\nu}{d\theta} \qquad (1.227)$$

as the Lagrangian function, the Euler–Lagrange equations (1.208) become

$$\frac{d}{d\theta}\left(\sum_{\mu=0}^{3} g_{\mu\lambda}\frac{dx^\mu}{d\theta}\right) = \tfrac{1}{2}\sum_{\mu=0}^{3}\sum_{\nu=0}^{3} \frac{\partial g_{\mu\nu}}{\partial x^\lambda}\frac{dx^\mu}{d\theta}\frac{dx^\nu}{d\theta} \qquad (1.228)$$

which we may rewrite in the form

$$\frac{d^2 x^\rho}{d\theta^2} + \sum_{\mu=0}^{3}\sum_{\nu=0}^{3} \Gamma^\rho_{\mu\nu} \frac{dx^\mu}{d\theta}\frac{dx^\nu}{d\theta} = 0. \qquad (1.229)$$

Since $ds = 0$ along the path of the light signal, we must have also

$$\sum_{\mu=0}^{3}\sum_{\nu=0}^{3} g_{\mu\nu}\frac{dx^\mu}{d\theta}\frac{dx^\nu}{d\theta} = 0. \qquad (1.230)$$

Equations (1.229) and (1.230) together determine the path of a light ray *in vacuo*.

CHAPTER 2

Optics, Wave Mechanics and Quantum Mechanics

In Chapter 1 we investigated the variational principles of dynamics from the classical non-relativistic aspect as well as from the viewpoint of the theory of relativity. We now turn our attention to the application of the variational principle to optics and establish Fermat's principle of least time, showing that there exists a close relationship between this principle and the principle of least action in dynamics. This then leads us to a general discussion of the analogy between optics and dynamics as seen through the similarity between the equation of geometrical optics and the Hamilton–Jacobi equation, followed by a development of the equations of wave mechanics from the extension to dynamics of the relationship between the equations of geometrical optics and wave motion. Finally, we give a short survey of the mathematical foundations of quantum mechanics and show that it is possible to establish an analogue of Hamilton's principle which provides the quantum dynamical equations of motion in a Lagrangian form.

The fundamental principle of geometrical optics is Huygens' principle and so we begin this chapter by studying its consequences.

2.1 Huygens' principle

This principle asserts that each point of a wave front behaves as a light source emitting a spherical wave with the speed of light u in the medium within which the light signal is being propagated, and that the superposition of these waves produces a progressive wave front. If the *refractive index* of the medium is $n(\mathbf{r})$ at the point with position vector $\mathbf{r}$ then

$$u = \frac{c}{n}, \tag{2.1}$$

where c is the speed of light *in vacuo*.

Suppose that there is a light disturbance at time $t = 0$ over an arbitrary basic surface. At the end of an infinitesimal time interval dt the light disturbance will have spread to an adjacent surface, the distance between the pair of surfaces in the direction of the normal at any given point of the basic surface being

$$ds = \frac{c}{n} dt. \tag{2.2}$$

In successive intervals of time dt the light disturbance will spread to further surfaces, each pair of adjacent surfaces being separated by the distance $(c/n) dt$.

We now introduce a function $\chi(\mathbf{r})$ which has a constant value over each one of these surfaces, equal to the time taken by the light disturbance to travel to the given surface from the basic surface over which χ is chosen to vanish. Then we have

$$dt = |\nabla \chi| \, ds \tag{2.3}$$

from which it follows by the use of (2.2) that

$$(\nabla \chi)^2 = \frac{n^2}{c^2}. \tag{2.4}$$

This is the fundamental differential equation of geometrical optics. It is closely similar to the equation

$$(\nabla S)^2 = 2m(E - V) \tag{2.5}$$

for Hamilton's characteristic function S associated with a particle of mass m moving under the influence of a potential $V(\mathbf{r})$, into which it may be easily transformed by taking a refractive index given by

$$\frac{n}{c} \sim \sqrt{\{2m(E - V)\}}. \tag{2.6}$$

OPTICS, WAVE MECHANICS AND QUANTUM MECHANICS

Thus we see that there exists a close resemblance between dynamics and geometrical optics.

2.2 Fermat's principle of least time

We now consider the path of a light ray which passes from a point P to a point Q. Denoting P by P_0 and Q by P_n we let $P_0 P_1 \ldots P_r P_{r+1} \ldots P_n$ be the actual path of the light ray and $P_0 P_1' \ldots P_r' P_{r+1}' \ldots P_n$ be any adjacent path having the same terminal points. Also we choose the points P_r and P_r' so that they lie on a common wave surface given by $\chi = t_r$, where t_r is the time taken by the light disturbance to reach this surface from the surface $\chi = 0$ which contains P.

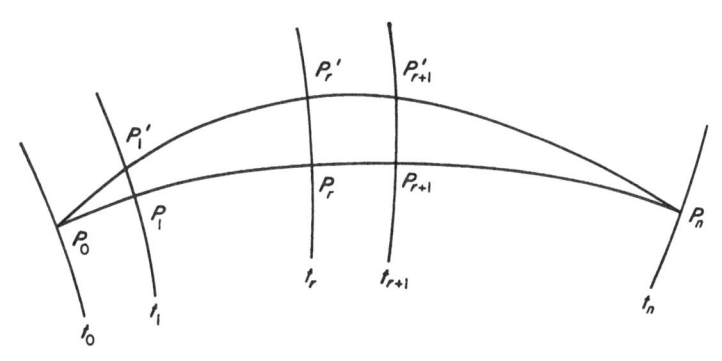

Fig. 1. Actual path $P_0 P_1 \ldots P_r P_{r+1} \ldots P_n$ of a light ray, and an adjacent path $P_0 P_1' \ldots P_r' P_{r+1}' \ldots P_n$ between the same terminal points $P \equiv P_0$ and $Q \equiv P_n$, employed in establishing Fermat's principle of least time. The points P_r and P_r' lie on a common wave surface given by $\chi = t_r$.

The actual path element $P_r P_{r+1}$ is normal to the wave surface $\chi = t_r$ whereas in general the adjacent path element $P_r' P_{r+1}'$ will not be. If θ_r is the angle which $P_r' P_{r+1}'$ makes with the normal to the wave surface, the time taken by the light ray to travel from P_r' to P_{r+1}' will be

$$\frac{t_{r+1} - t_r}{\cos \theta_r}$$

which is necessarily greater than or equal to the time $t_{r+1} - t_r$

taken by the light disturbance to travel along the actual path element $P_r P_{r+1}$. It follows that the time taken by the light to traverse the actual path from P to Q is less than or equal to that for any adjacent path. This is *Fermat's principle of least time* which we may write in the form

$$\delta \int_P^Q \frac{ds}{u} \geq 0. \qquad (2.7)$$

However, if a wave surface degenerates into a point or line at some point of the actual path of a light ray, there will be a region surrounding this point where the adjacent path will not intersect the wave surfaces, in which case the above proof fails because θ becomes complex and the minimum condition no longer applies. Since the length of the adjacent path differs from the length of the actual path of the light ray by a quantity of the second order of smallness, it follows that the minimum condition can be replaced in this case by the requirement that the time difference is of the second order of small quantities. Neglecting such quantities we obtain

$$\delta \int_P^Q \frac{ds}{u} = 0 \qquad (2.8)$$

or

$$\delta \int_P^Q \frac{n}{c} ds = 0. \qquad (2.9)$$

If we make the substitution (2.6) again, Fermat's principle (2.9) converts into the Jacobi form of the action principle

$$\delta \int_P^Q \sqrt{\{2m(E - V)\}}\, ds = 0, \qquad (2.10)$$

which reinforces the analogy between dynamics and geometrical optics.

In the previous section we defined the function χ to be the time taken by the light disturbance to travel from an arbitrary basic wave surface, given by $\chi = 0$, to a second wave surface. If P is any point of this basic surface, we may write χ in the form

$$\chi = \int_P^Q \frac{n}{c} ds, \qquad (2.11)$$

where the path of integration connecting the two points P and Q is normal to the wave surfaces.

Referring to equation (2.5) we see that Hamilton's characteristic function S may be expressed analogously as an action integral

$$S = \int_P^Q \sqrt{\{2m(E - V)\}}\, ds, \qquad (2.12)$$

where the integration is performed along the actual trajectory of a particle having total energy E whose first point P lies on an arbitrarily chosen basic surface. If we regard S as a function of position of the end point Q, we see that S vanishes over the basic surface and that the surfaces of constant S correspond to surfaces of constant action.

2.3 Wave motion

The general equation for a classical type of wave motion has the form

$$\nabla^2 \Psi = \frac{1}{u^2} \frac{\partial^2 \Psi}{\partial t^2}, \qquad (2.13)$$

where the scalar wave function Ψ describing the motion depends upon the time t as well as upon the position in space given by the vector $\mathbf{r}$.

For a monochromatic wave having *frequency* ν we may put

$$\Psi(\mathbf{r}, t) = \psi(\mathbf{r})\, e^{-2\pi i \nu t} \qquad (2.14)$$

where the spacial wave function $\psi(\mathbf{r})$ satisfies the equation

$$\nabla^2 \psi + \frac{4\pi^2 \nu^2}{u^2} \psi = 0. \qquad (2.15)$$

To solve this equation we make the substitution

$$\psi = A\, e^{2\pi i \phi}. \qquad (2.16)$$

For a homogeneous medium, u is independent of position, in which case the amplitude A can be taken as constant and then

$$\phi = \frac{\nu}{u} \hat{\mathbf{n}}.\mathbf{r}, \qquad (2.17)$$

where $\hat{\mathbf{n}}$ is a unit vector in the direction of propagation of the wave. Hence

$$\Psi(\mathbf{r}, t) = A\, e^{i\Phi(\mathbf{r},\,t)} \tag{2.18}$$

where Φ is the *phase*, defined according to the formula

$$\Phi(\mathbf{r}, t) = 2\pi\nu\left(\frac{\hat{\mathbf{n}}\cdot\mathbf{r}}{u} - t\right). \tag{2.19}$$

Consequently the planes of constant phase move with speed u which we therefore call the *phase velocity*.

Let us now suppose that u is a function of position. On substituting (2.16) into equation (2.15), we obtain

$$\nabla^2 A + 4\pi i \nabla A \cdot \nabla\phi + 2\pi i A \nabla^2 \phi - 4\pi^2 A (\nabla\phi)^2 + \frac{4\pi^2 \nu^2}{u^2} A = 0$$

and since A and ϕ are both real this means that

$$\nabla^2 A + 4\pi^2 A \left\{\frac{\nu^2}{u^2} - (\nabla\phi)^2\right\} = 0. \tag{2.20}$$

At high frequencies ν we have

$$\frac{\nabla^2 A}{4\pi^2 A} \ll \frac{\nu^2}{u^2} \tag{2.21}$$

which is equivalent to supposing that the properties of the medium in which the waves are being propagated do not change significantly in a distance of the order of the *wave length*

$$\lambda = \frac{u}{\nu} \tag{2.22}$$

which is small when ν is large. With this assumption (2.20) reduces to the approximate equation

$$(\nabla\phi)^2 = \frac{\nu^2}{u^2}. \tag{2.23}$$

On setting $\phi = \nu\chi$ this becomes the fundamental equation of geometrical optics (2.4) which we have already shown also follows as a consequence of Huygens' principle.

2.4 Schrödinger equation

In this section we shall carry the analogy between optics and mechanics a stage further by supposing that the Hamilton–Jacobi equation in dynamics results from making a small wavelength approximation to an appropriate wave equation describing a material particle.

We consider a free particle having mass m and energy E moving in a straight line with speed v in the direction of the unit vector $\hat{\mathbf{n}}$. From equations (1.177) and (1.181) we see that Hamilton's principal function for the particle is given by

$$W(\mathbf{r}, t) = S(\mathbf{r}) - Et,$$

where

$$S(\mathbf{r}) = mv\hat{\mathbf{n}}.\mathbf{r}$$

is Hamilton's characteristic function.

We now associate with this particle a wave propagated in the direction of $\hat{\mathbf{n}}$ having phase

$$\Phi(\mathbf{r}, t) = 2\pi(\phi - \nu t), \tag{2.24}$$

where

$$\phi = \frac{\hat{\mathbf{n}}.\mathbf{r}}{\lambda}, \tag{2.25}$$

ν being the frequency and λ the wave length. From the similarity between the equation

$$(\nabla S)^2 = 2mE$$

for Hamilton's characteristic function and the equation of geometrical optics (2.23), we are led to put

$$W = \frac{h}{2\pi} \Phi, \tag{2.26}$$

where h is a constant of proportionality. It then follows that

$$E = h\nu \tag{2.27}$$

and

$$\lambda = \frac{h}{mv}. \tag{2.28}$$

h is known as *Planck's constant* and λ as the *de Broglie wave length*.

Suppose now that the particle is moving under the influence of a field with potential $V(\mathbf{r})$. Again assuming that W and Φ are related by equation (2.26), we obtain

$$S = h\phi. \tag{2.29}$$

But

$$(\nabla S)^2 = 2m(E - V)$$

and hence, using (2.23), we get

$$\frac{v^2}{u^2} = \frac{2m(E - V)}{h^2}. \tag{2.30}$$

Therefore the wave equation (2.15) becomes

$$\nabla^2 \psi + \frac{8\pi^2 m}{h^2} (E - V)\psi = 0 \tag{2.31}$$

which is known as *Schrödinger's equation*, the wave equation of quantum mechanics. Thus we see that Schrödinger's equation bears the same relation to the equation for Hamilton's characteristic function as the scalar wave equation of optics bears to the equation of geometrical optics.

If Planck's constant h is allowed to approach zero, ν becomes large and then we revert to classical mechanics, as can be readily verified by pursuing a similar argument to that used in section 2.3 to derive the equation of geometrical optics from the scalar wave equation.

Because of the relation (2.27) between the frequency ν and the energy E, equation (2.14) for the total wave function takes the form

$$\Psi(\mathbf{r}, t) = \psi(\mathbf{r}) \, e^{-iEt/\hbar} \tag{2.32}$$

and so (2.31) may be written as

$$-\frac{\hbar^2}{2m} \nabla^2 \Psi + V\Psi = i\hbar \frac{\partial \Psi}{\partial t}, \tag{2.33}$$

where $h/2\pi$ is denoted by $\hbar$. Equation (2.33) is the time-dependent form of the Schrödinger equation. If H denotes the operator obtained from the Hamiltonian function

$$H(\mathbf{r}, \mathbf{p}) = \frac{1}{2m}\mathbf{p}^2 + V(\mathbf{r})$$

for a particle by making the substitution

$$\mathbf{p} = -i\hbar\nabla, \quad (2.34)$$

we may express the time-dependent Schrödinger equation in the form

$$H\Psi = i\hbar\frac{\partial \Psi}{\partial t} \quad (2.35)$$

while the time-independent Schrödinger equation (2.31) becomes

$$H\psi = E\psi. \quad (2.36)$$

The same procedure may be adopted to obtain the Schrödinger equation for a many particle system. Thus if

$$H(\mathbf{r}_1, \ldots, \mathbf{r}_N; \mathbf{p}_1, \ldots, \mathbf{p}_N) = \sum_{i=1}^{N} \frac{1}{2m_i}\mathbf{p}_i^2 + V(\mathbf{r}_1, \ldots, \mathbf{r}_N) \quad (2.37)$$

is the classical Hamiltonian function for a system composed of N particles having masses m_i, position vectors $\mathbf{r}_i$ and momenta $\mathbf{p}_i$, the Hamiltonian operator

$$H = -\sum_{i=1}^{N} \frac{\hbar^2}{2m_i}\nabla_i^2 + V(\mathbf{r}_1, \ldots, \mathbf{r}_N) \quad (2.38)$$

can be obtained by putting

$$\mathbf{p}_i = -i\hbar\nabla_i \quad (2.39)$$

in (2.37), in which case equations (2.35) and (2.36) are once again the appropriate time-dependent and time-independent Schrödinger equations for the system.

Substituting

$$\psi = A\, e^{iS/\hbar} \quad (2.40)$$

into (2.36) and making the small wavelength approximation that

A is a slowly varying function of position, we get

$$\sum_{i=1}^{N} \frac{1}{2m_i} (\nabla_i S)^2 + V = E \qquad (2.41)$$

which is just the equation (1.169) for Hamilton's characteristic function S corresponding to the classical Hamiltonian function (2.37) for a many particle system.

2.5 Klein–Gordon equation

We have shown in section 1.13 that relative to an observer situated at the origin of an inertial frame of reference S measuring time by means of a clock at rest in S, the total energy E and the momentum $\mathbf{p}$ of a free particle having rest mass m_0 are related by the formula

$$E = c\sqrt{(m_0^2 c^2 + \mathbf{p}^2)}.$$

Since the Hamiltonian function H is equal to the total energy, we have

$$H = c\sqrt{(m_0^2 c^2 + \mathbf{p}^2)} \qquad (2.42)$$

and so the Hamilton–Jacobi equation for a relativistic free particle is

$$\frac{\partial W}{\partial t} + c\sqrt{\{m_0^2 c^2 + (\nabla W)^2\}} = 0$$

which may be rewritten in the form

$$(\nabla W)^2 - \frac{1}{c^2}\left(\frac{\partial W}{\partial t}\right)^2 + m_0^2 c^2 = 0. \qquad (2.43)$$

By proceeding in this way we have regained the equation (1.225) which we derived in section 1.21 by employing a rather different approach. This equation is the classical approximation to the *Klein-Gordon equation*

$$\nabla^2 \Psi - \frac{1}{c^2} \frac{\partial^2 \Psi}{\partial t^2} = \frac{m_0^2 c^2}{\hbar^2} \Psi \qquad (2.44)$$

from which it may be obtained by substituting

$$\Psi = A\, e^{iW/\hbar} \qquad (2.45)$$

and allowing Planck's constant h to tend to zero.

OPTICS, WAVE MECHANICS AND QUANTUM MECHANICS

A comparison between (2.43) and (2.44) shows that the Hamilton–Jacobi equation (2.43) can be readily transformed into the wave equation by replacing ∇W by $-i\hbar\nabla$ and $\partial W/\partial t$ by $-i\hbar(\partial/\partial t)$ and operating on the wave function Ψ.

In the presence of an electromagnetic field characterized by a scalar potential V and vector potential $\mathbf{A}$, the Hamiltonian function of a particle is again equal to the total energy E and so, from (1.135) we get

$$H = c\sqrt{\left\{m_0^2 c^2 + \left(\mathbf{p} - \frac{e}{c}\mathbf{A}\right)^2\right\}} + eV. \qquad (2.46)$$

The corresponding Hamilton–Jacobi equation is therefore

$$\frac{\partial W}{\partial t} + eV + c\sqrt{\left\{m_0^2 c^2 + \left(\nabla W - \frac{e}{c}\mathbf{A}\right)^2\right\}} = 0$$

which we may rewrite as

$$\left(\nabla W - \frac{e}{c}\mathbf{A}\right)^2 - \frac{1}{c^2}\left(\frac{\partial W}{\partial t} + eV\right)^2 + m_0^2 c^2 = 0. \qquad (2.47)$$

This is the classical approximation to the equation

$$\left[\left(\frac{\partial}{\partial x} - \frac{ie}{\hbar c}A_x\right)^2 + \left(\frac{\partial}{\partial y} - \frac{ie}{\hbar c}A_y\right)^2 + \left(\frac{\partial}{\partial z} - \frac{ie}{\hbar c}A_z\right)^2 - \frac{1}{c^2}\left(\frac{\partial}{\partial t} + \frac{ie}{\hbar}V\right)^2\right]\Psi = \frac{m_0^2 c^2}{\hbar^2}\Psi. \qquad (2.48)$$

2.6 Dirac equation

The time-dependent Schrödinger equation (2.33) which describes nonrelativistic particles is of the first order in the time derivative but is of the second order in the space derivatives. To obtain a relativistic generalization of this equation, we used the classical Hamiltonian function (2.42) for a relativistic particle which yielded the Klein–Gordon equation (2.44) as shown in the previous section. However this equation involves second order space and time derivatives and therefore cannot be regarded as a satisfactory relativistic generalization of the Schrödinger equation. The correct equation for a relativistic particle should be of the first order in both the space and time derivatives and hence requires a Hamiltonian function which is linear in the momenta. Such a Hamiltonian function may be written in the form

$$H = -c\boldsymbol{\alpha}\cdot\mathbf{p} - \beta m_0 c^2. \tag{2.49}$$

The corresponding Hamilton–Jacobi equation is accordingly

$$\frac{\partial W}{\partial t} - c\boldsymbol{\alpha}\cdot\nabla W - \beta m_0 c^2 = 0 \tag{2.50}$$

which is the classical approximation to the *Dirac equation*

$$\left(i\hbar\frac{\partial}{\partial t} - i\hbar c\boldsymbol{\alpha}\cdot\nabla + \beta m_0 c^2\right)\Psi = 0. \tag{2.51}$$

The quantities $\boldsymbol{\alpha}$ and β are to be determined by requiring that the solution of the Dirac equation must also be a solution of the Klein–Gordon equation. Multiplying (2.51) on the left by

$$i\hbar\frac{\partial}{\partial t} + i\hbar c\boldsymbol{\alpha}\cdot\nabla - \beta m_0 c^2,$$

and letting $\alpha_x, \alpha_y, \alpha_z$ be the Cartesian components of $\boldsymbol{\alpha}$ we obtain

$$\begin{aligned}\Bigg[-\hbar^2\frac{\partial^2}{\partial t^2} & \\ + \hbar^2 c^2\bigg\{\alpha_x^2\frac{\partial^2}{\partial x^2} &+ \alpha_y^2\frac{\partial^2}{\partial y^2} + \alpha_z^2\frac{\partial^2}{\partial z^2} + (\alpha_x\alpha_y + \alpha_y\alpha_x)\frac{\partial^2}{\partial x\,\partial y} \\ + (\alpha_y\alpha_z + \alpha_z\alpha_y)\frac{\partial^2}{\partial y\,\partial z} &+ (\alpha_z\alpha_x + \alpha_x\alpha_z)\frac{\partial^2}{\partial z\,\partial x}\bigg\} \\ + i\hbar m_0 c^3\bigg\{(\alpha_x\beta + \beta\alpha_x)\frac{\partial}{\partial x} &+ (\alpha_y\beta + \beta\alpha_y)\frac{\partial}{\partial y} + (\alpha_z\beta + \beta\alpha_z)\frac{\partial}{\partial z}\bigg\}\Bigg]\Psi \\ &= m_0^2 c^4 \beta^2 \Psi\end{aligned}$$

which agrees with (2.44) if

$$\alpha_x^2 = \alpha_y^2 = \alpha_z^2 = \beta^2 = 1,$$
$$\alpha_x\alpha_y + \alpha_y\alpha_x = \alpha_y\alpha_z + \alpha_z\alpha_y = \alpha_z\alpha_x + \alpha_x\alpha_z = 0, \quad (2.52)$$
$$\alpha_x\beta + \beta\alpha_x = \alpha_y\beta + \beta\alpha_y = \alpha_z\beta + \beta\alpha_z = 0.$$

Evidently α_x, α_y, α_z and β cannot be numbers. However, we can represent them by matrices. The matrices of lowest rank which satisfy all the conditions (2.52) are the 4×4 matrices

$$\alpha_x = \begin{pmatrix} 0 & 0 & 0 & 1 \\ 0 & 0 & 1 & 0 \\ 0 & 1 & 0 & 0 \\ 1 & 0 & 0 & 0 \end{pmatrix}, \quad \alpha_y = \begin{pmatrix} 0 & 0 & 0 & -i \\ 0 & 0 & i & 0 \\ 0 & -i & 0 & 0 \\ i & 0 & 0 & 0 \end{pmatrix},$$

$$\alpha_z = \begin{pmatrix} 0 & 0 & 1 & 0 \\ 0 & 0 & 0 & -1 \\ 1 & 0 & 0 & 0 \\ 0 & -1 & 0 & 0 \end{pmatrix}, \quad \beta = \begin{pmatrix} 1 & 0 & 0 & 0 \\ 0 & 1 & 0 & 0 \\ 0 & 0 & -1 & 0 \\ 0 & 0 & 0 & -1 \end{pmatrix}. \quad (2.53)$$

The solution of equation (2.51) must now be expressed in the form of a 4-component column matrix

$$\Psi = \begin{pmatrix} \Psi_1 \\ \Psi_2 \\ \Psi_3 \\ \Psi_4 \end{pmatrix}. \quad (2.54)$$

In the presence of an electromagnetic field determined by a scalar potential V and a vector potential $\mathbf{A}$, it is clear that the Dirac equation takes the form

$$\left[i\hbar\left(\frac{\partial}{\partial t} + \frac{ie}{\hbar}V\right) - i\hbar c\boldsymbol{\alpha}\cdot\left(\nabla - \frac{ie}{\hbar c}\mathbf{A}\right) + \beta m_0 c^2\right]\Psi = 0, \quad (2.55)$$

in analogous fashion to the case of the Klein–Gordon equation discussed in the previous section.

2.7 Quantum mechanics

Our final considerations in the present chapter are concerned with the role played by Hamilton's principle in quantum mechanics. However, before proceeding to the examination of this principle, we shall need to give a brief summary of the mathematical foundations of quantum mechanics to which we shall now devote our attention.

State vectors and linear operators

We begin by associating a *state vector*, which we shall denote by $|a\rangle$, with each state of a quantum dynamical system. We shall suppose that if this vector is multiplied by any nonvanishing complex number c, the resulting vector $c|a\rangle$ will correspond to the same state. Then the *principle of superposition of states* can be expressed mathematically by requiring that if a given state arises from the superposition of a number of other states associated with vectors $|a_1\rangle, |a_2\rangle, \ldots, |a_m\rangle$, its state vector $|a\rangle$ can be written as the linear combination

$$|a\rangle = \sum_{i=1}^{m} c_i |a_i\rangle \qquad (2.56)$$

where the coefficients c_i are certain complex numbers.

Corresponding to each state vector $|a\rangle$ is a *complex conjugate vector* denoted by $\langle a|$ which is associated with the same state of the dynamical system. We shall suppose that the complex conjugate vector of (2.56) is

$$\sum_{i=1}^{m} \bar{c}_i \langle a_i|.$$

For any pair of vectors $|a\rangle$ and $|b\rangle$ we can define a number, written $\langle b | a \rangle$, which is called the *scalar product* of $\langle b|$ and $|a\rangle$. We shall take the scalar product to satisfy the law

$$\langle b| \left\{ \sum_{i=1}^{m} c_i |a_i\rangle \right\} = \sum_{i=1}^{m} c_i \langle b | a_i \rangle \qquad (2.57)$$

as well as the relation

$$\langle b | a \rangle = \overline{\langle a | b \rangle} \tag{2.58}$$

from which it can be inferred at once that $\langle a | a \rangle$ is a real number.

We next introduce the concept of a *linear operator* ω which transforms any state vector $|a\rangle$ into another state vector $\omega |a\rangle$. The linearity of this operator ensures that for any set of arbitrary vectors $|a_i\rangle$ and complex numbers c_i we have

$$\omega \left\{ \sum_{i=1}^{m} c_i |a_i\rangle \right\} = \sum_{i=1}^{m} c_i \omega |a_i\rangle. \tag{2.59}$$

The complex conjugate vector of $\omega |a\rangle$ is written $\langle a| \omega^*$ where ω^* is called the *Hermitian adjoint operator* of ω.

If we now replace $|b\rangle$ by $\omega |b\rangle$ in (2.58) we obtain the important formula

$$\langle b| \omega^* |a\rangle = \overline{\langle a| \omega |b\rangle} \tag{2.60}$$

relating the operator ω and its Hermitian adjoint ω^*. Further let us replace $\langle a|$ by $\langle a| \lambda$ and $|b\rangle$ by $\mu |b\rangle$ in (2.58) where λ and μ are any two linear operators. Then we get

$$\langle b| \mu^*\lambda^* |a\rangle = \overline{\langle a| \lambda\mu |b\rangle}$$

which, on making use of the relation (2.60) with $\omega = \lambda\mu$, may be rewritten as

$$\langle b| \mu^*\lambda^* |a\rangle = \langle b| (\lambda\mu)^* |a\rangle$$

and so we arrive at the result

$$\mu^*\lambda^* = (\lambda\mu)^*. \tag{2.61}$$

Observables

In quantum mechanics a measurable physical quantity, or *observable* as it has been termed by Dirac, is represented by a *linear Hermitian operator* α. Since we are given that this operator is Hermitian, that is self-adjoint, we have

$$\alpha^* = \alpha. \tag{2.62}$$

Associated with the operator $\boldsymbol{\alpha}$ is the *eigenvalue equation*

$$\boldsymbol{\alpha} |\alpha\rangle = \alpha |\alpha\rangle \qquad (2.63)$$

where α is the *eigenvalue* corresponding to the *eigenvector* solution $|\alpha\rangle$. Then if the dynamical system is in the state associated with the eigenvector $|\alpha\rangle$, a measurement of the physical quantity represented by $\boldsymbol{\alpha}$ will give the eigenvalue α as the result.

In order that the above physical interpretation of the eigenvalue α should be a plausible one, it is necessary to show that α is a real number. To do this we consider the adjoint form of the eigenvalue equation (2.63):

$$\langle\alpha| \boldsymbol{\alpha}^* = \langle\alpha| \bar{\alpha}. \qquad (2.64)$$

Using the Hermitian property (2.62) of the operator $\boldsymbol{\alpha}$, taking the scalar product with $\langle\alpha|$ on both sides of equation (2.63) and with $|\alpha\rangle$ on both sides of (2.64), we find that $\bar{\alpha} = \alpha$ which shows that the eigenvalue α is indeed a real number. We may now rewrite the adjoint equation (2.64) as

$$\langle\alpha| \boldsymbol{\alpha} = \langle\alpha| \alpha. \qquad (2.65)$$

Another important result which follows from the Hermitian property of $\boldsymbol{\alpha}$ can be obtained by replacing $\boldsymbol{\omega}$ by $\boldsymbol{\alpha}$ in formula (2.60). We get

$$\langle b| \boldsymbol{\alpha} |a\rangle = \overline{\langle a| \boldsymbol{\alpha} |b\rangle}. \qquad (2.66)$$

The orthogonality property of eigenvectors $|\alpha\rangle$ and $|\alpha'\rangle$ can be established by taking the scalar product with $\langle\alpha'|$ on both sides of (2.63) and taking the scalar product with $|\alpha\rangle$ on both sides of the eigenvalue equation

$$\langle\alpha'| \boldsymbol{\alpha} = \langle\alpha'| \alpha'.$$

This gives

$$\langle\alpha'| \boldsymbol{\alpha} |\alpha\rangle = \alpha\langle\alpha' | \alpha\rangle = \alpha'\langle\alpha' | \alpha\rangle$$

from which we see that

$$(\alpha - \alpha')\langle\alpha' | \alpha\rangle = 0$$

yielding the *orthogonality* property embodied in the formula

$$\langle \alpha' \mid \alpha \rangle = 0$$

provided the eigenvalues α and α' are different. Since it is usual to normalize all the eigenvectors to unity so that $\langle \alpha \mid \alpha \rangle = 1$, we arrive at the *orthonormality* relation

$$\langle \alpha' \mid \alpha \rangle = \delta_{\alpha'\alpha} \tag{2.67}$$

where $\delta_{\alpha'\alpha}$ vanishes when $\alpha' \neq \alpha$ and is equal to 1 when $\alpha' = \alpha$. In the case of continuous eigenvalues this is replaced by

$$\langle \alpha' \mid \alpha \rangle = \delta(\alpha' - \alpha) \tag{2.68}$$

where $\delta(x)$ is the Dirac function which satisfies

$$\delta(x) = 0 \quad (x \neq 0)$$

and

$$\int_{-\infty}^{\infty} \delta(x)\, dx = 1,$$

so that

$$\int_{-\infty}^{\infty} \delta(x) f(x)\, dx = f(0) \tag{2.69}$$

for any continuous function $f(x)$.

Next we consider two linear operators **α** and **β** which *commute* so that we have

$$\boldsymbol{\alpha\beta} = \boldsymbol{\beta\alpha}. \tag{2.70}$$

Then $\boldsymbol{\alpha\beta} \mid \alpha \rangle = \boldsymbol{\beta\alpha} \mid \alpha \rangle = \alpha \boldsymbol{\beta} \mid \alpha \rangle$ and so $\boldsymbol{\beta} \mid \alpha \rangle$ is an eigenvector of the operator **α** associated with eigenvalue α. If there is just one eigenvector $\mid \alpha \rangle$ with eigenvalue α, it follows at once that $\boldsymbol{\beta} \mid \alpha \rangle$ is proportional to $\mid \alpha \rangle$ and so we see that $\mid \alpha \rangle$ must also be an eigenvector of the operator **β**. If there are m eigenvectors $\mid \alpha 1 \rangle, \mid \alpha 2 \rangle, \ldots, \mid \alpha m \rangle$ of the operator **α** all associated with the same eigenvalue α, which is then said to be m-fold *degenerate*, m linear combinations of these can be determined which are simultaneously eigenvectors of the commuting operators **α** and **β**. More generally, it can be readily verified that any number of mutually commuting operators possess a set of simultaneous eigenvectors, which means therefore that observables represented by mutually commuting operators are *compatible* in the sense that measurements of the physical quantities associated with them can be performed simultaneously. Moreover, although we will not do so here, it can be established that the set of simultaneous eigenvectors is *complete*, so that an arbitrary vector can be expressed as a linear combination of these eigenvectors.

Suppose now that we have a *complete set of n commuting Hermitian operators* α_r ($r = 1, \ldots, n$), that is a set of commuting operators for which there is just one *simultaneous eigenvector* $|\alpha_1 \alpha_2 \ldots \alpha_n\rangle$, which we will abbreviate by $|\alpha_r\rangle$, associated with any set of eigenvalues α_r ($r = 1, \ldots, n$).

Since the simultaneous eigenvectors form a complete set, we can express an arbitrary vector $|a\rangle$ in the form

$$|a\rangle = \int |\alpha_r\rangle \, d\alpha_r \langle \alpha_r | a \rangle \qquad (2.71)$$

where the set of numbers $\langle \alpha_r | a \rangle$ is called the *representative* of $|a\rangle$ and $\int d\alpha_r$ indicates an integration and summation over all the eigenvalues $\alpha_1, \alpha_2, \ldots, \alpha_n$. Let β_s be a second set of commuting Hermitian operators in which case we may write also

$$|a\rangle = \int |\beta_s\rangle \, d\beta_s \langle \beta_s | a \rangle, \qquad (2.72)$$

the two representations associated with the different sets of operators being related by

$$\langle \alpha_r | a \rangle = \int \langle \alpha_r | \beta_s \rangle \, d\beta_s \langle \beta_s | a \rangle \qquad (2.73)$$

where $\langle \alpha_r | \beta_s \rangle$ is called the *transformation function*. If we now introduce a third set of commuting Hermitian operators γ_t we see that

$$\langle \alpha_r | \gamma_t \rangle = \int \langle \alpha_r | \beta_s \rangle \, d\beta_s \langle \beta_s | \gamma_t \rangle \qquad (2.74)$$

which is known as the *multiplicative composition law* of transformation functions.

OPTICS, WAVE MECHANICS AND QUANTUM MECHANICS 69

Next we come to the notion of *probability amplitude*. We assume that if we make a large number of measurements of a physical quantity described by a Hermitian operator α on a dynamical system in a state associated with the vector $|a\rangle$, the average of all the results will be

$$\langle a| \, \alpha \, |a\rangle, \tag{2.75}$$

the state vector $|a\rangle$ being normalized so that $\langle a \mid a \rangle = 1$. The numerical quantity $\langle a| \, \alpha \, |a\rangle$ is called the *expectation value* of α. Suppose that the dynamical system is entirely characterized by the operator α. Then we have

$$\langle a| \, \alpha \, |a\rangle = \int \langle a| \, \alpha \, |\alpha\rangle \, d\alpha \langle \alpha \mid a \rangle = \int \alpha \, |\langle \alpha \mid a \rangle|^2 \, d\alpha \tag{2.76}$$

and so we see that we may regard $|\langle \alpha \mid a \rangle|^2$ as the probability that a measurement of the observable associated with the operator α yields the number α as the result. In general when we have a set of commuting Hermitian operators α_r characterizing a dynamical system, the probability of making simultaneous measurements of the associated observables which yield the values α_r ($r = 1, \ldots, n$) is

$$|\langle \alpha_r \mid a \rangle|^2 \tag{2.77}$$

where $|a\rangle$ is the normalized state vector of the system. The numbers $\langle \alpha_r \mid a \rangle$ forming the representative of $|a\rangle$ are often referred to as *probability amplitudes*.

Schrödinger equation

We now direct our attention to the evolution with time of an undisturbed quantum dynamical system having state vector $|a, t\rangle$ which depends upon the time t. We suppose that the state vectors at times t_1 and t_2 are related by the transformation

$$|a, t_2\rangle = U |a, t_1\rangle \tag{2.78}$$

where U is a linear operator which is independent of a and depends only on t_1 and t_2. Moreover, we shall assume that U is a *unitary operator* satisfying the relation

$$U^*U = I \qquad (2.79)$$

where I is the unit operator. Then we see that

$$\langle a, t_2 \mid a, t_2 \rangle = \langle a, t_1 \mid U^*U \mid a, t_1 \rangle = \langle a, t_1 \mid a, t_1 \rangle$$

from which it follows that $\langle a, t \mid a, t \rangle$ does not change with the time.

Let us now allow t_2 to approach close to t_1 and set $t_2 - t_1 = \delta t$ where δt is an infinitesimal time interval. Then if H is an Hermitian operator we see that

$$U = I - \frac{i}{\hbar} H \, \delta t \qquad (2.80)$$

is a unitary operator since its Hermitian adjoint is given by

$$U^* = I + \frac{i}{\hbar} H \, \delta t,$$

and thus the relation (2.79) is satisfied to the first order in δt. Proceeding to the limit of vanishing δt we now find that (2.78) becomes

$$i\hbar \frac{d}{dt} \mid a, t \rangle = H \mid a, t \rangle \qquad (2.81)$$

for the case of general t, which is just the Schrödinger equation of motion if we identify the Hermitian operator H with the total energy of the system. The introduction of a complete set of commuting operators α_r enables us to rewrite the Schrödinger equation (2.81) as

$$i\hbar \frac{d}{dt} \langle \alpha_r \mid a, t \rangle = \langle \alpha_r \mid H \mid a, t \rangle. \qquad (2.82)$$

Next suppose that the quantum dynamical system has a classical analogue which can be specified by generalized coordinates q_r ($r = 1, \ldots, n$) together with their canonical momenta p_r ($r = 1, \ldots, n$) and possesses a Hamiltonian function $H(q_r, p_r)$.

Then we may characterize the quantum dynamical system by a set of commuting operators q_r ($r = 1, \ldots, n$) associated with the configuration of the system. If $|q_r\rangle$ is the simultaneous eigenvector for the operators q_r ($r = 1, \ldots, n$) corresponding to the set of eigenvalues q_r ($r = 1, \ldots, n$), we can introduce a *coordinate representation* of the state vector $|a, t\rangle$ given by the representative $\langle q_r | a, t \rangle$ which we may denote by the continuous function $\Psi_a(q_r, t)$. Moreover, it can be shown that the use of the coordinate representation enables us to put

$$H |a, t\rangle = \int |q_r'\rangle \, dq_r' H(q_r', -i\hbar \, \partial/\partial q_r') \langle q_r' | a, t \rangle$$

where the differential operator $H(q_r, -i\hbar \, \partial/\partial q_r)$ is obtained by setting $p_r = -i\hbar \, \partial/\partial q_r$ in the classical Hamiltonian function $H(q_r, p_r)$. Now using the orthonormality condition

$$\langle q_r | q_r' \rangle = \delta(q_r - q_r')$$

we see that equation (2.82) can be written as

$$i\hbar \frac{\partial}{\partial t} \Psi_a(q_r, t) = H(q_r, -i\hbar \, \partial/\partial q_r) \Psi_a(q_r, t) \qquad (2.83)$$

which is just a generalized form of the Schrödinger equation for the wave function $\Psi_a(q_r, t)$. For the particular case of a dynamical system composed of a single particle whose spacial position is characterized by the operator $\mathbf{r}$ having eigenvectors denoted by $|\mathbf{r}\rangle$, the representative $\langle \mathbf{r} | a, t \rangle$ is a solution of the single particle Schrödinger wave equation

$$i\hbar \frac{\partial}{\partial t} \Psi(\mathbf{r}, t) = H(\mathbf{r}, -i\hbar \nabla) \Psi(\mathbf{r}, t) \qquad (2.84)$$

where $H(\mathbf{r}, -i\hbar \nabla)$ is the differential operator obtained by replacing the momentum $\mathbf{p}$ by $-i\hbar \nabla$ in the classical Hamiltonian function $H(\mathbf{r}, \mathbf{p})$. Equation (2.84) is of course the same as the wave equation (2.35) established in section 2.4 by employing the optico-mechanical analogy.

Unitary transformations

Our next concern is with the transformation

$$\alpha_{r2} = U\alpha_{r1}U^{-1} \tag{2.85}$$

from a set of commuting Hermitian operators α_{r1} to a set of commuting Hermitian operators α_{r2}, where U^{-1} is the reciprocal operator of the linear operator U so that

$$U^{-1}U = I. \tag{2.86}$$

It can be readily verified that if $|\alpha_{r1}\rangle$ is a simultaneous eigenvector of the set of operators α_{r1} associated with a set of eigenvalues α_{r1}, then $U|\alpha_{r1}\rangle$ is a simultaneous eigenvector of the set of operators α_{r2} associated with the original set of eigenvalues α_{r1}. Thus the two sets of operators α_{r1} and α_{r2} possess the same eigenvalue spectrum.

Now suppose that $U^* = U^{-1}$ so that the relation (2.79) is satisfied and therefore U is a unitary operator. Then (2.85) is referred to as a *unitary transformation*. Since in this case

$$\langle \alpha_{r2}| = \langle \alpha_{r1}| U^* = \langle \alpha_{r1}| U^{-1},$$

it follows that the transformation function $\langle \alpha'_{r2} | \alpha_{r1} \rangle$ may be written in the form

$$\langle \alpha'_{r2} | \alpha_{r1} \rangle = \langle \alpha'_{r1}| U^{-1} |\alpha_{r1}\rangle. \tag{2.87}$$

If F is an infinitesimal Hermitian operator, then

$$U = I - \frac{i}{\hbar} F \tag{2.88}$$

is a unitary operator since its Hermitian adjoint is given by

$$U^* = I + \frac{i}{\hbar} F = U^{-1} \tag{2.89}$$

to the first order in F. Let us now apply the unitary transformation generated by F to a set of commuting Hermitian operators α_r. If

$\delta\alpha_r$ is the infinitesimal change in α_r due to the application of this transformation, we have

$$\alpha_r + \delta\alpha_r = \left(I - \frac{i}{\hbar}F\right)\alpha_r\left(I + \frac{i}{\hbar}F\right)$$

and so

$$i\hbar\,\delta\alpha_r = F\alpha_r - \alpha_r F \qquad (2.90)$$

to the first order of small quantities.

The change in the representative $\langle\alpha_r\,|\,a\rangle$ when the set of operators α_r is subjected to the infinitesimal unitary transformation generated by the operator F, is given by

$$\delta\langle\alpha_r\,|\,a\rangle = \langle\alpha_r + \delta\alpha_r\,|\,a\rangle - \langle\alpha_r\,|\,a\rangle.$$

But

$$\langle\alpha_r + \delta\alpha_r| = \langle\alpha_r|\,U^* = \langle\alpha_r| + \frac{i}{\hbar}\langle\alpha_r|\,F$$

and so we see that

$$\delta\langle\alpha_r\,|\,a\rangle = \frac{i}{\hbar}\langle\alpha_r|\,F\,|a\rangle \qquad (2.91)$$

which we may rewrite in the alternative form

$$\delta\langle\alpha_r\,|\,a\rangle = \frac{i}{\hbar}\int\langle\alpha_r|\,F\,|\alpha'_r\rangle\,d\alpha'_r\langle\alpha'_r\,|\,a\rangle. \qquad (2.92)$$

Lastly we suppose that the two sets of commuting Hermitian operators α_r and β_s are subjected to infinitesimal unitary transformations generated by the Hermitian operators F_α and F_β respectively. Then the change in the transformation function $\langle\alpha_r\,|\,\beta_s\rangle$ is given by

$$\delta\langle\alpha_r\,|\,\beta_s\rangle = \frac{i}{\hbar}\langle\alpha_r|\,F_\alpha - F_\beta\,|\beta_s\rangle. \qquad (2.93)$$

2.8 Hamilton's principle in quantum mechanics

Having concluded our short survey of the mathematical formalism of quantum mechanics we are now in a position to investigate the role of Hamilton's principle in quantum theory.

We consider the time development of a quantum dynamical system characterized by a set of commuting Hermitian linear operators α_r, and make the supposition that the operator sets α_{r1} and α_{r2}, corresponding to the times t_1 and t_2 respectively, are related by the unitary transformation

$$\alpha_{r2} = U_{21}\alpha_{r1}U_{21}^{-1}$$
$$|\alpha_{r2}\rangle = U_{21}|\alpha_{r1}\rangle \tag{2.94}$$

where U_{21} is a unitary operator. Then the eigenvalue spectrum of the set of operators α_r does not change with the time t so that $\alpha_{r2} = \alpha_{r1}$.

Let us now introduce an infinitesimal variation in the transformation function $\langle \alpha'_{r2} | \alpha_{r1} \rangle$ arising from alterations in the operator sets α_{r1}, α_{r2} and in the times t_1, t_2. Since such a variation produces a corresponding infinitesimal change in the operator U_{21}^{-1}, we have from equation (2.87) that

$$\delta\langle \alpha'_{r2} | \alpha_{r1} \rangle = \langle \alpha'_{r1} | \delta U_{21}^{-1} | \alpha_{r1} \rangle \tag{2.95}$$

to the first order of small quantities. Remembering that the operator U_{21} is unitary so that $U_{21}^{-1} = U_{21}^*$, it follows that $i\, U_{21} \delta U_{21}^{-1}$ is Hermitian because

$$i\, U_{21}\, \delta U_{21}^* = -i(\delta U_{21}) U_{21}^* = (i\, U_{21}\, \delta U_{21}^*)^*$$

as a consequence of formula (2.61), and so we may set

$$\delta U_{21}^{-1} = \frac{i}{\hbar} U_{21}^{-1}\, \delta W_{21} \tag{2.96}$$

where δW_{21} is an infinitesimal Hermitian operator. Hence the variation in the transformation function may be expressed as

$$\delta\langle \alpha'_{r2} | \alpha_{r1} \rangle = \frac{i}{\hbar} \langle \alpha'_{r2} | \delta W_{21} | \alpha_{r1} \rangle. \tag{2.97}$$

Now from the multiplicative composition law (2.74) we have

$$\langle \alpha''_{r3} | \alpha_{r1} \rangle = \int \langle \alpha''_{r3} | \alpha'_{r2} \rangle\, d\alpha'_{r2} \langle \alpha'_{r2} | \alpha_{r1} \rangle \tag{2.98}$$

and so, taking a variation of the transformation functions, it follows that

$$\langle \alpha_{r3}''| \delta W_{31} |\alpha_{r1}\rangle = \int \langle \alpha_{r3}''| \delta W_{32} |\alpha_{r2}'\rangle \, d\alpha_{r2}' \langle \alpha_{r2}' | \alpha_{r1}\rangle$$
$$+ \int \langle \alpha_{r3}'' | \alpha_{r2}'\rangle \, d\alpha_{r2}' \langle \alpha_{r2}'| \delta W_{21} |\alpha_{r1}\rangle$$

and therefore

$$\delta W_{31} = \delta W_{32} + \delta W_{21}. \tag{2.99}$$

We can satisfy this additive composition law by assuming that

$$W_{21} = \int_{t_1}^{t_2} L(\boldsymbol{\alpha}_r, \dot{\boldsymbol{\alpha}}_r) \, dt \tag{2.100}$$

where $L(\boldsymbol{\alpha}_r, \dot{\boldsymbol{\alpha}}_r)$ can be regarded as a Lagrangian function of the operators $\boldsymbol{\alpha}_r$ and their derivatives $\dot{\boldsymbol{\alpha}}_r$ with respect to the time t, so that W_{21} becomes the analogue of Hamilton's integral (1.162) in classical mechanics. We can now rewrite (2.97) in the form

$$\delta \langle \alpha_{r2}' | \alpha_{r1}\rangle = \frac{i}{\hbar} \langle \alpha_{r2}'| \delta \int_{t_1}^{t_2} L \, dt \, |\alpha_{r1}\rangle \tag{2.101}$$

which is the dynamical principle obtained by Schwinger.[1]

If the change in the transformation function is the result of infinitesimal variations in α_{r1}, t_1 and α_{r2}, t_2 due to infinitesimal unitary transformations generated by Hermitian operators F_1 and F_2 respectively, we see from (2.93) that

$$\delta \langle \alpha_{r2}' | \alpha_{r1}\rangle = \frac{i}{\hbar} \langle \alpha_{r2}'| F_2 - F_1 |\alpha_{r1}\rangle \tag{2.102}$$

and so we obtain

$$\delta W_{21} = F_2 - F_1 \tag{2.103}$$

which implies that δW_{21} does not depend upon variations occurring at times t within the interval from t_1 to t_2 but only upon the variations at the terminal points. Now choosing only those variations for which $F_2 = F_1$ leads to the *operator analogue of Hamilton's principle*:

$$\delta W_{21} = 0. \tag{2.104}$$

To derive the equations of motion we take infinitesimal variations at each time t of the operators $\alpha_r(t)$ given by

$$\alpha_r(t) \to \alpha_r(t) + \delta\alpha_r(t), \qquad (2.105)$$

and infinitesimal changes in the times t_1, t_2 at the terminal points given by

$$t_1 \to t_1 + \Delta t_1, \qquad t_2 \to t_2 + \Delta t_2. \qquad (2.106)$$

Taking care to preserve the correct order of the operators in L, the resulting alteration in Hamilton's integral (2.100) becomes

$$\delta W_{21} = \int_{t_1}^{t_2} \sum_{r=1}^{n} \delta\alpha_r \left\{ \frac{\partial L}{\partial \alpha_r} - \frac{d}{dt}\left(\frac{\partial L}{\partial \dot\alpha_r}\right)\right\} dt + [F(t)]_{t_1}^{t_2} \qquad (2.107)$$

where

$$F(t) = L\,\Delta t + \sum_{r=1}^{n} \delta\alpha_r \frac{\partial L}{\partial \dot\alpha_r} \qquad (t = t_1, t_2), \qquad (2.108)$$

the supposition being made that the commutation properties of the $\delta\alpha_r$ and the structure of the Lagrangian operator L permit the $\delta\alpha_r$ to be moved without the brackets in formula (2.107). Now applying the principle (2.103), we obtain the operator equations of motion for the quantum dynamical system in the Lagrangian form

$$\frac{d}{dt}\left(\frac{\partial L}{\partial \dot\alpha_r}\right) = \frac{\partial L}{\partial \alpha_r}. \qquad (2.109)$$

Thus we see that a quantum mechanical description of a dynamical system can be given which is analogous to that introduced by Hamilton in classical mechanics. It involves the stationary property of an integral over time of an appropriate Lagrangian function of the operators characterizing the dynamical system, and leads to the equations of motion in Lagrangian form.[2]

References

1. Schwinger, J., *Phys. Rev.*, **82**, 914 (1951).
2. See Feynman, R. P., *Rev. Mod. Phys.*, **20**, 367 (1948) for an alternative treatment to the one given here.

CHAPTER 3

Field Equations

The first two chapters of this book have dealt with the variational principles of dynamics and optics, and the relationship between them, as well as with some of their more important consequences. We now direct our attention to the wider problem of obtaining variational principles for the field equations of mathematical physics. In the first instance we shall investigate the problem from a quite general point of view, employing both a Lagrangian and a Hamiltonian formalism, and subsequently turn our attention to the treatment of specific cases such as, for example, the electromagnetic field equations and the various field equations of wave mechanics.

3.1 Lagrangian formalism

We consider a field whose n *components* are the real functions $\psi_\sigma(x_\mu)$ ($\sigma = 1, \ldots, n$), where the x_μ ($\mu = 1, \ldots, m$) are a set of m coordinates specifying location in space and time. To derive the field equations, we introduce a *Lagrangian density*

$$\mathscr{L}\left(x_\mu, \psi_\sigma, \frac{\partial \psi_\sigma}{\partial x_\mu}\right) \tag{3.1}$$

which is a function depending upon the field components ψ_σ and their derivatives $\partial \psi_\sigma / \partial x_\mu$ as well as upon the coordinates x_μ, and then define the integral

$$I = \int_T \mathscr{L}\left(x_\mu, \psi_\sigma, \frac{\partial \psi_\sigma}{\partial x_\mu}\right) d\tau, \tag{3.2}$$

where the domain of integration T is a hypervolume in the m-dimensional space determined by the coordinates x_μ.

We now choose a variation of the field components

$$\psi_\sigma(x_\mu) \to \psi_\sigma(x_\mu) + \epsilon\eta_\sigma(x_\mu), \tag{3.3}$$

where the $\eta_\sigma(x_\mu)$ are any set of linearly independent functions of the x_μ which vanish over the boundary hypersurface S of the region T of integration. Then the resulting change in I is given by

$$\delta I = \int_T \mathscr{L}\left(x_\mu, \psi_\sigma + \epsilon\eta_\sigma, \frac{\partial\psi_\sigma}{\partial x_\mu} + \epsilon\frac{\partial\eta_\sigma}{\partial x_\mu}\right) d\tau \\ - \int_T \mathscr{L}\left(x_\mu, \psi_\sigma, \frac{\partial\psi_\sigma}{\partial x_\mu}\right) d\tau \tag{3.4}$$

which we may rewrite in the form

$$\delta I = \epsilon \int_T \sum_{\sigma=1}^n \left\{\frac{\partial\mathscr{L}}{\partial\psi_\sigma}\eta_\sigma + \sum_{\mu=1}^m \frac{\partial\mathscr{L}}{\partial(\partial\psi_\sigma/\partial x_\mu)}\frac{\partial\eta_\sigma}{\partial x_\mu}\right\} d\tau \tag{3.5}$$

to the first order in the parameter ϵ.

Now the generalized form of Green's theorem enables us to put

$$\int_T \sum_{\mu=1}^m \frac{\partial\mathscr{L}}{\partial(\partial\psi_\sigma/\partial x_\mu)}\frac{\partial\eta_\sigma}{\partial x_\mu} d\tau \\ = \int_S \eta_\sigma \sum_{\mu=1}^m \frac{\partial\mathscr{L}}{\partial(\partial\psi_\sigma/\partial x_\mu)} l_\mu \, dS - \int_T \eta_\sigma \sum_{\mu=1}^m \frac{\partial}{\partial x_\mu}\frac{\partial\mathscr{L}}{\partial(\partial\psi_\sigma/\partial x_\mu)} d\tau, \tag{3.6}$$

where the l_μ ($\mu = 1, \ldots, m$) are the direction cosines of the outdrawn normal to the hypersurface S at any point, and so

$$\delta I = \epsilon \int_T \sum_{\sigma=1}^n \eta_\sigma \left\{\frac{\partial\mathscr{L}}{\partial\psi_\sigma} - \sum_{\mu=1}^m \frac{\partial}{\partial x_\mu}\frac{\partial\mathscr{L}}{\partial(\partial\psi_\sigma/\partial x_\mu)}\right\} d\tau \tag{3.7}$$

since the η_σ all vanish over S.

If the integral I is assumed to be stationary so that $\delta I = 0$, it follows that

$$\frac{\partial\mathscr{L}}{\partial\psi_\sigma} = \sum_{\mu=1}^m \frac{\partial}{\partial x_\mu}\frac{\partial\mathscr{L}}{\partial(\partial\psi_\sigma/\partial x_\mu)} \qquad (\sigma = 1, \ldots, n). \tag{3.8}$$

FIELD EQUATIONS

These are the Euler–Lagrange equations corresponding to the Lagrangian density (3.1) and they provide the field equations.

We now specialize to the case for which $m = 4$ with x_1, x_2, x_3 being the coordinates of a point in space with position vector $\mathbf{r}$ and x_4 being the time t. We choose a Lagrangian density

$$\mathscr{L}(t, \psi_\sigma, \nabla\psi_\sigma, \dot\psi_\sigma) \qquad (3.9)$$

which is a function of the time t, the field components $\psi_\sigma(\mathbf{r}, t)$ and their gradients $\nabla\psi_\sigma$ and derivatives $\dot\psi_\sigma$ with respect to the time. Next we introduce the integral

$$I = \int_{t_1}^{t_2} dt \int_V \mathscr{L}(t, \psi_\sigma, \nabla\psi_\sigma, \dot\psi_\sigma)\, d\mathbf{r}, \qquad (3.10)$$

where the range of integration is the volume V and the time interval from t_1 to t_2. We then take a variation of the field components given by

$$\psi_\sigma(\mathbf{r}, t) \to \psi_\sigma(\mathbf{r}, t) + \delta\psi_\sigma(\mathbf{r}, t) \qquad (3.11)$$

such that the small changes $\delta\psi_\sigma(\mathbf{r}, t)$ vanish over the boundary surface S of the volume V and satisfy the terminal conditions

$$\delta\psi_\sigma(\mathbf{r}, t_1) = \delta\psi_\sigma(\mathbf{r}, t_2) = 0. \qquad (3.12)$$

These changes in the field components are analogous to the increments $\epsilon\eta_\sigma$ introduced at the beginning of this section. As a consequence of this variation, the integral I is altered by the amount

$$\delta I = \int_{t_1}^{t_2} dt \int_V \mathscr{L}(t, \psi_\sigma + \delta\psi_\sigma, \nabla\psi_\sigma + \delta\nabla\psi_\sigma, \dot\psi_\sigma + \delta\dot\psi_\sigma)\, d\mathbf{r}$$
$$- \int_{t_1}^{t_2} dt \int_V \mathscr{L}(t, \psi_\sigma, \nabla\psi_\sigma, \dot\psi_\sigma)\, d\mathbf{r} \qquad (3.13)$$

which may be expressed to the first order of small quantities as

$$\delta I = \int_{t_1}^{t_2} dt \int_V \sum_{\sigma=1}^{n} \left\{ \frac{\partial \mathscr{L}}{\partial \psi_\sigma} \delta\psi_\sigma \right.$$
$$\left. + \sum_{k=1}^{3} \frac{\partial \mathscr{L}}{\partial(\partial\psi_\sigma/\partial x_k)} \delta\!\left(\frac{\partial\psi_\sigma}{\partial x_k}\right) + \frac{\partial \mathscr{L}}{\partial \dot\psi_\sigma} \delta\dot\psi_\sigma \right\} d\mathbf{r}. \qquad (3.14)$$

By Green's theorem we have

$$\int_V \sum_{k=1}^{3} \frac{\partial \mathscr{L}}{\partial(\partial \psi_\sigma/\partial x_k)} \frac{\partial}{\partial x_k}(\delta \psi_\sigma)\, d\mathbf{r}$$
$$= \int_S \delta \psi_\sigma \sum_{k=1}^{3} \frac{\partial \mathscr{L}}{\partial(\partial \psi_\sigma/\partial x_k)} l_k\, dS - \int_V \delta \psi_\sigma \sum_{k=1}^{3} \frac{\partial}{\partial x_k} \frac{\partial \mathscr{L}}{\partial(\partial \psi_\sigma/\partial x_k)}\, d\mathbf{r}, \tag{3.15}$$

where the l_k are the direction cosines of the outdrawn normal to the boundary surface S at any point, and an integration by parts gives us

$$\int_{t_1}^{t_2} \frac{\partial \mathscr{L}}{\partial \dot\psi_\sigma} \frac{\partial}{\partial t}(\delta \psi_\sigma)\, dt = \left[\delta \psi_\sigma \frac{\partial \mathscr{L}}{\partial \dot\psi_\sigma}\right]_{t_1}^{t_2} - \int_{t_1}^{t_2} \delta \psi_\sigma \frac{\partial}{\partial t}\left(\frac{\partial \mathscr{L}}{\partial \dot\psi_\sigma}\right) dt. \tag{3.16}$$

Since the $\delta\psi_\sigma(\mathbf{r}, t)$ vanish over the surface S and at times $t = t_1, t_2$ we obtain

$$\delta I = \int_{t_1}^{t_2} dt \int_V \sum_{\sigma=1}^{n} \delta \psi_\sigma \left\{ \frac{\partial \mathscr{L}}{\partial \psi_\sigma} - \sum_{k=1}^{3} \frac{\partial}{\partial x_k} \frac{\partial \mathscr{L}}{\partial(\partial \psi_\sigma/\partial x_k)} - \frac{\partial}{\partial t}\left(\frac{\partial \mathscr{L}}{\partial \dot\psi_\sigma}\right) \right\} d\mathbf{r}. \tag{3.17}$$

Hence if I is stationary for arbitrary independent variations $\delta\psi_\sigma$, it follows that

$$\frac{\partial \mathscr{L}}{\partial \psi_\sigma} - \sum_{k=1}^{3} \frac{\partial}{\partial x_k} \frac{\partial \mathscr{L}}{\partial(\partial \psi_\sigma/\partial x_k)} - \frac{\partial}{\partial t}\left(\frac{\partial \mathscr{L}}{\partial \dot\psi_\sigma}\right) = 0 \tag{3.18}$$

which are the Euler–Lagrange equations.

Introducing a Lagrangian function L given by

$$L = \int_V \mathscr{L}(t, \psi_\sigma, \nabla\psi_\sigma, \dot\psi_\sigma)\, d\mathbf{r}, \tag{3.19}$$

the stationary property of I may be expressed in the form

$$\delta \int_{t_1}^{t_2} L\, dt = 0 \tag{3.20}$$

which is analogous to Hamilton's principle in dynamics.

FIELD EQUATIONS 81

The analogy to classical mechanics can be carried a stage further by consideration of

$$\delta L = \int_V \sum_{\sigma=1}^{n} \left[\left\{ \frac{\partial \mathscr{L}}{\partial \psi_\sigma} - \sum_{k=1}^{3} \frac{\partial}{\partial x_k} \frac{\partial \mathscr{L}}{\partial(\partial \psi_\sigma/\partial x_k)} \right\} \delta\psi_\sigma + \frac{\partial \mathscr{L}}{\partial \dot\psi_\sigma} \delta\dot\psi_\sigma \right] d\mathbf{r}. \quad (3.21)$$

This may be rewritten in the form

$$\delta L = \int_V \sum_{\sigma=1}^{n} \left[\frac{\delta L}{\delta \psi_\sigma} \delta\psi_\sigma + \frac{\delta L}{\delta \dot\psi_\sigma} \delta\dot\psi_\sigma \right] d\mathbf{r} \quad (3.22)$$

by introducing the *functional derivatives* of L with respect to ψ_σ and $\dot\psi_\sigma$ given by

$$\frac{\delta L}{\delta \psi_\sigma} = \frac{\partial \mathscr{L}}{\partial \psi_\sigma} - \sum_{k=1}^{3} \frac{\partial}{\partial x_k} \frac{\partial \mathscr{L}}{\partial(\partial \psi_\sigma/\partial x_k)} \quad (3.23)$$

and

$$\frac{\delta L}{\delta \dot\psi_\sigma} = \frac{\partial \mathscr{L}}{\partial \dot\psi_\sigma} - \sum_{k=1}^{3} \frac{\partial}{\partial x_k} \frac{\partial \mathscr{L}}{\partial(\partial \dot\psi_\sigma/\partial x_k)}, \quad (3.24)$$

the terms involving the $\{\partial \mathscr{L}/\partial(\partial \dot\psi_\sigma/\partial x_k)\}$ on the right hand side of (3.24) being zero here because $\mathscr{L}$ does not depend upon the spacial derivatives of $\dot\psi_\sigma$.

The Euler–Lagrange equations (3.18) may then be put into the form

$$\frac{\partial}{\partial t} \frac{\delta L}{\delta \dot\psi_\sigma} = \frac{\delta L}{\delta \psi_\sigma} \quad (\sigma = 1, \ldots, n) \quad (3.25)$$

which are analogous to Lagrange's equations in dynamics if we regard the field component ψ_σ at a particular point of space as one of the generalized coordinates q_r of a dynamical system having an infinite number of degrees of freedom.

3.2 Hamiltonian formalism

We now introduce a Hamiltonian function

$$H = \int_V \mathscr{H} \, d\mathbf{r}, \quad (3.26)$$

where the *Hamiltonian density* $\mathscr{H}$ is defined by the formula

$$\mathscr{H} = \sum_{\sigma=1}^{n} \pi_\sigma \dot{\psi}_\sigma - \mathscr{L} \tag{3.27}$$

with the *canonical momentum density* π_σ corresponding to the field component ψ_σ being given by

$$\pi_\sigma = \frac{\partial \mathscr{L}}{\partial \dot{\psi}_\sigma}. \tag{3.28}$$

The change in H arising from independent increments in the field components ψ_σ and their time derivatives $\dot{\psi}_\sigma$ or, alternatively, in the ψ_σ and the momentum densities π_σ is given by

$$\delta H = \int_V \sum_{\sigma=1}^{n} \left(\pi_\sigma \, \delta \dot{\psi}_\sigma + \dot{\psi}_\sigma \, \delta \pi_\sigma - \frac{\delta L}{\delta \psi_\sigma} \delta \psi_\sigma - \frac{\delta L}{\delta \dot{\psi}_\sigma} \delta \dot{\psi}_\sigma \right) d\mathbf{r}$$

which may be put in the form

$$\delta H = \int_V \sum_{\sigma=1}^{n} (\dot{\psi}_\sigma \, \delta \pi_\sigma - \dot{\pi}_\sigma \, \delta \psi_\sigma) \, d\mathbf{r} \tag{3.29}$$

using equations (3.25) and (3.28).

If we now regard $\mathscr{H}$ as a function of the field components ψ_σ, the canonical momentum densities π_σ, the spacial derivatives $\partial \psi_\sigma / \partial x_k$ and the time t, we obtain

$$\delta H = \int_V \sum_{\sigma=1}^{n} \left[\frac{\partial \mathscr{H}}{\partial \psi_\sigma} \delta \psi_\sigma + \frac{\partial \mathscr{H}}{\partial \pi_\sigma} \delta \pi_\sigma + \sum_{k=1}^{3} \frac{\partial \mathscr{H}}{\partial(\partial \psi_\sigma / \partial x_k)} \delta\left(\frac{\partial \psi_\sigma}{\partial x_k}\right) \right] d\mathbf{r}$$
$$= \int_V \sum_{\sigma=1}^{n} \left[\left\{ \frac{\partial \mathscr{H}}{\partial \psi_\sigma} - \sum_{k=1}^{3} \frac{\partial}{\partial x_k} \frac{\partial \mathscr{H}}{\partial(\partial \psi_\sigma / \partial x_k)} \right\} \delta \psi_\sigma + \frac{\partial \mathscr{H}}{\partial \pi_\sigma} \delta \pi_\sigma \right] d\mathbf{r}$$
$$\tag{3.30}$$

using Green's theorem and the boundary conditions imposed on $\delta \psi_\sigma$ which we choose to be the same as those introduced in the Lagrangian formalism discussed in the previous section. This may be expressed in the form

$$\delta H = \int_V \sum_{\sigma=1}^{n} \left[\frac{\delta H}{\delta \psi_\sigma} \delta \psi_\sigma + \frac{\delta H}{\delta \pi_\sigma} \delta \pi_\sigma \right] d\mathbf{r}, \tag{3.31}$$

FIELD EQUATIONS 83

where
$$\frac{\delta H}{\delta \psi_\sigma} = \frac{\partial \mathcal{H}}{\partial \psi_\sigma} - \sum_{k=1}^{3} \frac{\partial}{\partial x_k} \frac{\partial \mathcal{H}}{\partial(\partial \psi_\sigma / \partial x_k)} \quad (3.32)$$

and
$$\frac{\delta H}{\delta \pi_\sigma} = \frac{\partial \mathcal{H}}{\partial \pi_\sigma} - \sum_{k=1}^{3} \frac{\partial}{\partial x_k} \frac{\partial \mathcal{H}}{\partial(\partial \pi_\sigma / \partial x_k)} \quad (3.33)$$

are the functional derivatives of H with respect to ψ_σ and π_σ. Since $\mathcal{H}$ does not depend upon the spacial derivatives of π_σ in the present case, the second term on the right-hand side of (3.33) vanishes here.

By comparing (3.31) with (3.29) we now see that

$$\dot{\psi}_\sigma = \frac{\delta H}{\delta \pi_\sigma}, \qquad \dot{\pi}_\sigma = -\frac{\delta H}{\delta \psi_\sigma} \quad (3.34)$$

which closely resemble Hamilton's equations (1.107) in classical dynamics.

We conclude this section by examining the condition for the Hamiltonian function H to be conserved. Taking the differential coefficient of H with respect to time, we obtain

$$\frac{dH}{dt} = \int_V \left[\sum_{\sigma=1}^{n} \left\{ \frac{\delta H}{\delta \psi_\sigma} \dot{\psi}_\sigma + \frac{\delta H}{\delta \pi_\sigma} \dot{\pi}_\sigma \right\} + \frac{\partial \mathcal{H}}{\partial t} \right] d\mathbf{r}$$

and so it follows from the field equations (3.34) that

$$\frac{dH}{dt} = \int_V \frac{\partial \mathcal{H}}{\partial t} d\mathbf{r}. \quad (3.35)$$

Hence H is constant if the Hamiltonian density $\mathcal{H}$ does not depend explicitly on the time t, for then we must have

$$\frac{dH}{dt} = 0. \quad (3.36)$$

3.3 Laplace's equation

The first example of a field equation which we shall discuss in this chapter is *Laplace's equation*

$$\nabla^2 \psi = 0, \quad (3.37)$$

where

$$\nabla^2 = \sum_{k=1}^{3} \frac{\partial^2}{\partial x_k^2} \qquad (3.38)$$

is *Laplace's operator*. This is a case for which there is only a single field component $\psi(\mathbf{r})$ depending upon position in space but not on the time.

We choose for the Lagrangian density the function

$$\mathscr{L} = \tfrac{1}{2}(\nabla\psi)^2 = \tfrac{1}{2} \sum_{k=1}^{3} \left(\frac{\partial\psi}{\partial x_k}\right)^2. \qquad (3.39)$$

Since $\mathscr{L}$ does not depend directly on the field component ψ, the Euler–Lagrange equation (3.18) takes the form

$$\sum_{k=1}^{3} \frac{\partial}{\partial x_k} \frac{\partial \mathscr{L}}{\partial(\partial\psi/\partial x_k)} = 0 \qquad (3.40)$$

which yields

$$\sum_{k=1}^{3} \frac{\partial^2 \psi}{\partial x_k^2} = 0.$$

This is just Laplace's equation (3.37).

Irrotational motion of an incompressible fluid

An illustrative example of Laplace's equation occurs in hydrodynamics in the form of the equation of continuity for the irrotational motion of an incompressible fluid. This equation arises from conservation of mass considerations in the following manner. If ρ is the density of the fluid at any point, the rate of change of mass enclosed within a volume V bounded by a fixed closed surface S is given by

$$\int_V \frac{\partial \rho}{\partial t} \, d\mathbf{r}. \qquad (3.41)$$

This must be equal to the rate at which mass flows into the volume V which is readily seen to be

$$-\oint_S \rho \mathbf{v} \cdot \mathbf{n} \, dS, \qquad (3.42)$$

FIELD EQUATIONS

where **v** is the velocity of the fluid at any point and **n** is a unit vector in the direction of the outdrawn normal to the surface S.

Using Gauss's divergence theorem, (3.42) can be put into the form

$$-\int_V \text{div}(\rho\mathbf{v})\, d\tau \tag{3.43}$$

and so equating (3.41) and (3.43) we get

$$\int_V \left\{\frac{\partial \rho}{\partial t} + \text{div}(\rho\mathbf{v})\right\} d\tau = 0. \tag{3.44}$$

Since the volume V is quite arbitrary it must follow that

$$\frac{\partial \rho}{\partial t} + \text{div}(\rho\mathbf{v}) = 0 \tag{3.45}$$

which is referred to as the *continuity equation*.

In the present section we are concerned with the motion of an incompressible fluid for which the density ρ is constant in space and time and then (3.45) becomes

$$\text{div}\,\mathbf{v} = 0. \tag{3.46}$$

The quantity

$$\boldsymbol{\omega} = \tfrac{1}{2}\,\text{curl}\,\mathbf{v} \tag{3.47}$$

is called the *vorticity*. If it vanishes everywhere within a given region, the fluid is said to be *irrotational* within this region. The reason for this nomenclature comes from the fact that if an element of the fluid has angular velocity $\boldsymbol{\omega}$, its velocity at a point with position vector **r** referred to a base point 0 on the axis of rotation is given by

$$\mathbf{v} = \mathbf{v}_0 + \boldsymbol{\omega} \times \mathbf{r}$$

where $\mathbf{v}_0$ is the velocity of 0, so that

$$\text{curl}\,\mathbf{v} = \nabla \times (\boldsymbol{\omega} \times \mathbf{r}) = 2\boldsymbol{\omega}.$$

For an irrotational fluid we have

$$\text{curl}\,\mathbf{v} = 0 \tag{3.48}$$

and then it follows that the velocity vector can be expressed in the form

$$\mathbf{v} = -\nabla\phi, \tag{3.49}$$

where ϕ is a scalar function of position called the *velocity potential*. Substituting into (3.46) we get Laplace's equation

$$\nabla^2\phi = 0. \tag{3.50}$$

This equation may also be derived from a variational principle by employing a suitable Lagrangian density in the following fashion. The kinetic energy per unit volume of fluid is given by

$$\mathscr{T} = \tfrac{1}{2}\rho \mathbf{v}^2 = \tfrac{1}{2}\rho(\nabla\phi)^2 \tag{3.51}$$

and since the potential energy is constant for an incompressible fluid, the Lagrangian density is just

$$\mathscr{L} = \tfrac{1}{2}\rho(\nabla\phi)^2 \tag{3.52}$$

which gives rise to (3.50) as the Euler–Lagrange equation.

As the total kinetic energy of the fluid

$$L = \int \mathscr{L} \, d\mathbf{r} \tag{3.53}$$

is positive definite, it follows that the stationary property of L is equivalent to the total kinetic energy being a minimum for the actual motion of the fluid as compared to all other motions which are consistent with the boundary conditions.

3.4 Poisson's equation

If we now choose the Lagrangian density to have the form

$$\mathscr{L} = \tfrac{1}{2}(\nabla\psi)^2 + f\psi = \tfrac{1}{2}\sum_{k=1}^{3}\left(\frac{\partial\psi}{\partial x_k}\right)^2 + f\psi, \tag{3.54}$$

where the single field component $\psi(\mathbf{r})$ is independent of the time

FIELD EQUATIONS

and f is an arbitrary function of position, the Euler–Lagrange equation (3.18) may be written

$$\sum_{k=1}^{3} \frac{\partial}{\partial x_k} \frac{\partial \mathscr{L}}{\partial(\partial \psi/\partial x_k)} = \frac{\partial \mathscr{L}}{\partial \psi} \tag{3.55}$$

giving

$$\sum_{k=1}^{3} \frac{\partial^2 \psi}{\partial x_k^2} = f, \tag{3.56}$$

that is

$$\nabla^2 \psi = f \tag{3.57}$$

which is known as *Poisson's equation*.

We shall consider two examples of this equation which arise in electrostatics and in Newtonian gravitation.

Electrostatics

Coulomb's law states that the electric field intensity due to a particle of charge e is given by

$$\mathbf{E} = \frac{e}{r^3} \mathbf{r}, \tag{3.58}$$

where $\mathbf{r}$ is the position vector of the field point referred to the charged particle as origin. This may be rewritten in the form

$$\mathbf{E} = -\nabla V, \tag{3.59}$$

where

$$V = \frac{e}{r} \tag{3.60}$$

is the electrostatic potential due to the charge e.

For a continuous distribution of charge having density ρ the electrostatic potential is given by

$$V = \int \frac{\rho \, d\mathbf{r}}{r}, \tag{3.61}$$

where r is the distance from a typical volume element $d\mathbf{r}$ to the field point.

Now *Gauss's theorem of total normal intensity* states that

$$\oint_S \mathbf{E}.d\mathbf{S} = 4\pi Q, \tag{3.62}$$

where

$$Q = \int \rho \, d\mathbf{r} \tag{3.63}$$

is the total charge contained within an arbitrary closed surface S. Hence, by Gauss's divergence theorem, we have

$$\int (\text{div } \mathbf{E} - 4\pi\rho) \, d\mathbf{r} = 0, \tag{3.64}$$

the region of integration being the volume bounded by S. It follows that

$$\text{div } \mathbf{E} = 4\pi\rho \tag{3.65}$$

which is just one of the Maxwell equations. Substituting for $\mathbf{E}$ using (3.59), we obtain Poisson's equation in the form

$$\nabla^2 V = -4\pi\rho. \tag{3.66}$$

In order to derive equation (3.66) by employing a variational principle, we consider the work done against the electrostatic forces in assembling a finite distribution of charge. This is called the *mutual potential energy* of the distribution and is given by

$$W = \tfrac{1}{2} \int \rho V \, d\mathbf{r}. \tag{3.67}$$

Using Maxwell's equation (3.65) it may be rewritten as

$$W = \frac{1}{8\pi} \int V \text{ div } \mathbf{E} \, d\mathbf{r}$$

and hence

$$W = \frac{1}{8\pi} \int \text{div } (V\mathbf{E}) \, d\mathbf{r} - \frac{1}{8\pi} \int (\nabla V).\mathbf{E} \, d\mathbf{r}$$

$$= \frac{1}{8\pi} \oint_S V\mathbf{E}.d\mathbf{S} + \frac{1}{8\pi} \int (\nabla V)^2 \, d\mathbf{r}$$

by Gauss's divergence theorem. If we allow the surface S to tend to infinite distance away from the distribution of charge, the

FIELD EQUATIONS

surface integral will provide a vanishing contribution to the mutual potential energy and then we get

$$W = \frac{1}{8\pi} \int (\nabla V)^2 \, d\mathbf{r}, \qquad (3.68)$$

where the volume integration is now extended over all space. Combining (3.67) and (3.68) we see that we may rewrite the mutual potential energy in the form

$$W = \int \left\{ \rho V - \frac{1}{8\pi} (\nabla V)^2 \right\} d\mathbf{r}. \qquad (3.69)$$

Taking $-W$ as the Lagrangian function with V as the field component, the Lagrangian density is accordingly

$$\mathscr{L} = \frac{1}{8\pi} (\nabla V)^2 - \rho V \qquad (3.70)$$

in which case the Euler–Lagrange equation becomes Poisson's equation (3.66).

Since we know from (3.68) that W is positive definite it follows that the stationary property of the Lagrangian function corresponds to the mutual potential energy of the distribution of charge being a minimum for the actual solution of Poisson's equation.

Newtonian gravitation

Our second example of Poisson's equation is obtained from Newtonian gravitation. According to Newton's law of gravitation the field intensity due to a particle of mass m is given by

$$\mathbf{F} = -\frac{\gamma m}{r^3} \mathbf{r}, \qquad (3.71)$$

where $\mathbf{r}$ is the position vector of the field point relative to the particle and γ is the constant of gravitation. This may be rewritten in the form

$$\mathbf{F} = -\nabla V, \qquad (3.72)$$

where

$$V = -\frac{\gamma m}{r} \qquad (3.73)$$

is the potential of the field.

For a continuous distribution of matter with mass density ρ the potential is given by

$$V = -\gamma \int \frac{\rho\, d\mathbf{r}}{r}, \qquad (3.74)$$

where r is the distance from the volume element $d\mathbf{r}$ to the field point. Thus we see that Newtonian gravitation is completely analogous to the case of electrostatics treated in the preceding section. The application of Gauss's theorem of total normal intensity provides Poisson's equation in the form

$$\nabla^2 V = 4\pi\gamma\rho. \qquad (3.75)$$

Further, the potential energy of the distribution of matter is given by

$$\begin{aligned} W &= \tfrac{1}{2} \int \rho V\, d\mathbf{r} = -\frac{1}{8\pi\gamma} \int (\nabla V)^2\, d\mathbf{r} \\ &= \int \left\{ \rho V + \frac{1}{8\pi\gamma} (\nabla V)^2 \right\} d\mathbf{r}. \end{aligned} \qquad (3.76)$$

Finally, taking W for the Lagrangian function so that

$$\mathscr{L} = \frac{1}{8\pi\gamma}(\nabla V)^2 + \rho V \qquad (3.77)$$

is the Lagrangian density, yields Poisson's equation (3.75) on applying the variational principle.

3.5 Scalar wave equation

In this section we are concerned with the derivation of the scalar wave equation

$$\nabla^2 \psi = \frac{1}{c^2} \frac{\partial^2 \psi}{\partial t^2} \qquad (3.78)$$

from the stationary property of an appropriate Lagrangian function. To this end it is first convenient to denote the Cartesian coordinates of a point in space by x_1, x_2, x_3 and to put $x_4 = ict$. Introducing the *d'Alembertian operator*

$$\Box^2 = \sum_{\mu=1}^{4} \frac{\partial^2}{\partial x_\mu^2}, \qquad (3.79)$$

FIELD EQUATIONS

we may rewrite the scalar wave equation (3.78) in the form

$$\Box^2 \psi = 0. \tag{3.80}$$

Rather than finding the Lagrangian density which yields (3.80) directly, we investigate the more general problem of a scalar field which is a function of position in an m-dimensional space having $x_1, x_2, \ldots, x_m$ as coordinates. If the Lagrangian density is chosen to depend only upon the derivatives $\partial \psi / \partial x_\mu$ and not on the field component ψ, the Euler–Lagrange equation (3.8) takes the form

$$\sum_{\mu=1}^{m} \frac{\partial}{\partial x_\mu} \frac{\partial \mathscr{L}}{\partial (\partial \psi / \partial x_\mu)} = 0. \tag{3.81}$$

It is at once evident that if we put

$$\mathscr{L} = \tfrac{1}{2} \sum_{\mu=1}^{m} \left(\frac{\partial \psi}{\partial x_\mu} \right)^2 \tag{3.82}$$

equation (3.81) becomes

$$\sum_{\mu=1}^{m} \frac{\partial^2 \psi}{\partial x_\mu^2} = 0. \tag{3.83}$$

If we now particularize to the case for which $m = 4$, the field equation (3.83) reduces to (3.80) and consequently to the scalar wave equation (3.78), while the Lagrangian density may be expressed as

$$\mathscr{L} = \tfrac{1}{2} (\Box \psi)^2, \tag{3.84}$$

where $\Box$ is the four dimensional gradient operator.

Vibrating string

An important example of a dynamical system which is described by a scalar wave equation is the vibrating string. Before treating the case of the continuous string, we consider a system consisting of N identical particles, each of mass m, attached at equal intervals a apart to a light string of length $l = (N + 1)a$, the ith particle being at a distance $x_i = ia$ from one end of the string. If the tension in the string is P and the displacement $y_i(t)$ of the ith particle at time t from its equilibrium position is small, then the element of string between the ith and the $(i + 1)$th particle is extended by an amount

$$\sqrt{\{a^2 + (y_{i+1} - y_i)^2\}} - a \cong \frac{1}{2a}(y_{i+1} - y_i)^2$$

and so, to the second order of small quantities, the potential energy of the string is given by

$$V = \frac{P}{2a} \sum_{i=0}^{N} (y_{i+1} - y_i)^2. \quad (3.85)$$

Since the total kinetic energy of the system of particles is just

$$T = \frac{m}{2} \sum_{i=1}^{N} \dot{y}_i^2, \quad (3.86)$$

the Lagrange equation for the ith particle

$$\frac{d}{dt}\frac{\partial T}{\partial \dot{y}_i} - \frac{\partial T}{\partial y_i} = -\frac{\partial V}{\partial y_i} \quad (3.87)$$

yields as the equation of motion

$$m\ddot{y}_i = \frac{P}{a}(y_{i+1} - 2y_i + y_{i-1}). \quad (3.88)$$

We are now in a position to make the transition to a continuous uniform string of length l having mass ρ per unit length. This can be achieved by letting the number N of particles become large and the mass m of each particle and the interval a become small in such a manner that $m/a \to \rho$ and $(N + 1)a = l$. If $y(x, t)$ is the displacement of the string at time t at the point whose distance from one end of the string is x, we have

$$\lim_{a \to 0} \frac{1}{a}(y_{i+1} - y_i) = \left(\frac{\partial y}{\partial x}\right)_{x=x_i} \quad (3.89)$$

and

$$\lim_{a \to 0} \frac{1}{a^2}(y_{i+1} - 2y_i + y_{i-1}) = \left(\frac{\partial^2 y}{\partial x^2}\right)_{x=x_i} \quad (3.90)$$

Hence it follows from (3.88) that

$$\frac{\partial^2 y}{\partial x^2} = \frac{1}{c^2}\frac{\partial^2 y}{\partial t^2}, \tag{3.91}$$

where $c = \sqrt{(P/\rho)}$ is the phase velocity. This is the wave equation for a vibrating string.

We also see from (3.85) and (3.86) that in the limit of the continuous string the potential energy takes the form

$$V = \frac{P}{2}\int_0^l \left(\frac{\partial y}{\partial x}\right)^2 dx \tag{3.92}$$

and the kinetic energy becomes

$$T = \tfrac{1}{2}\int_0^l \rho \left(\frac{\partial y}{\partial t}\right)^2 dx. \tag{3.93}$$

These expressions for the energies can also be obtained by investigating the case of the continuous string directly. We consider a non-uniform string having line density ρ which is a function of the coordinate x. If the displacement of the string from its equilibrium configuration is $y(x, t)$ at the point x and at the time t, the kinetic energy of an element of length δx of the string is

$$\tfrac{1}{2}\rho\left(\frac{\partial y}{\partial t}\right)^2 \delta x. \tag{3.94}$$

Now an element of the string of length δx is extended due to the displacement of the string by an amount

$$\left[\sqrt{\left\{1 + \left(\frac{\partial y}{\partial x}\right)^2\right\}} - 1\right]\delta x \simeq \tfrac{1}{2}\left(\frac{\partial y}{\partial x}\right)^2 \delta x$$

and so the potential energy of this element of string is

$$\tfrac{1}{2}P\left(\frac{\partial y}{\partial x}\right)^2 \delta x. \tag{3.95}$$

On integrating (3.95) and (3.94) over the length of the string, the formulae (3.92) and (3.93) are regained respectively.

We now see that the Lagrangian density for the vibrating string is given by

$$\mathscr{L} = \tfrac{1}{2}\left\{\rho\left(\frac{\partial y}{\partial t}\right)^2 - P\left(\frac{\partial y}{\partial x}\right)^2\right\} \tag{3.96}$$

which, on using the Euler–Lagrange equation

$$\frac{\partial}{\partial x}\frac{\partial \mathscr{L}}{\partial(\partial y/\partial x)} + \frac{\partial}{\partial t}\frac{\partial \mathscr{L}}{\partial(\partial y/\partial t)} = 0, \tag{3.97}$$

results in the equation of motion of the string in the form (3.91).

It is also of interest to consider the Hamiltonian formalism for the vibrating string. Introducing the momentum density

$$\pi = \frac{\partial \mathscr{L}}{\partial(\partial y/\partial t)} = \rho\frac{\partial y}{\partial t}, \tag{3.98}$$

we may express the Hamiltonian density in the form

$$\mathscr{H} = \pi\frac{\partial y}{\partial t} - \mathscr{L} = \tfrac{1}{2}\left\{\rho\left(\frac{\partial y}{\partial t}\right)^2 + P\left(\frac{\partial y}{\partial x}\right)^2\right\} \tag{3.99}$$

which is just the energy density of the string. Rewriting (3.99) as

$$\mathscr{H} = \frac{1}{2\rho}\pi^2 + \frac{P}{2}\left(\frac{\partial y}{\partial x}\right)^2 \tag{3.100}$$

we find that Hamilton's equations (3.34) regenerate the formula (3.98) for the momentum density and the wave equation (3.91).

Vibrating membrane

In the preceding subsection we dealt with the case of the vibrating string which is described by a one-dimensional wave equation. We now consider the analogous case of the vibrations of a stretched membrane which however are determined by a two-dimensional equation. Suppose that the membrane has mass ρ per unit area and that it is stretched at tension P. Suppose further that when the membrane is in equilibrium, it lies in the xy plane of a Cartesian frame of reference. If the displacement of the membrane from its equilibrium configuration is $z(x, y, t)$ at the point of the xy plane

FIELD EQUATIONS

with coordinates x, y and at time t, the kinetic energy of an element $\delta x \, \delta y$ of the membrane is given by

$$\tfrac{1}{2}\rho\left(\frac{\partial z}{\partial t}\right)^2 \delta x \, \delta y$$

while the potential energy of the element is

$$\tfrac{1}{2}P\left\{\left(\frac{\partial z}{\partial x}\right)^2 + \left(\frac{\partial z}{\partial y}\right)^2\right\} \delta x \, \delta y.$$

Hence the Lagrangian density is

$$\mathscr{L} = \tfrac{1}{2}\left[\rho\left(\frac{\partial z}{\partial t}\right)^2 - P\left\{\left(\frac{\partial z}{\partial x}\right)^2 + \left(\frac{\partial z}{\partial y}\right)^2\right\}\right] \quad (3.101)$$

which yields the Euler–Lagrange equation

$$\frac{\partial^2 z}{\partial x^2} + \frac{\partial^2 z}{\partial y^2} = \frac{1}{c^2}\frac{\partial^2 z}{\partial t^2}, \quad (3.102)$$

where $c = \sqrt{(P/\rho)}$ is the phase velocity. This is the wave equation for a vibrating membrane.

The Hamiltonian density is given by

$$\mathscr{H} = \pi \frac{\partial z}{\partial t} - \mathscr{L}, \quad (3.103)$$

where

$$\pi = \frac{\partial \mathscr{L}}{\partial(\partial z/\partial t)} = \rho \frac{\partial z}{\partial t} \quad (3.104)$$

is the momentum density. It follows that

$$\mathscr{H} = \tfrac{1}{2}\left[\rho\left(\frac{\partial z}{\partial t}\right)^2 + P\left\{\left(\frac{\partial z}{\partial x}\right)^2 + \left(\frac{\partial z}{\partial y}\right)^2\right\}\right] \quad (3.105)$$

which is the energy density of the membrane. Further, if we express the Hamiltonian density as

$$\mathscr{H} = \frac{1}{2\rho}\pi^2 + \frac{P}{2}\left\{\left(\frac{\partial z}{\partial x}\right)^2 + \left(\frac{\partial z}{\partial y}\right)^2\right\}, \quad (3.106)$$

Hamilton's equations (3.34) may be employed to derive the wave equation (3.102) as well as regaining the formula (3.104) for the momentum density.

Sound waves

The last example of a scalar wave equation which we shall discuss here is the three dimensional case associated with waves in a compressible fluid. We begin by deriving the *Euler equation of motion* for the fluid. To do this we consider the fluid contained within a closed surface S. If $\mathbf{v}$ is the fluid velocity and ρ is the density of the fluid, the rate of change of momentum is

$$\int \rho \frac{d\mathbf{v}}{dt} d\mathbf{r},$$

where the integration is over the volume bounded by S. Also, if $\mathbf{F}$ is the external force per unit mass of fluid and p is the pressure within the fluid, the total force acting on the fluid is

$$\int \rho \mathbf{F} \, d\mathbf{r} - \oint_S p\mathbf{n} \, dS,$$

where $\mathbf{n}$ is a unit vector in the direction of the outdrawn normal to the boundary surface S. Since

$$\oint_S p\mathbf{n} \, dS = \int \nabla p \, d\mathbf{r}$$

it follows that

$$\int \rho \frac{d\mathbf{v}}{dt} d\mathbf{r} = \int (\rho \mathbf{F} - \nabla p) \, d\mathbf{r}$$

by Newton's second law of motion. Hence, since the volume of integration is arbitrary,

$$\frac{d\mathbf{v}}{dt} = \mathbf{F} - \frac{1}{\rho} \nabla p \qquad (3.107)$$

which is the equation of fluid flow due to Euler.

Now

$$\frac{d\mathbf{v}}{dt} = \frac{\partial \mathbf{v}}{\partial t} + \mathbf{v} \cdot \nabla \mathbf{v}$$

and since

$$\mathbf{v} \cdot \nabla \mathbf{v} = \tfrac{1}{2} \nabla \mathbf{v}^2 - \mathbf{v} \times \operatorname{curl} \mathbf{v},$$

FIELD EQUATIONS

we see that in the case of irrotational motion for which curl $\mathbf{v} = 0$, the Euler equation takes the form

$$\frac{\partial \mathbf{v}}{\partial t} + \tfrac{1}{2}\nabla \mathbf{v}^2 = \mathbf{F} - \frac{1}{\rho}\nabla p. \tag{3.108}$$

If the density ρ is a function of the pressure p, we may introduce the function

$$P = \int_{p_0}^{p} \frac{dp}{\rho}, \tag{3.109}$$

where p_0 is the average pressure at a given point of the fluid, and then we have

$$\nabla P = \frac{1}{\rho}\nabla p.$$

Supposing that the external forces are negligible and putting

$$\mathbf{v} = -\nabla \phi, \tag{3.110}$$

where ϕ is the velocity potential, the Euler equation (3.108) becomes

$$\nabla\left(P + \tfrac{1}{2}v^2 - \frac{\partial \phi}{\partial t}\right) = 0$$

and so

$$P + \tfrac{1}{2}v^2 - \frac{\partial \phi}{\partial t} = 0, \tag{3.111}$$

since the arbitrary function of t which arises as a result of the integration may be absorbed into the potential ϕ without affecting the velocity $\mathbf{v}$.

For sound waves it may be assumed that the oscillations of the fluid are small in which case terms of the second order in v may be neglected and then the equation (3.111) reduces to

$$P = \frac{\partial \phi}{\partial t}. \tag{3.112}$$

If we denote the average density of the fluid by ρ_0, the actual density at any time may be written in the form

$$\rho = \rho_0(1 + s), \tag{3.113}$$

where s is called the *condensation*. Hence the pressure may be expressed as

$$p = p_0 + (\rho - \rho_0)\left(\frac{dp}{d\rho}\right)_0 + \cdots \cong p_0 + \rho_0 c^2 s, \quad (3.114)$$

where

$$c^2 = \left(\frac{dp}{d\rho}\right)_0. \quad (3.115)$$

Also we may write

$$P = \int_{\rho_0}^{\rho} \frac{dp}{d\rho} \frac{d\rho}{\rho} \cong \left(\frac{dp}{d\rho}\right)_0 \int_{\rho_0}^{\rho} \frac{d\rho}{\rho}$$

and so, to the first order in s, we have

$$P \cong c^2 \ln \frac{\rho}{\rho_0} = c^2 \ln(1 + s) \cong c^2 s$$

which, together with (3.112), yields the relation

$$s = \frac{1}{c^2} \frac{\partial \phi}{\partial t}. \quad (3.116)$$

We now consider a given mass element of the fluid which occupies a volume τ_0 when its density is ρ_0, and which has volume

$$\tau = \tau_0/(1 + s) \cong \tau_0(1 - s) \quad (3.117)$$

when its density is ρ. The work done in compressing the volume from τ_0 to τ is

$$-\int_{\tau_0}^{\tau} p \, d\tau = \tau_0 \int_0^s p \, ds$$

which, on using (3.114), becomes

$$\tau_0 \int_0^s (p_0 + \rho_0 c^2 s) \, ds = \tau_0 p_0 s + \tfrac{1}{2}\tau_0 \rho_0 c^2 s^2$$
$$= (\tau_0 - \tau)p_0 + \tfrac{1}{2}\tau_0 \rho_0 c^2 s^2. \quad (3.118)$$

This is the potential energy of the fluid within the volume element τ_0. Integrating over the total constant volume occupied by the fluid, we see that the first term of (3.118) gives a vanishing contribution to the potential energy which therefore takes the form

$$V = \tfrac{1}{2} \int \rho_0 c^2 s^2 \, d\mathbf{r} = \tfrac{1}{2} \int \frac{\rho_0}{c^2} \left(\frac{\partial \phi}{\partial t}\right)^2 d\mathbf{r}. \qquad (3.119)$$

Since the kinetic energy of the gas is

$$T = \tfrac{1}{2} \int \rho_0 (\nabla \phi)^2 \, d\mathbf{r} \qquad (3.120)$$

it follows that the Lagrangian density is given by

$$\mathscr{L} = \tfrac{1}{2}\rho_0 \left\{ (\nabla \phi)^2 - \frac{1}{c^2}\left(\frac{\partial \phi}{\partial t}\right)^2 \right\}. \qquad (3.121)$$

The Euler–Lagrange equation then yields the equation for sound waves:

$$\nabla^2 \phi = \frac{1}{c^2} \frac{\partial^2 \phi}{\partial t^2}. \qquad (3.122)$$

Referring to the formulae (3.120) and (3.119), we see that the spacial derivatives of the field component ϕ are associated with the kinetic energy of the fluid while the time derivative of ϕ is associated with the potential energy, which is the reverse situation to that found to exist for the vibrating string and membrane.

The Hamiltonian density for wave motion in a compressible fluid has the form

$$\mathscr{H} = \pi \frac{\partial \phi}{\partial t} - \mathscr{L}, \qquad (3.123)$$

where the canonical momentum conjugate to ϕ, given by the formula

$$\pi = \frac{\partial \mathscr{L}}{\partial (\partial \phi / \partial t)} = -\frac{\rho_0}{c^2} \frac{\partial \phi}{\partial t} = -\rho_0 s, \qquad (3.124)$$

is proportional to the condensation and not to the velocity of the fluid.

Substituting (3.124) into (3.123) gives

$$\mathscr{H} = -\tfrac{1}{2}\rho_0 \left\{ (\nabla \phi)^2 + \frac{1}{c^2}\left(\frac{\partial \phi}{\partial t}\right)^2 \right\} \qquad (3.125)$$

from which it follows that $-\mathcal{H}$ is equal to the energy density of the fluid.

Lastly to obtain the wave equation (3.122) by the use of Hamilton's equations (3.34), we require to express the Hamiltonian density in terms of π rather than $\partial\phi/\partial t$. It can be readily verified that

$$\mathcal{H} = -\tfrac{1}{2}\rho_0(\nabla\phi)^2 - \frac{c^2}{2\rho_0}\pi^2. \qquad (3.126)$$

3.6 Subsidiary conditions

We now return to the Lagrangian formalism discussed in section 3.1. In a number of instances the stationary property of the integral

$$I = \int_T \mathscr{L}\left(x_\mu, \psi_\sigma, \frac{\partial\psi_\sigma}{\partial x_\mu}\right) d\tau$$

is subject to certain subsidiary conditions being satisfied. Let us suppose that these subsidiary conditions can be expressed in the form

$$\int_T \mathscr{G}_i\left(x_\mu, \psi_\sigma, \frac{\partial\psi_\sigma}{\partial x_\mu}\right) d\tau = c_i \qquad (i = 1, \ldots, p), \qquad (3.127)$$

where $\mathscr{G}_1, \ldots, \mathscr{G}_p$ are functions of the coordinates x_μ, the field components ψ_σ and their derivatives $\partial\psi_\sigma/\partial x_\mu$, and $c_1, \ldots, c_p$ are constants, p being less than the number n of field components.

To solve this modified variational problem, we introduce a set of p undetermined quantities $\lambda_1, \ldots, \lambda_p$ which we may call Lagrange multipliers, and employ the stationary property of the integral

$$I' = \int_T \left(\mathscr{L} + \sum_{i=1}^{p} \lambda_i \mathscr{G}_i\right) d\tau \qquad (3.128)$$

with respect to the variation of the field components given by

$$\psi_\sigma(x_\mu) \to \psi_\sigma(x_\mu) + \epsilon\eta_\sigma(x_\mu),$$

where the η_σ are any set of functions of the x_μ which vanish over

FIELD EQUATIONS 101

the boundary hypersurface S of the region T and are consistent with the auxiliary conditions (3.127) remaining satisfied. It then follows that

$$\int_T \sum_{\sigma=1}^n \eta_\sigma \left\{ \frac{\partial \mathscr{L}'}{\partial \psi_\sigma} - \sum_{\mu=1}^m \frac{\partial}{\partial x_\mu} \frac{\partial \mathscr{L}'}{\partial(\partial \psi_\sigma/\partial x_\mu)} \right\} d\tau = 0, \quad (3.129)$$

where

$$\mathscr{L}' = \mathscr{L} + \sum_{i=1}^p \lambda_i \mathscr{G}_i \quad (3.130)$$

is a modified Lagrangian density.

In the case under consideration only $n - p$ of the functions $\eta_\sigma(x_\mu)$ are completely independent since the auxiliary conditions have to be obeyed. However, we can choose the p Lagrange multipliers λ_i so that the coefficients of all the functions η_σ vanish within the integral (3.129) by following a similar procedure to that used on pp. 15 and 16, in which case we obtain the Euler–Lagrange equations for the variational problem in the form

$$\frac{\partial \mathscr{L}'}{\partial \psi_\sigma} = \sum_{\mu=1}^m \frac{\partial}{\partial x_\mu} \frac{\partial \mathscr{L}'}{\partial(\partial \psi_\sigma/\partial x_\mu)} \quad (\sigma = 1, \ldots, n). \quad (3.131)$$

Together with the subsidiary conditions (3.127) these provide the field equations for the components ψ_σ as well as determining the values of the Lagrange multipliers λ_i.

3.7 Helmholtz equation

Let us now proceed to the examination of the solutions of the scalar wave equation

$$\nabla^2 \Psi = \frac{1}{c^2} \frac{\partial^2 \Psi}{\partial t^2} \quad (3.132)$$

which have the separable form

$$\Psi(\mathbf{r}, t) = \psi(\mathbf{r}) T(t) \quad (3.133)$$

composed of the product of a function ψ, depending on spacial coordinates only, and a function T, depending on the time only.

Substituting (3.133) into equation (3.132) we obtain

$$\frac{1}{\psi} \nabla^2 \psi = \frac{1}{c^2} \frac{1}{T} \frac{d^2 T}{dt^2}. \tag{3.134}$$

Since the left-hand side of this equation is a function of the spacial coordinates only and the right-hand side is a function of the time only, both sides must be independent of $\mathbf{r}$ and t and consequently equal to a constant. If we choose this constant to be $-k^2$ we get

$$\frac{d^2 T}{dt^2} = -k^2 c^2 T \tag{3.135}$$

whose general solution can be expressed in the form of a sine wave with frequency $\nu = kc/2\pi$, and

$$\nabla^2 \psi + k^2 \psi = 0 \tag{3.136}$$

which is known as the *Helmholtz equation*. The permissible values of k are determined by the boundary conditions imposed on the function ψ.

The Helmholtz equation (3.136) can be derived from a variational principle by choosing the Lagrangian density to be

$$\mathscr{L} = \tfrac{1}{2}(\nabla \psi)^2 \tag{3.137}$$

and taking the normalization

$$\int_V \psi^2(\mathbf{r}) \, d\mathbf{r} = 1 \tag{3.138}$$

as a subsidiary condition, the region of integration being the volume V.

Introducing a Lagrange multiplier λ, the integral

$$I' = \int_V \tfrac{1}{2}\{(\nabla \psi)^2 - \lambda \psi^2\} \, d\mathbf{r} \tag{3.139}$$

is stationary with respect to small variations which are consistent with the normalization condition (3.138) and which vanish over the boundary surface of V if

$$\sum_{k=1}^{3} \frac{\partial}{\partial x_k} \frac{\partial \mathscr{L}'}{\partial(\partial \psi/\partial x_k)} = \frac{\partial \mathscr{L}'}{\partial \psi}, \tag{3.140}$$

FIELD EQUATIONS

where the modified Lagrangian density $\mathscr{L}'$ is given by

$$\mathscr{L}' = \tfrac{1}{2}\{(\nabla\psi)^2 - \lambda\psi^2\} = \tfrac{1}{2}\left\{\sum_{k=1}^{3}\left(\frac{\partial\psi}{\partial x_k}\right)^2 - \lambda\psi^2\right\}. \quad (3.141)$$

The Euler–Lagrange equation therefore takes the form

$$\sum_{k=1}^{3}\frac{\partial^2\psi}{\partial x_k^2} + \lambda\psi = 0$$

which is just the Helmholtz equation (3.136) with $k^2 = \lambda$.

3.8 Sturm–Liouville equation

In spherical polar coordinates r, θ, ϕ the Helmholtz equation

$$\nabla^2\psi + k^2\psi = 0$$

assumes the form

$$\frac{1}{r^2}\frac{\partial}{\partial r}\left(r^2\frac{\partial\psi}{\partial r}\right) + \frac{1}{r^2\sin\theta}\frac{\partial}{\partial\theta}\left(\sin\theta\frac{\partial\psi}{\partial\theta}\right)$$
$$+ \frac{1}{r^2\sin^2\theta}\frac{\partial^2\psi}{\partial\phi^2} + k^2\psi = 0. \quad (3.142)$$

If we separate the variables by substituting

$$\psi(r,\theta,\phi) = R(r)S(\theta,\phi), \quad (3.143)$$

we obtain

$$\frac{1}{R}\frac{d}{dr}\left(r^2\frac{dR}{dr}\right) + k^2 r^2 = -\frac{1}{S}\left\{\frac{1}{\sin\theta}\frac{\partial}{\partial\theta}\left(\sin\theta\frac{\partial S}{\partial\theta}\right) + \frac{1}{\sin^2\theta}\frac{\partial^2 S}{\partial\phi^2}\right\}$$

and, since the left-hand side of this equation is a function of r only and the right-hand side is a function of θ and ϕ only, both sides may be put equal to a constant α so that we get

$$\frac{d}{dr}\left(r^2\frac{dR}{dr}\right) + (k^2 r^2 - \alpha)R = 0 \quad (3.144)$$

and

$$\frac{1}{\sin\theta}\frac{\partial}{\partial\theta}\left(\sin\theta\frac{\partial S}{\partial\theta}\right) + \frac{1}{\sin^2\theta}\frac{\partial^2 S}{\partial\phi^2} + \alpha S = 0. \quad (3.145)$$

Separating the variables further by putting
$$S(\theta, \phi) = \Theta(\theta)\Phi(\phi) \tag{3.146}$$
we obtain
$$\frac{d^2\Phi}{d\phi^2} + \beta\Phi = 0 \tag{3.147}$$
and
$$\frac{1}{\sin\theta}\frac{d}{d\theta}\left(\sin\theta\frac{d\Theta}{d\theta}\right) + \left(\alpha - \frac{\beta}{\sin^2\theta}\right)\Theta = 0, \tag{3.148}$$
β being the constant of separation.

Equation (3.147) has a single valued solution within the range $0 \leq \phi < 2\pi$ if $\beta = m^2$ where m is an integer, in which case we have
$$\Phi(\phi) = A\,e^{im\phi} + B\,e^{-im\phi}, \tag{3.149}$$
where A and B are constants. Defining the new variable $\mu = \cos\theta$ and writing $\Theta(\theta) = P(\mu)$ we see that (3.148) becomes
$$\frac{d}{d\mu}\left\{(1-\mu^2)\frac{dP}{d\mu}\right\} + \left(\alpha - \frac{m^2}{1-\mu^2}\right)P = 0. \tag{3.150}$$

In general the solutions of this equation are infinite at $\mu = \pm 1$. However if we put $\alpha = l(l+1)$, where l is a positive integer or zero, one of its solutions is bounded for $-1 \leq \mu \leq 1$. This bounded solution is denoted by $P_l^m(\mu)$ and is called an *associated Legendre function*, being of the form of a polynomial of order $l - |m|$ in μ times the function $(1-\mu^2)^{(\frac{1}{2})|m|}$.

The three equations (3.144), (3.147) and (3.150) are all of the *Sturm–Liouville* type
$$\frac{d}{dx}\left\{p(x)\frac{d\psi}{dx}\right\} + \{q(x) + \lambda r(x)\}\psi = 0, \tag{3.151}$$
where λ is the separation constant. $p(x)$ and $r(x)$ are algebraic functions of x having a finite number of zeros and poles. The points at which $p(x)$ vanishes are singular points of equation (3.151) and the range of definition $a \leq x \leq b$ must not include any of these singular points except at the end points $x = a, b$. In addition the function $q(x)$ must not have a singularity at any point within the range of definition, apart from the end points.

FIELD EQUATIONS 105

The Sturm–Liouville equation may be derived by supposing that the integral

$$I = \int_a^b \tfrac{1}{2}\left\{p(x)\left(\frac{d\psi}{dx}\right)^2 - q(x)\psi^2\right\} dx \qquad (3.152)$$

is stationary subject to the subsidiary condition

$$\int_a^b r(x)\psi^2 \, dx = \text{constant} \qquad (3.153)$$

being satisfied. Thus, introducing the modified Lagrangian density

$$\mathcal{L}' = \tfrac{1}{2}p(x)\left(\frac{d\psi}{dx}\right)^2 - \tfrac{1}{2}\{q(x) + \lambda r(x)\}\psi^2, \qquad (3.154)$$

where λ is a Lagrange multiplier, and using the Euler–Lagrange equation

$$\frac{d}{dx}\frac{\partial \mathcal{L}'}{\partial(\partial\psi/\partial x)} - \frac{\partial \mathcal{L}'}{\partial \psi} = 0, \qquad (3.155)$$

(3.151) follows immediately.

3.9 Electromagnetic field

So far in this chapter we have confined our attention to examples of fields described by a single component, that is to *scalar fields*. We now consider a case of a *vector field*, the electromagnetic field, which is governed by Maxwell's equations for the electric field vector **E** and the magnetic field vector **H**. In the presence of an electric charge density ρ and an electric current density **j**, these assume the form

$$\text{curl } \mathbf{E} + \frac{1}{c}\frac{\partial \mathbf{H}}{\partial t} = 0, \qquad (3.156)$$

$$\text{div } \mathbf{H} = 0, \qquad (3.157)$$

$$\text{curl } \mathbf{H} - \frac{1}{c}\frac{\partial \mathbf{E}}{\partial t} = \frac{4\pi}{c}\mathbf{j}, \qquad (3.158)$$

and

$$\text{div } \mathbf{E} = 4\pi\rho. \qquad (3.159)$$

It then follows in the way explained on p. 11 that we may write
$$\mathbf{H} = \operatorname{curl} \mathbf{A} \tag{3.160}$$
and
$$\mathbf{E} = -\nabla V - \frac{1}{c}\frac{\partial \mathbf{A}}{\partial t}, \tag{3.161}$$

where $\mathbf{A}$ and V are vector and scalar potentials. Substituting expressions (3.160) and (3.161) for $\mathbf{H}$ and $\mathbf{E}$ into equations (3.158) and (3.159), we obtain

$$\nabla^2 \mathbf{A} - \frac{1}{c^2}\frac{\partial^2 \mathbf{A}}{\partial t^2} - \nabla\left(\operatorname{div} \mathbf{A} + \frac{1}{c}\frac{\partial V}{\partial t}\right) = -\frac{4\pi}{c}\mathbf{j} \tag{3.162}$$

and

$$\nabla^2 V + \frac{1}{c}\frac{\partial}{\partial t}\operatorname{div} \mathbf{A} = -4\pi\rho, \tag{3.163}$$

using the formula
$$\operatorname{curl}\operatorname{curl} \mathbf{A} = \operatorname{grad}\operatorname{div} \mathbf{A} - \nabla^2 \mathbf{A}.$$

Now the potentials are not uniquely defined since the electric and magnetic field vectors $\mathbf{E}$ and $\mathbf{H}$ remain unchanged if $\mathbf{A}$ and V are replaced by
$$\mathbf{A}' = \mathbf{A} + \nabla\chi \tag{3.164}$$
and
$$V' = V - \frac{1}{c}\frac{\partial \chi}{\partial t}, \tag{3.165}$$

where χ is an arbitrary function of position and time, the transformation from the potentials $\mathbf{A}$ and V to $\mathbf{A}'$ and V' given by equations (3.164) and (3.165) being called a *gauge transformation*. Inverting the transformation and starting with any pair of potentials $\mathbf{A}'$ and V', we see that $\mathbf{A}$ and V will satisfy the auxiliary condition
$$\operatorname{div} \mathbf{A} + \frac{1}{c}\frac{\partial V}{\partial t} = 0 \tag{3.166}$$

if χ is chosen so that
$$\nabla^2\chi - \frac{1}{c^2}\frac{\partial^2\chi}{\partial t^2} = \operatorname{div} \mathbf{A}' + \frac{1}{c}\frac{\partial V'}{\partial t}, \tag{3.167}$$

FIELD EQUATIONS

in which case (3.162) and (3.163) simplify to the wave equations

$$\nabla^2 \mathbf{A} - \frac{1}{c^2}\frac{\partial^2 \mathbf{A}}{\partial t^2} = -\frac{4\pi}{c}\mathbf{j}, \quad (3.168)$$

$$\nabla^2 V - \frac{1}{c^2}\frac{\partial^2 V}{\partial t^2} = -4\pi\rho. \quad (3.169)$$

Our next step is to rewrite our equations in four-tensor notation. To do this we introduce a four-potential ψ_μ such that

$$\psi_1 = A_x, \quad \psi_2 = A_y, \quad \psi_3 = A_z, \quad \psi_4 = iV \quad (3.170)$$

and a four-current j_μ such that

$$j_1 = j_x, \quad j_2 = j_y, \quad j_3 = j_z, \quad j_4 = ic\rho. \quad (3.171)$$

Then the auxiliary condition (3.166) may be expressed in the form

$$\sum_{\mu=1}^{4} \frac{\partial \psi_\mu}{\partial x_\mu} = 0 \quad (3.172)$$

and the wave equations (3.168) and (3.169) may be written as

$$\sum_{\mu=1}^{4} \frac{\partial^2 \psi_\nu}{\partial x_\mu^2} = -\frac{4\pi}{c} j_\nu \quad (\nu = 1, 2, 3, 4), \quad (3.173)$$

where x_1, x_2, x_3 are the Cartesian coordinates x, y, z of a point in space and $x_4 = ict$.

Since the current density may be put into the form

$$\mathbf{j} = \rho\mathbf{v}, \quad (3.174)$$

where $\mathbf{v}$ is the velocity of the electric charge, conservation of charge yields the continuity equation

$$\frac{\partial \rho}{\partial t} + \text{div } \mathbf{j} = 0 \quad (3.175)$$

which is analogous to the continuity equation (3.45) derived in the case of fluid flow. Using the four-vector notation (3.171) this may be rewritten in the form

$$\sum_{\mu=1}^{4} \frac{\partial j_\mu}{\partial x_\mu} = 0. \quad (3.176)$$

We may also express Maxwell's equations in four-tensor notation by denoting the components of the electromagnetic field by the antisymmetric tensor $F_{\mu\nu}$, where $F_{\mu\mu} = 0$ ($\mu = 1, 2, 3, 4$),

$$iE_x = F_{41} = -F_{14}, \quad iE_y = F_{42} = -F_{24}, \quad iE_z = F_{43} = -F_{34} \tag{3.177}$$

and

$$H_x = F_{23} = -F_{32}, \quad H_y = F_{31} = -F_{13}, \quad H_z = F_{12} = -F_{21}. \tag{3.178}$$

Then equations (3.160) and (3.161) can be written as

$$F_{\mu\nu} = \frac{\partial \psi_\nu}{\partial x_\mu} - \frac{\partial \psi_\mu}{\partial x_\nu} \quad (\mu, \nu = 1, 2, 3, 4) \tag{3.179}$$

and (3.158) and (3.159) are equivalent to

$$\sum_{\mu=1}^{4} \frac{\partial F_{\mu\nu}}{\partial x_\mu} = -\frac{4\pi}{c} j_\nu \quad (\nu = 1, 2, 3, 4). \tag{3.180}$$

Further, introducing the third order tensor given by

$$t_{\mu\nu\lambda} = \frac{\partial F_{\mu\nu}}{\partial x_\lambda}, \tag{3.181}$$

we see that

$$t_{\mu\nu\lambda} + t_{\lambda\mu\nu} + t_{\nu\lambda\mu} = 0 \tag{3.182}$$

is equivalent to the remaining two Maxwell's equations (3.156) and (3.157).

Let us now consider the derivation of the wave equation (3.173) for the four potentials ψ_ν by means of a variational principle. Choosing the Lagrangian density to have the form

$$\mathscr{L} = \frac{1}{c} \sum_{\mu=1}^{4} j_\mu \psi_\mu - \frac{1}{16\pi} \sum_{\mu=1}^{4} \sum_{\nu=1}^{4} \left(\frac{\partial \psi_\nu}{\partial x_\mu} - \frac{\partial \psi_\mu}{\partial x_\nu} \right)^2 - \frac{1}{8\pi} \left(\sum_{\mu=1}^{4} \frac{\partial \psi_\mu}{\partial x_\mu} \right)^2, \tag{3.183}$$

the Euler–Lagrange equation

$$\frac{\partial \mathscr{L}}{\partial \psi_\nu} - \sum_{\mu=1}^{4} \frac{\partial}{\partial x_\mu} \frac{\partial \mathscr{L}}{\partial (\partial \psi_\nu / \partial x_\mu)} = 0$$

corresponding to the variation of the component ψ_ν, yields

$$\frac{4\pi}{c}j_\nu + \sum_{\mu=1}^{4}\frac{\partial}{\partial x_\mu}\left(\frac{\partial \psi_\nu}{\partial x_\mu} - \frac{\partial \psi_\mu}{\partial x_\nu}\right) + \frac{\partial}{\partial x_\nu}\left(\sum_{\mu=1}^{4}\frac{\partial \psi_\mu}{\partial x_\mu}\right) = 0$$

which is readily seen to be equivalent to equation (3.173). However, on making the substitution (3.179) we get

$$\sum_{\mu=1}^{4}\frac{\partial F_{\mu\nu}}{\partial x_\mu} + \frac{\partial}{\partial x_\nu}\left(\sum_{\mu=1}^{4}\frac{\partial \psi_\mu}{\partial x_\mu}\right) = -\frac{4\pi}{c}j_\nu$$

which is the same as equation (3.180), or Maxwell's equations (3.158) and (3.159), only if

$$\Lambda = \sum_{\mu=1}^{4}\frac{\partial \psi_\mu}{\partial x_\mu} \qquad (3.184)$$

is independent of the coordinates x_ν. By taking

$$\Lambda = 0, \qquad \frac{\partial \Lambda}{\partial t} = 0 \qquad (3.185)$$

at time $t = 0$ and noting that Λ must satisfy the equation

$$\nabla^2 \Lambda = \frac{1}{c^2}\frac{\partial^2 \Lambda}{\partial t^2}, \qquad (3.186)$$

we obtain $\partial^n \Lambda/\partial t^n = 0$ for all n at time $t = 0$ from which it follows by using Taylor's theorem that Λ must vanish for all values of the time t. Hence the auxiliary condition (3.172) is satisfied and only those solutions of equation (3.173) are selected which correspond to the electromagnetic field. Then the Lagrangian density (3.183) becomes

$$\begin{aligned}\mathscr{L} &= \frac{1}{c}\sum_{\mu=1}^{4}j_\mu\psi_\mu - \frac{1}{16\pi}\sum_{\mu=1}^{4}\sum_{\nu=1}^{4}(F_{\mu\nu})^2 \\ &= \frac{1}{c}\mathbf{A}\cdot\mathbf{j} - \rho V + \frac{1}{8\pi}(\mathbf{E}^2 - \mathbf{H}^2).\end{aligned} \qquad (3.187)$$

To obtain the Hamiltonian density $\mathscr{H}$ we require the canonical momenta π_ν conjugate to the ψ_ν. They are given by

$$\pi_\nu = \frac{\partial \mathscr{L}}{\partial \dot{\psi}_\nu} = \frac{1}{ic} \frac{\partial \mathscr{L}}{\partial(\partial \psi_\nu/\partial x_4)}$$
$$= \frac{1}{4\pi ic}\left\{\frac{\partial \psi_4}{\partial x_\nu} - \frac{\partial \psi_\nu}{\partial x_4} - \delta_{\nu 4}\sum_{\mu=1}^{4}\frac{\partial \psi_\mu}{\partial x_\mu}\right\} \qquad (3.188)$$

and so

$$\mathscr{H} = \sum_{\nu=1}^{4}\pi_\nu \dot{\psi}_\nu - \mathscr{L}$$
$$= \frac{1}{16\pi}\sum_{j=1}^{3}\sum_{k=1}^{3}\left(\frac{\partial \psi_j}{\partial x_k} - \frac{\partial \psi_k}{\partial x_j}\right)^2 + \frac{1}{8\pi}\sum_{k=1}^{3}\left(\frac{\partial \psi_k}{\partial x_4} - \frac{\partial \psi_4}{\partial x_k}\right)^2$$
$$+ \frac{1}{4\pi}\sum_{k=1}^{3}\frac{\partial \psi_k}{\partial x_4}\left(\frac{\partial \psi_4}{\partial x_k} - \frac{\partial \psi_k}{\partial x_4}\right) + \frac{1}{8\pi}\left(\sum_{\mu=1}^{4}\frac{\partial \psi_\mu}{\partial x_\mu}\right)^2$$
$$- \frac{1}{4\pi}\frac{\partial \psi_4}{\partial x_4}\left(\sum_{\nu=1}^{4}\frac{\partial \psi_\nu}{\partial x_\nu}\right) - \frac{1}{c}\sum_{\mu=1}^{4}j_\mu \psi_\mu,$$

that is

$$\mathscr{H} = \frac{1}{8\pi}(\mathbf{H}^2 - \mathbf{E}^2) - \frac{1}{4\pi c}\frac{\partial \mathbf{A}}{\partial t}\cdot \mathbf{E} + \frac{1}{8\pi}\left(\text{div } \mathbf{A} + \frac{1}{c}\frac{\partial V}{\partial t}\right)^2$$
$$- \frac{1}{4\pi c}\frac{\partial V}{\partial t}\left(\text{div } \mathbf{A} + \frac{1}{c}\frac{\partial V}{\partial t}\right) + \rho V - \frac{1}{c}\mathbf{A}\cdot\mathbf{j}. \quad (3.189)$$

Using equation (3.161) we see that expression (3.189) for $\mathscr{H}$ may be rewritten in the form

$$\mathscr{H} = \frac{1}{8\pi}(\mathbf{E}^2 + \mathbf{H}^2) + \frac{1}{4\pi}\mathbf{E}\cdot\nabla V$$
$$+ \frac{1}{8\pi}\left\{(\text{div }\mathbf{A})^2 - \frac{1}{c^2}\left(\frac{\partial V}{\partial t}\right)^2\right\} + \rho V - \frac{1}{c}\mathbf{A}\cdot\mathbf{j} \quad (3.190)$$

and so, on applying the auxiliary condition (3.166) and using the Maxwell equation (3.159), the Hamiltonian function becomes

$$H = \int \mathscr{H}\, d\mathbf{r} = \int\left[\frac{1}{8\pi}(\mathbf{E}^2 + \mathbf{H}^2) + \frac{1}{4\pi}\text{div}(V\mathbf{E}) - \frac{1}{c}\mathbf{A}\cdot\mathbf{j}\right]d\mathbf{r}. \quad (3.191)$$

But

$$\int \text{div}(V\mathbf{E})\, d\mathbf{r} = \int V\mathbf{E}\cdot d\mathbf{S}$$

FIELD EQUATIONS

which vanishes if the potentials fall off sufficiently fast at large distances, and then we have

$$H = \int \left[\frac{1}{8\pi} (\mathbf{E}^2 + \mathbf{H}^2) - \frac{1}{c} \mathbf{A} \cdot \mathbf{j} \right] d\mathbf{r}. \tag{3.192}$$

To apply the Hamiltonian formalism described in section 3.2 we must now express $\mathscr{H}$ in terms of the momenta π_ν and eliminate $\partial V/\partial t$. From (3.188) we see that

$$\pi_k = -\frac{1}{4\pi c} E_k \quad (k = 1, 2, 3), \tag{3.193}$$

where E_1, E_2, E_3 denote the Cartesian components of the electric field, and

$$\pi_4 = \frac{i}{4\pi c} \left(\text{div } \mathbf{A} + \frac{1}{c} \frac{\partial V}{\partial t} \right), \tag{3.194}$$

so that (3.190) may be written in the form

$$\mathscr{H} = 2\pi c^2 \sum_{\mu=1}^{4} \pi_\mu^2 + \frac{1}{16\pi} \sum_{j=1}^{3} \sum_{k=1}^{3} \left(\frac{\partial \psi_j}{\partial x_k} - \frac{\partial \psi_k}{\partial x_j} \right)^2$$
$$+ ic \sum_{k=1}^{3} \pi_k \frac{\partial \psi_4}{\partial x_k} - ic\pi_4 \sum_{k=1}^{3} \frac{\partial \psi_k}{\partial x_k} - \frac{1}{c} \sum_{\mu=1}^{4} j_\mu \psi_\mu. \tag{3.195}$$

Then the Hamilton equation

$$\frac{\partial \psi_\mu}{\partial t} = \frac{\partial \mathscr{H}}{\partial \pi_\mu} \tag{3.196}$$

yields expression (3.188) for the momentum density, while the Hamilton equation

$$\frac{\partial \pi_\mu}{\partial t} = -\frac{\partial \mathscr{H}}{\partial \psi_\mu} + \sum_{k=1}^{3} \frac{\partial}{\partial x_k} \frac{\partial \mathscr{H}}{\partial (\partial \psi_\mu/\partial x_k)} \tag{3.197}$$

gives rise to Maxwell's equations (3.158) and (3.159) provided we satisfy the auxiliary condition (3.166) by putting $\pi_4 = 0$ in the field equations.

3.10 Diffusion equation

In this section we are concerned with diffusion phenomena and in particular with the diffusion of a fluid through another fluid and with the flow of heat through a conducting medium. In the former case, if ρ is the mass of the diffusing fluid per unit volume, the direction of flow at any point P is normal to the surface of constant ρ at that point and the rate of flow across unit area of that surface is proportional to the gradient of ρ at P. Hence we may put

$$\mathbf{F} = -k\nabla\rho, \tag{3.198}$$

where $\mathbf{F}$ has the direction of the flow of fluid and magnitude equal to the mass of fluid crossing unit area per unit time and k is a constant called the *coefficient of diffusion*. Since the continuity equation

$$\frac{\partial \rho}{\partial t} + \operatorname{div} \mathbf{F} = 0 \tag{3.199}$$

must also be satisfied by the diffusing fluid, we see that ρ satisfies

$$\frac{\partial \rho}{\partial t} = k\nabla^2\rho \tag{3.200}$$

which is known as the *diffusion equation*. It applies not only to the flow of an actual fluid but also to other diffusion phenomena such as the motion of electrons and ions through a gas, the diffusion of neutrons through matter* and the conduction of heat through an isotropic medium.

In the last mentioned example the flow of heat at any point P of the medium is normal to the isothermal surface through that point and the rate of flow of heat per unit area is proportional to the gradient of the temperature θ at P. Hence we may represent the flow of heat by the vector

$$\mathbf{F} = -k\nabla\theta, \tag{3.201}$$

---

* The determination of asymptotic neutron densities in certain neutron diffusion problems by the use of a variational principle has been investigated by R. E. Marshak.[1]

FIELD EQUATIONS

where k is called the *thermal conductivity*. If m is the mass per unit volume of the conducting medium and C is the *specific heat*, that is, the amount of heat absorbed by unit mass of the body per unit rise in temperature, we see that the rate of absorption of heat by the matter within an arbitrary volume V bounded by a closed surface S is

$$\int_V mC \frac{\partial \theta}{\partial t} d\mathbf{r}.$$

But the rate at which heat is flowing out through S is

$$\oint_S \mathbf{F}.d\mathbf{S} = \int_V \text{div } \mathbf{F} \, d\mathbf{r}$$

by Gauss's divergence theorem, and so

$$mC \frac{\partial \theta}{\partial t} + \text{div } \mathbf{F} = 0, \qquad (3.202)$$

which has the same form as the continuity equation (3.199). It follows that

$$\frac{\partial \theta}{\partial t} = \kappa \nabla^2 \theta, \qquad (3.203)$$

where $\kappa = k/mC$ is called the *diffusivity*. Equation (3.203) is just the *equation for heat conduction* and is analogous to the diffusion equation (3.200).

Since the total energy of a diffusing system is not constant, we are unable to treat the problem of diffusion by the Lagrangian formalism without some modification. In order to produce a constant Hamiltonian function we introduce an imaginary system which takes up the energy lost by the actual system. This can be done by choosing for the Lagrangian density

$$\mathscr{L} = -k\nabla\psi^\dagger.\nabla\psi - \tfrac{1}{2}(\psi^\dagger\dot\psi - \dot\psi\psi^\dagger), \qquad (3.204)$$

where ψ is the field component describing the actual system and $\psi^\dagger$ is the field component for the compensating system. Then independent variations of $\psi^\dagger$ and ψ give rise to the Euler–Lagrange equations

$$\frac{\partial \psi}{\partial t} = k\nabla^2 \psi \qquad (3.205)$$

and

$$\frac{\partial \psi^\dagger}{\partial t} = -k\nabla^2 \psi^\dagger, \qquad (3.206)$$

where (3.205) is just the diffusion equation derived previously in this section for the particular cases of fluid flow and heat flow, while (3.206) is the equation for the imaginary system which absorbs all the energy lost by the actual system. We see at once that

$$\psi^\dagger(t) = \psi(-t) \qquad (3.207)$$

so that the compensating system can be regarded as the reflection of the actual system with respect to time.

Defining the canonical momenta

$$\pi = \frac{\partial \mathscr{L}}{\partial \dot\psi} = -\tfrac{1}{2}\psi^\dagger \qquad (3.208)$$

and

$$\pi^\dagger = \frac{\partial \mathscr{L}}{\partial \dot\psi^\dagger} = \tfrac{1}{2}\psi, \qquad (3.209)$$

we see that the Hamiltonian density takes the form

$$\mathscr{H} = \pi\dot\psi + \pi^\dagger\dot\psi^\dagger - \mathscr{L} = k\nabla\psi^\dagger \cdot \nabla\psi. \qquad (3.210)$$

Hence the Hamiltonian function is given by

$$H = k \int \nabla\psi^\dagger \cdot \nabla\psi \, d\mathbf{r} \qquad (3.211)$$

and so, using Green's theorem and assuming that there is no diffusion out of the boundary surface, we find that

$$\frac{dH}{dt} = -k \int (\psi^\dagger \nabla^2 \psi + \psi \nabla^2 \psi^\dagger) \, d\mathbf{r} = 0 \qquad (3.212)$$

from which it follows at once that the Hamiltonian function is independent of the time.

3.11 Complex field components

In all the foregoing discussion in this chapter we have supposed that the field components are real functions of the spacial co-ordinates x_1, x_2, x_3 and the time t. We now turn our attention to those cases for which the field components ψ_σ are complex and may be written in the form

$$\psi_\sigma = \frac{1}{\sqrt{2}}(\psi_\sigma^{(1)} + i\psi_\sigma^{(2)}) \qquad (\sigma = 1,\ldots,n), \qquad (3.213)$$

where $\psi_\sigma^{(1)}$ and $\psi_\sigma^{(2)}$ are real functions. Assuming that $\psi_\sigma^{(1)}$ and $\psi_\sigma^{(2)}$ are all entirely independent of each other we may conveniently express the Lagrangian density for the field as

$$\mathscr{L}(t, \psi_\sigma, \bar{\psi}_\sigma, \nabla\psi_\sigma, \nabla\bar{\psi}_\sigma, \dot{\psi}_\sigma, \dot{\bar{\psi}}_\sigma), \qquad (3.214)$$

where $\bar{\psi}_\sigma$ is the complex conjugate of ψ_σ. Independent variations of the functions ψ_σ and $\bar{\psi}_\sigma$ then lead to the Euler–Lagrange equations

$$\frac{\partial\mathscr{L}}{\partial\psi_\sigma} - \sum_{k=1}^{3}\frac{\partial}{\partial x_k}\frac{\partial\mathscr{L}}{\partial(\partial\psi_\sigma/\partial x_k)} - \frac{\partial}{\partial t}\frac{\partial\mathscr{L}}{\partial\dot{\psi}_\sigma} = 0 \qquad (3.215)$$

and

$$\frac{\partial\mathscr{L}}{\partial\bar{\psi}_\sigma} - \sum_{k=1}^{3}\frac{\partial}{\partial x_k}\frac{\partial\mathscr{L}}{\partial(\partial\bar{\psi}_\sigma/\partial x_k)} - \frac{\partial}{\partial t}\frac{\partial\mathscr{L}}{\partial\dot{\bar{\psi}}_\sigma} = 0 \qquad (3.216)$$

by following an analogous procedure to that described in section 3.1.

Associated with the complex field we can introduce an electric charge density ρ defined by the expression

$$\rho = -i\epsilon \sum_{\sigma=1}^{n}\left(\frac{\partial\mathscr{L}}{\partial\dot{\psi}_\sigma}\psi_\sigma - \frac{\partial\mathscr{L}}{\partial\dot{\bar{\psi}}_\sigma}\bar{\psi}_\sigma\right) \qquad (3.217)$$

and an electric current vector density **j** with components defined by

$$j_k = -i\epsilon \sum_{\sigma=1}^{n}\left\{\frac{\partial\mathscr{L}}{\partial(\partial\psi_\sigma/\partial x_k)}\psi_\sigma - \frac{\partial\mathscr{L}}{\partial(\partial\bar{\psi}_\sigma/\partial x_k)}\bar{\psi}_\sigma\right\} \qquad (k=1,2,3), \qquad (3.218)$$

where ϵ is a real number. Then we have

$$\frac{\partial \rho}{\partial t} + \text{div}\,\mathbf{j} = -i\epsilon \sum_{\sigma=1}^{n} \left[\left\{ \frac{\partial \mathscr{L}}{\partial \psi_\sigma} \psi_\sigma + \sum_{k=1}^{3} \frac{\partial \mathscr{L}}{\partial(\partial \psi_\sigma/\partial x_k)} \frac{\partial \psi_\sigma}{\partial x_k} + \frac{\partial \mathscr{L}}{\partial \dot\psi_\sigma} \dot\psi_\sigma \right\} \right.$$
$$\left. - \left\{ \frac{\partial \mathscr{L}}{\partial \bar\psi_\sigma} \bar\psi_\sigma + \sum_{k=1}^{3} \frac{\partial \mathscr{L}}{\partial(\partial \bar\psi_\sigma/\partial x_k)} \frac{\partial \bar\psi_\sigma}{\partial x_k} + \frac{\partial \mathscr{L}}{\partial \dot{\bar\psi}_\sigma} \dot{\bar\psi}_\sigma \right\} \right]$$

making use of the field equations (3.215) and (3.216).

At this stage we suppose that the Lagrangian density (3.214) is invariant under the gauge transformation

$$\psi_\sigma \to \psi_\sigma e^{i\alpha}, \qquad \bar\psi_\sigma \to \bar\psi_\sigma e^{-i\alpha}, \qquad (3.219)$$

where α is a real number. Now the infinitesimal variation

$$\psi_\sigma \to \psi_\sigma(1 + i\alpha), \qquad \bar\psi_\sigma \to \bar\psi_\sigma(1 - i\alpha)$$

results in a change in $\mathscr{L}$ of amount

$$i\alpha \sum_{\sigma=1}^{n} \left[\left\{ \frac{\partial \mathscr{L}}{\partial \psi_\sigma} \psi_\sigma + \sum_{k=1}^{3} \frac{\partial \mathscr{L}}{\partial(\partial \psi_\sigma/\partial x_k)} \frac{\partial \psi_\sigma}{\partial x_k} + \frac{\partial \mathscr{L}}{\partial \dot\psi_\sigma} \dot\psi_\sigma \right\} \right.$$
$$\left. - \left\{ \frac{\partial \mathscr{L}}{\partial \bar\psi_\sigma} \bar\psi_\sigma + \sum_{k=1}^{3} \frac{\partial \mathscr{L}}{\partial(\partial \bar\psi_\sigma/\partial x_k)} \frac{\partial \bar\psi_\sigma}{\partial x_k} + \frac{\partial \mathscr{L}}{\partial \dot{\bar\psi}_\sigma} \dot{\bar\psi}_\sigma \right\} \right]$$

which must therefore vanish because of the invariant property of the Lagrangian density under the transformation (3.219). It then follows from this result, which is a particular case of a theorem due to Noether[2] on invariant problems in the calculus of variations, that

$$\frac{\partial \rho}{\partial t} + \text{div}\,\mathbf{j} = 0 \qquad (3.220)$$

and so the continuity equation is satisfied by ρ and $\mathbf{j}$. Of course in the case of a real field, both ρ and $\mathbf{j}$ vanish and the above analysis no longer applies.

We conclude this section by considering the Hamiltonian density $\mathscr{H}$. Introducing the momenta

$$\pi_\sigma = \frac{\partial \mathscr{L}}{\partial \dot\psi_\sigma}, \qquad \bar\pi_\sigma = \frac{\partial \mathscr{L}}{\partial \dot{\bar\psi}_\sigma} \qquad (3.221)$$

FIELD EQUATIONS

canonically conjugate to ψ_σ and $\bar{\psi}_\sigma$ respectively, and the momenta

$$\pi_\sigma^{(1)} = \frac{\partial \mathscr{L}}{\partial \dot{\psi}_\sigma^{(1)}}, \quad \pi_\sigma^{(2)} = \frac{\partial \mathscr{L}}{\partial \dot{\psi}_\sigma^{(2)}} \qquad (3.222)$$

canonically conjugate to $\psi_\sigma^{(1)}$ and $\psi_\sigma^{(2)}$ respectively, we see that

$$\pi_\sigma^{(1)} = \frac{1}{\sqrt{2}} (\pi_\sigma + \bar{\pi}_\sigma) \qquad (3.223)$$

and

$$\pi_\sigma^{(2)} = \frac{i}{\sqrt{2}} (\pi_\sigma - \bar{\pi}_\sigma). \qquad (3.224)$$

Hence we have for the Hamiltonian density

$$\mathscr{H} = \sum_{\sigma=1}^{n} \{\pi_\sigma^{(1)} \dot{\psi}_\sigma^{(1)} + \pi_\sigma^{(2)} \dot{\psi}_\sigma^{(2)}\} - \mathscr{L} = \sum_{\sigma=1}^{n} \left\{ \pi_\sigma \dot{\psi}_\sigma + \bar{\pi}_\sigma \dot{\bar{\psi}}_\sigma \right\} - \mathscr{L}. \qquad (3.225)$$

Finally it should be noted that unless the Lagrangian density $\mathscr{L}$ is real, $\bar{\pi}_\sigma$ will not be the complex conjugate of π_σ.

3.12 Schrödinger equation

An important example of a field with a complex component ψ is provided by the Schrödinger equation

$$-\frac{\hbar^2}{2m} \nabla^2 \psi + V(\mathbf{r}, t)\psi = i\hbar \frac{\partial \psi}{\partial t} \qquad (3.226)$$

for a particle of mass m moving under the action of a potential $V(\mathbf{r}, t)$. This equation can be readily derived from a variational principle by taking for the Lagrangian density

$$\mathscr{L} = -\frac{\hbar^2}{2m} \nabla \bar{\psi} \cdot \nabla \psi - V(\mathbf{r}, t)\bar{\psi}\psi + \tfrac{1}{2}i\hbar(\bar{\psi}\dot{\psi} - \psi\dot{\bar{\psi}}) \qquad (3.227)$$

which depends on both ψ and its conjugate $\bar{\psi}$. Independent variations of $\bar{\psi}$ and ψ yield the Schrödinger equation (3.226) and its conjugate equation

$$-\frac{\hbar^2}{2m} \nabla^2 \bar{\psi} + V(\mathbf{r}, t)\bar{\psi} = -i\hbar \frac{\partial \bar{\psi}}{\partial t}. \qquad (3.228)$$

Defining the momenta canonically conjugate to ψ and $\bar{\psi}$ by

$$\pi = \frac{\partial \mathscr{L}}{\partial \dot{\psi}} = \tfrac{1}{2} i\hbar\bar{\psi} \qquad (3.229)$$

and

$$\bar{\pi} = \frac{\partial \mathscr{L}}{\partial \dot{\bar{\psi}}} = -\tfrac{1}{2} i\hbar\psi, \qquad (3.230)$$

we see that the Hamiltonian density becomes

$$\begin{aligned}\mathscr{H} &= \pi\dot{\psi} + \bar{\pi}\dot{\bar{\psi}} - \mathscr{L} \\ &= \frac{\hbar^2}{2m}\nabla\bar{\psi}\cdot\nabla\psi + V(\mathbf{r},t)\bar{\psi}\psi.\end{aligned} \qquad (3.231)$$

If we rewrite the Hamiltonian density in the form

$$\mathscr{H} = \frac{1}{i\hbar}\left\{\frac{\hbar^2}{2m}(\nabla\pi\cdot\nabla\psi - \nabla\bar{\pi}\cdot\nabla\bar{\psi}) + V(\mathbf{r},t)(\pi\psi - \bar{\pi}\bar{\psi})\right\} \qquad (2.232)$$

and use Hamilton's equations

$$\dot{\psi} = \frac{\delta H}{\delta \pi}, \quad \dot{\pi} = -\frac{\delta H}{\delta \psi}; \quad \dot{\bar{\psi}} = \frac{\delta H}{\delta \bar{\pi}}, \quad \dot{\bar{\pi}} = -\frac{\delta H}{\delta \bar{\psi}}, \qquad (3.233)$$

we regain the Schrödinger equation (3.226) and its conjugate equation (3.228), remembering that the complete expressions for the functional derivatives $\delta H/\delta\pi$ and $\delta H/\delta\bar{\pi}$ must be employed here since $\mathscr{H}$ depends upon the gradients of the canonical momenta in the present case.

In wave mechanics the wave function ψ is interpreted in such a manner that $\bar{\psi}\psi\,d\mathbf{r}$ is equal to the probability of observing the particle at time t within the volume element $d\mathbf{r}$ surrounding the point with position vector $\mathbf{r}$. Hence $\bar{\psi}\psi$ is referred to as the *probability density*. Putting $\epsilon = e/\hbar$ into expression (3.217) we get

$$\rho = -\frac{ie}{\hbar}\left(\frac{\partial \mathscr{L}}{\partial \dot{\psi}}\psi - \frac{\partial \mathscr{L}}{\partial \dot{\bar{\psi}}}\bar{\psi}\right) = e\bar{\psi}\psi \qquad (3.234)$$

FIELD EQUATIONS 119

for the charge density, e being the charge of the particle, in accord with the above interpretation. Now

$$\frac{\partial}{\partial t}\int \bar{\psi}\psi\, d\mathbf{r} = \int \left(\bar{\psi}\frac{\partial \psi}{\partial t} + \psi\frac{\partial \bar{\psi}}{\partial t}\right) d\mathbf{r}$$

$$= \frac{i\hbar}{2m}\int (\bar{\psi}\nabla^2\psi - \psi\nabla^2\bar{\psi})\, d\mathbf{r}$$

$$= \frac{i\hbar}{2m}\int \nabla\cdot(\bar{\psi}\nabla\psi - \psi\nabla\bar{\psi})\, d\mathbf{r}$$

and so

$$\int \left(\frac{\partial \rho}{\partial t} + \operatorname{div}\mathbf{j}\right) d\mathbf{r} = 0, \quad (3.235)$$

where

$$\mathbf{j} = -\frac{ie\hbar}{2m}(\bar{\psi}\nabla\psi - \psi\nabla\bar{\psi}) \quad (3.236)$$

is the electric current vector density whose components may be expressed in terms of the Lagrangian density (3.227) by the formula

$$j_k = -\frac{ie}{\hbar}\left\{\frac{\partial \mathscr{L}}{\partial(\partial\psi/\partial x_k)}\psi - \frac{\partial \mathscr{L}}{\partial(\partial\bar{\psi}/\partial x_k)}\bar{\psi}\right\}, \quad (3.237)$$

which comes from substituting $\epsilon = e/\hbar$ into (3.218).

The continuity equation

$$\frac{\partial \rho}{\partial t} + \operatorname{div}\mathbf{j} = 0 \quad (3.238)$$

relating the charge and current densities follows at once from (3.235) since the volume of integration is arbitrary.

3.13 Klein–Gordon equation

A *neutral scalar meson* having rest mass m_0 may be described by a single real field component ψ satisfying the Klein–Gordon equation

$$\nabla^2\psi - \frac{1}{c^2}\frac{\partial^2\psi}{\partial t^2} = \frac{m_0^2 c^2}{\hbar^2}\psi \quad (3.239)$$

discussed previously in section 2.5. This equation can be derived from a variational principle by choosing for the Lagrangian density

$$\begin{aligned}\mathscr{L} &= -\frac{\hbar^2}{2m_0}(\nabla\psi)^2 + \frac{\hbar^2}{2m_0 c^2}\dot\psi^2 - \frac{m_0 c^2}{2}\psi^2 \\ &= -\frac{\hbar^2}{2m_0}\left\{\sum_{\mu=1}^{4}\left(\frac{\partial\psi}{\partial x_\mu}\right)^2 + \frac{m_0^2 c^2}{\hbar^2}\psi^2\right\}, \quad (3.240)\end{aligned}$$

where x_1, x_2, x_3 are the Cartesian coordinates x, y, z respectively and $x_4 = ict$, since the Euler–Lagrange equation

$$\sum_{\mu=1}^{4}\frac{\partial}{\partial x_\mu}\frac{\partial\mathscr{L}}{\partial(\partial\psi/\partial x_\mu)} = \frac{\partial\mathscr{L}}{\partial\psi}$$

yields the field equation

$$\sum_{\mu=1}^{4}\frac{\partial^2\psi}{\partial x_\mu^2} = \frac{m_0^2 c^2}{\hbar^2}\psi \quad (3.241)$$

which is just the Klein–Gordon equation (3.239).

Defining the momentum canonically conjugate to ψ by

$$\pi = \frac{\partial\mathscr{L}}{\partial\dot\psi} = \frac{\hbar^2}{m_0 c^2}\dot\psi, \quad (3.242)$$

the Hamiltonian density becomes

$$\begin{aligned}\mathscr{H} &= \pi\dot\psi - \mathscr{L} \\ &= \frac{m_0 c^2}{2\hbar^2}\pi^2 + \frac{\hbar^2}{2m_0}(\nabla\psi)^2 + \frac{m_0 c^2}{2}\psi^2 \quad (3.243)\end{aligned}$$

from which it follows readily that Hamilton's equations

$$\dot\psi = \frac{\delta H}{\delta\pi}, \quad \dot\pi = -\frac{\delta H}{\delta\psi}$$

reproduce the formula (3.242) for the momentum density and the Klein–Gordon equation (3.239) respectively.

For the case of a *charged scalar meson* the field component ψ must be complex and so we take for the Lagrangian density

$$\begin{aligned}\mathscr{L} &= -\frac{\hbar^2}{2m_0}\nabla\bar\psi\cdot\nabla\psi + \frac{\hbar^2}{2m_0 c^2}\dot{\bar\psi}\dot\psi - \frac{m_0 c^2}{2}\bar\psi\psi \\ &= -\frac{\hbar^2}{2m_0}\left\{\sum_{\mu=1}^{4}\frac{\partial\bar\psi}{\partial x_\mu}\frac{\partial\psi}{\partial x_\mu} + \frac{m_0^2 c^2}{\hbar^2}\bar\psi\psi\right\}, \quad (3.244)\end{aligned}$$

rather than expression (3.240) appropriate to a neutral meson, which gives rise to the Euler–Lagrange equations

$$\sum_{\mu=1}^{4} \frac{\partial^2 \psi}{\partial x_\mu^2} = \frac{m_0^2 c^2}{\hbar^2} \psi \qquad (3.245)$$

and

$$\sum_{\mu=1}^{4} \frac{\partial^2 \bar{\psi}}{\partial x_\mu^2} = \frac{m_0^2 c^2}{\hbar^2} \bar{\psi}. \qquad (3.246)$$

These are just the Klein–Gordon equation and its conjugate equation.

Putting

$$\pi = \frac{\partial \mathscr{L}}{\partial \dot{\psi}} = \frac{\hbar^2}{2m_0 c^2} \dot{\bar{\psi}} \qquad (3.247)$$

and

$$\bar{\pi} = \frac{\partial \mathscr{L}}{\partial \dot{\bar{\psi}}} = \frac{\hbar^2}{2m_0 c^2} \dot{\psi}, \qquad (3.248)$$

the Hamiltonian density takes the form

$$\begin{aligned} \mathscr{H} &= \pi \dot{\psi} + \bar{\pi} \dot{\bar{\psi}} - \mathscr{L} \\ &= \frac{2m_0 c^2}{\hbar^2} \bar{\pi}\pi + \frac{\hbar^2}{2m_0} \nabla \bar{\psi} \cdot \nabla \psi + \frac{m_0 c^2}{2} \bar{\psi}\psi. \end{aligned} \qquad (3.249)$$

Using Hamilton's equations again we regenerate the definitions (3.247), (3.248) and the field equations (3.245), (3.246).

The Hamiltonian densities (3.243) and (3.249) are both positive definite which is a necessary condition for them to be equivalent to energy densities.

If e is the charge of the meson, the charge density ρ can be obtained from expression (3.217) by putting $\epsilon = e/\hbar$. We then get

$$\rho = -\frac{ie\hbar}{2m_0 c^2} (\dot{\bar{\psi}}\psi - \dot{\psi}\bar{\psi}) \qquad (3.250)$$

which can attain both positive and negative values, unlike the case of the Schrödinger equation treated in the previous section. On the other hand using expression (3.218) we obtain for the electric current density

$$\mathbf{j} = -\frac{ie\hbar}{2m_0}(\bar{\psi}\nabla\psi - \psi\nabla\bar{\psi}) \qquad (3.251)$$

which is identical to the expression (3.236) obtained for the Schrödinger equation. Now

$$\frac{\partial}{\partial t}\int(\dot{\bar{\psi}}\psi - \dot{\psi}\bar{\psi})\,d\mathbf{r} = \int\left(\psi\frac{\partial^2\bar{\psi}}{\partial t^2} - \bar{\psi}\frac{\partial^2\psi}{\partial t^2}\right)d\mathbf{r}$$

$$= c^2\int(\psi\nabla^2\bar{\psi} - \bar{\psi}\nabla^2\psi)\,d\mathbf{r}$$

$$= c^2\int\nabla\cdot(\psi\nabla\bar{\psi} - \bar{\psi}\nabla\psi)\,d\mathbf{r}$$

from which we see that

$$\int\left(\frac{\partial\rho}{\partial t} + \operatorname{div}\mathbf{j}\right)d\mathbf{r} = 0$$

and so the continuity equation is satisfied by ρ and $\mathbf{j}$. Of course this also follows as a consequence of the fact that the Lagrangian density is invariant under the transformation (3.219).

In the presence of an electromagnetic field determined by the scalar potential V and the vector potential $\mathbf{A}$, the Lagrangian function for the charged scalar meson requires modification. It now takes the form

$$\mathscr{L} = -\frac{\hbar^2}{2m_0}\left(\nabla\bar{\psi} + \frac{ie}{\hbar c}\mathbf{A}\bar{\psi}\right)\cdot\left(\nabla\psi - \frac{ie}{\hbar c}\mathbf{A}\psi\right)$$
$$+ \frac{\hbar^2}{2m_0 c^2}\left(\dot{\bar{\psi}} - \frac{ie}{\hbar}V\bar{\psi}\right)\left(\dot{\psi} + \frac{ie}{\hbar}V\psi\right) - \frac{m_0 c^2}{2}\bar{\psi}\psi \qquad (3.252)$$

which we may rewrite as

$$\mathscr{L} = -\frac{\hbar^2}{2m_0}\left\{\sum_{\mu=1}^{4}\left(\frac{\partial\bar{\psi}}{\partial x_\mu} + \frac{ie}{\hbar c}A_\mu\bar{\psi}\right)\left(\frac{\partial\psi}{\partial x_\mu} - \frac{ie}{\hbar c}A_\mu\psi\right) + \frac{m_0^2 c^2}{\hbar^2}\bar{\psi}\psi\right\}, \qquad (3.253)$$

where $A_1 = A_x$, $A_2 = A_y$, $A_3 = A_z$ and $A_4 = iV$. The Euler–Lagrange equations associated with independent variations of $\bar{\psi}$ and ψ are accordingly

$$\left\{\sum_{\mu=1}^{4}\left(\frac{\partial}{\partial x_\mu} - \frac{ie}{\hbar c}A_\mu\right)^2 - \frac{m_0^2 c^2}{\hbar^2}\right\}\psi = 0 \qquad (3.254)$$

and

$$\left\{\sum_{\mu=1}^{4}\left(\frac{\partial}{\partial x_\mu} + \frac{ie}{\hbar c}A_\mu\right)^2 - \frac{m_0^2 c^2}{\hbar^2}\right\}\bar{\psi} = 0 \qquad (3.255)$$

which are equivalent to equation (2.48) derived in section 2.5.

Introducing the canonical momenta

$$\pi = \frac{\partial \mathscr{L}}{\partial \dot{\psi}} = \frac{\hbar^2}{2m_0 c^2}\left(\dot{\bar{\psi}} - \frac{ie}{\hbar}V\bar{\psi}\right) \qquad (3.256)$$

and

$$\bar{\pi} = \frac{\partial \mathscr{L}}{\partial \dot{\bar{\psi}}} = \frac{\hbar^2}{2m_0 c^2}\left(\dot{\psi} + \frac{ie}{\hbar}V\psi\right), \qquad (3.257)$$

the Hamiltonian density takes the form

$$\begin{aligned}\mathscr{H} &= \pi\dot{\psi} + \bar{\pi}\dot{\bar{\psi}} - \mathscr{L} \\ &= \frac{2m_0 c^2}{\hbar^2}\bar{\pi}\pi + \frac{\hbar^2}{2m_0}\left(\nabla\bar{\psi} + \frac{ie}{\hbar c}\mathbf{A}\bar{\psi}\right)\cdot\left(\nabla\psi - \frac{ie}{\hbar c}\mathbf{A}\psi\right) \\ &\quad - \frac{ie}{\hbar}V(\pi\psi - \bar{\pi}\bar{\psi}) + \frac{m_0 c^2}{2}\bar{\psi}\psi. \end{aligned} \qquad (3.258)$$

Once again Hamilton's equations reproduce the field equations (3.254) and (3.255) as well as the formulae (3.256) and (3.257) for the momenta canonically conjugate to ψ and $\bar{\psi}$.

3.14 Vector meson field

In the preceding section we assumed that a meson particle can be completely represented by a single field component ψ. We now suppose that we require four complex field components ψ_ν to describe our particle, which we therefore call a *vector meson*, each component satisfying the Klein–Gordon equation

$$\sum_{\mu=1}^{4}\frac{\partial^2 \psi_\nu}{\partial x_\mu^2} = \frac{m_0^2 c^2}{\hbar^2}\psi_\nu \qquad (\nu = 1, 2, 3, 4). \qquad (3.259)$$

As for the case of the electromagnetic field, the components ψ_ν are not independent of each other but satisfy the auxiliary condition

$$\sum_{\nu=1}^{4} \frac{\partial \psi_\nu}{\partial x_\nu} = 0. \qquad (3.260)$$

If we choose for the Lagrangian density the function

$$\mathscr{L} = -\frac{\hbar^2}{2m_0}\left\{ \tfrac{1}{2} \sum_{\mu=1}^{4} \sum_{\nu=1}^{4} \left(\frac{\partial \bar{\psi}_\nu}{\partial x_\mu} - \frac{\partial \bar{\psi}_\mu}{\partial x_\nu}\right)\left(\frac{\partial \psi_\nu}{\partial x_\mu} - \frac{\partial \psi_\mu}{\partial x_\nu}\right) + \frac{m_0^2 c^2}{\hbar^2} \sum_{\nu=1}^{4} \bar{\psi}_\nu \psi_\nu \right\}, \qquad (3.261)$$

the Euler–Lagrange equations

$$\sum_{\mu=1}^{4} \frac{\partial}{\partial x_\mu} \frac{\partial \mathscr{L}}{\partial(\partial \bar{\psi}_\nu/\partial x_\mu)} = \frac{\partial \mathscr{L}}{\partial \bar{\psi}_\nu}$$

and

$$\sum_{\mu=1}^{4} \frac{\partial}{\partial x_\mu} \frac{\partial \mathscr{L}}{\partial(\partial \psi_\nu/\partial x_\mu)} = \frac{\partial \mathscr{L}}{\partial \psi_\nu}$$

associated with independent variations of $\bar{\psi}_\nu$ and ψ_ν furnish

$$\sum_{\mu=1}^{4} \frac{\partial}{\partial x_\mu} \left(\frac{\partial \psi_\nu}{\partial x_\mu} - \frac{\partial \psi_\mu}{\partial x_\nu}\right) = \frac{m_0^2 c^2}{\hbar^2} \psi_\nu \qquad (3.262)$$

and its conjugate equation. On differentiating (3.262) with respect to x_ν and summing over ν we arrive at the result

$$\frac{m_0^2 c^2}{\hbar^2} \sum_{\nu=1}^{4} \frac{\partial \psi_\nu}{\partial x_\nu} = \sum_{\mu=1}^{4} \sum_{\nu=1}^{4} \frac{\partial^2}{\partial x_\mu \partial x_\nu} \left(\frac{\partial \psi_\nu}{\partial x_\mu} - \frac{\partial \psi_\mu}{\partial x_\nu}\right) = 0 \qquad (3.263)$$

and so the auxiliary condition (3.260) follows automatically provided the mass m_0 of the meson is non-zero. Now making use of the auxiliary condition, we see that (3.262) reduces to the Klein–Gordon equation (3.259) as required.

If m_0 is zero we revert to the case of the electromagnetic field *in vacuo*, treated in section 3.9, for which the auxiliary condition no longer follows by virtue of equation (3.263) but must be imposed as an extra condition on the variational problem.

We now turn to the Hamiltonian formalism for the vector meson field. Defining the canonical momenta according to the formulae

$$\pi_\nu = \frac{\partial \mathscr{L}}{\partial \dot{\psi}_\nu} = \frac{1}{ic}\frac{\partial \mathscr{L}}{\partial(\partial \psi_\nu/\partial x_4)} = \frac{i\hbar^2}{2m_0 c}\left(\frac{\partial \bar{\psi}_\nu}{\partial x_4} - \frac{\partial \bar{\psi}_4}{\partial x_\nu}\right) \quad (\nu = 1,2,3,4)$$
(3.264)

and

$$\bar{\pi}_\nu = \frac{\partial \mathscr{L}}{\partial \dot{\bar{\psi}}_\nu} = \frac{1}{ic}\frac{\partial \mathscr{L}}{\partial(\partial \bar{\psi}_\nu/\partial x_4)} = \frac{i\hbar^2}{2m_0 c}\left(\frac{\partial \psi_\nu}{\partial x_4} - \frac{\partial \psi_4}{\partial x_\nu}\right) \quad (\nu = 1,2,3,4),$$
(3.265)

we see that

$$\dot{\psi}_k = \frac{2m_0 c^2}{\hbar^2}\bar{\pi}_k + ic\frac{\partial \psi_4}{\partial x_k} \quad (k = 1,2,3),$$
(3.266)

$$\dot{\bar{\psi}}_k = \frac{2m_0 c^2}{\hbar^2}\pi_k + ic\frac{\partial \bar{\psi}_4}{\partial x_k} \quad (k = 1,2,3),$$
(3.267)

and

$$\pi_4 = 0, \quad \bar{\pi}_4 = 0.$$
(3.268)

The Hamiltonian density is therefore given by

$$\mathscr{H} = \sum_{k=1}^{3}(\pi_k \dot{\psi}_k + \bar{\pi}_k \dot{\bar{\psi}}_k) - \mathscr{L}$$

$$= \frac{2m_0 c^2}{\hbar^2}\sum_{k=1}^{3}\bar{\pi}_k \pi_k + \frac{\hbar^2}{4m_0}\sum_{j=1}^{3}\sum_{k=1}^{3}\left(\frac{\partial \bar{\psi}_k}{\partial x_j} - \frac{\partial \bar{\psi}_j}{\partial x_k}\right)\left(\frac{\partial \psi_k}{\partial x_j} - \frac{\partial \psi_j}{\partial x_k}\right)$$

$$+ \tfrac{1}{2}m_0 c^2 \sum_{\nu=1}^{4}\bar{\psi}_\nu \psi_\nu + ic\sum_{k=1}^{3}\left(\pi_k \frac{\partial \psi_4}{\partial x_k} + \bar{\pi}_k \frac{\partial \bar{\psi}_4}{\partial x_k}\right). \quad (3.269)$$

In order to apply Hamilton's equations we need to eliminate the components ψ_4 and $\bar{\psi}_4$ from the Hamiltonian density because of the vanishing of their conjugate momenta. However, the derivatives of these components occur within expression (3.269) for $\mathscr{H}$ and so we add

$$-ic\sum_{k=1}^{3}\frac{\partial}{\partial x_k}(\pi_k \psi_4 + \bar{\pi}_k \bar{\psi}_4)$$

to $\mathscr{H}$ which is permissible because this term has the form of a divergence and does not alter the Hamiltonian function H. We now get for the Hamiltonian density

$$\mathscr{H} = \frac{2m_0 c^2}{\hbar^2} \sum_{k=1}^{3} \bar{\pi}_k \pi_k + \frac{\hbar^2}{4m_0} \sum_{j=1}^{3} \sum_{k=1}^{3} \left(\frac{\partial \bar{\psi}_k}{\partial x_j} - \frac{\partial \bar{\psi}_j}{\partial x_k}\right)\left(\frac{\partial \psi_k}{\partial x_j} - \frac{\partial \psi_j}{\partial x_k}\right)$$
$$+ \tfrac{1}{2} m_0 c^2 \sum_{\nu=1}^{4} \bar{\psi}_\nu \psi_\nu - ic \sum_{k=1}^{3} \left(\psi_4 \frac{\partial \pi_k}{\partial x_k} + \bar{\psi}_4 \frac{\partial \bar{\pi}_k}{\partial x_k}\right)$$

and since it follows from (3.262) that

$$\psi_4 = \frac{2i}{m_0 c} \sum_{k=1}^{3} \frac{\partial \bar{\pi}_k}{\partial x_k}, \qquad \bar{\psi}_4 = \frac{2i}{m_0 c} \sum_{k=1}^{3} \frac{\partial \pi_k}{\partial x_k},$$

we obtain finally

$$\mathscr{H} = \frac{2m_0 c^2}{\hbar^2} \sum_{k=1}^{3} \bar{\pi}_k \pi_k + \frac{2}{m_0} \sum_{j=1}^{3} \frac{\partial \bar{\pi}_j}{\partial x_j} \sum_{k=1}^{3} \frac{\partial \pi_k}{\partial x_k} + \tfrac{1}{2} m_0 c^2 \sum_{k=1}^{3} \bar{\psi}_k \psi_k$$
$$+ \frac{\hbar^2}{4m_0} \sum_{j=1}^{3} \sum_{k=1}^{3} \left(\frac{\partial \bar{\psi}_k}{\partial x_j} - \frac{\partial \bar{\psi}_j}{\partial x_k}\right)\left(\frac{\partial \psi_k}{\partial x_j} - \frac{\partial \psi_j}{\partial x_k}\right). \quad (3.270)$$

Hamilton's equations now give us equations (3.266) and (3.267) once more, and the field equation (3.259) as well as its conjugate, after application of the auxiliary condition (3.260).

3.15 Dirac equation

According to the relativistic theory of Dirac, briefly introduced in section 2.6, an electron having rest mass m_0 may be described by a column matrix ψ, having four components $\psi_1, \psi_2, \psi_3, \psi_4$, which satisfies the wave equation

$$\left(i\hbar \frac{\partial}{\partial t} - i\hbar c \boldsymbol{\alpha} \cdot \nabla + m_0 c^2 \beta\right)\psi = 0, \quad (3.271)$$

where $\alpha_x, \alpha_y, \alpha_z$ and β are the 4×4 matrices (2.53).

The Dirac equation (3.271) can be derived from a variational principle if the Lagrangian density is chosen to have the form

$$\mathscr{L} = \psi^* \left(i\hbar \frac{\partial}{\partial t} - i\hbar c \boldsymbol{\alpha} \cdot \nabla + m_0 c^2 \beta\right)\psi, \quad (3.272)$$

FIELD EQUATIONS

where ψ^* is the Hermitian adjoint of ψ and is accordingly a row matrix with elements $\bar{\psi}_1, \bar{\psi}_2, \bar{\psi}_3, \bar{\psi}_4$. Rewriting (3.272) as

$$\mathscr{L} = \sum_{\mu=1}^{4} \bar{\psi}_\mu \left(i\hbar \frac{\partial \psi_\mu}{\partial t} - i\hbar c \sum_{\nu=1}^{4} \boldsymbol{\alpha}_{\mu\nu} \cdot \nabla \psi_\nu + m_0 c^2 \sum_{\nu=1}^{4} \beta_{\mu\nu} \psi_\nu \right), \quad (3.273)$$

we see that the Euler–Lagrange equation associated with the variation of $\bar{\psi}_\mu$ is just

$$\frac{\partial \mathscr{L}}{\partial \bar{\psi}_\mu} = 0, \quad (3.274)$$

furnishing for the field equations

$$i\hbar \frac{\partial \psi_\mu}{\partial t} - i\hbar c \sum_{\nu=1}^{4} \boldsymbol{\alpha}_{\mu\nu} \cdot \nabla \psi_\nu + m_0 c^2 \sum_{\nu=1}^{4} \beta_{\mu\nu} \psi_\nu = 0 \quad (\mu = 1, 2, 3, 4) \quad (3.275)$$

which are equivalent to (3.271). The Euler–Lagrange equation corresponding to the variation of ψ_ν is

$$\frac{\partial \mathscr{L}}{\partial \psi_\nu} - \sum_{k=1}^{3} \frac{\partial}{\partial x_k} \frac{\partial \mathscr{L}}{\partial (\partial \psi_\nu / \partial x_k)} - \frac{\partial}{\partial t} \frac{\partial \mathscr{L}}{\partial (\partial \psi_\nu / \partial t)} = 0 \quad (3.276)$$

which provides us with

$$m_0 c^2 \sum_{\mu=1}^{4} \bar{\psi}_\mu \beta_{\mu\nu} + i\hbar c \sum_{\mu=1}^{4} \nabla \bar{\psi}_\mu \cdot \boldsymbol{\alpha}_{\mu\nu} - i\hbar \frac{\partial \bar{\psi}_\nu}{\partial t} = 0 \quad (\nu = 1, 2, 3, 4). \quad (3.277)$$

These are equivalent to

$$-i\hbar \frac{\partial \psi^*}{\partial t} + i\hbar c \nabla \psi^* \cdot \boldsymbol{\alpha} + m_0 c^2 \psi^* \beta = 0 \quad (3.278)$$

which is the conjugate of the Dirac equation (3.271).

Since the momenta conjugate to the $\bar{\psi}_\mu$, given by

$$\bar{\pi}_\mu = \frac{\partial \mathscr{L}}{\partial \dot{\bar{\psi}}_\mu}, \quad (3.279)$$

are all zero we must eliminate the $\bar{\psi}_\mu$ from the Lagrangian density before employing the Hamiltonian formalism. This can be done by putting

$$\pi_\mu = \frac{\partial \mathscr{L}}{\partial \dot\psi_\mu} = i\hbar \bar\psi_\mu \qquad (3.280)$$

for then

$$\mathscr{H} = \sum_{\mu=1}^{4} (\pi_\mu \dot\psi_\mu + \bar\pi_\mu \dot{\bar\psi}_\mu) - \mathscr{L}$$

$$= \sum_{\mu=1}^{4} \pi_\mu \left(c \sum_{\nu=1}^{4} \alpha_{\mu\nu} \cdot \nabla \psi_\nu + \frac{im_0 c^2}{\hbar} \sum_{\nu=1}^{4} \beta_{\mu\nu} \psi_\nu \right). \qquad (3.281)$$

Hamilton's equations

$$\dot\psi_\mu = \frac{\delta H}{\delta \pi_\mu}, \qquad \dot\pi_\mu = -\frac{\delta H}{\delta \psi_\mu}$$

now give us the Dirac equation (3.275) and its conjugate (3.277).

It is also of interest to derive expressions for the electric charge and electric current densities. From (3.217) with $\epsilon = e/\hbar$ we have

$$\rho = -\frac{ie}{\hbar} \sum_{\nu=1}^{4} \left(\frac{\partial \mathscr{L}}{\partial \dot\psi_\nu} \psi_\nu - \frac{\partial \mathscr{L}}{\partial \dot{\bar\psi}_\nu} \bar\psi_\nu \right) = e \sum_{\nu=1}^{4} \bar\psi_\nu \psi_\nu = e\psi^*\psi \qquad (3.282)$$

and from (3.218) we obtain

$$j_k = -\frac{ie}{\hbar} \sum_{\nu=1}^{4} \left\{ \frac{\partial \mathscr{L}}{\partial (\partial \psi_\nu / \partial x_k)} \psi_\nu - \frac{\partial \mathscr{L}}{\partial (\partial \bar\psi_\nu / \partial x_k)} \bar\psi_\nu \right\}$$

$$= -ec \sum_{\mu=1}^{4} \sum_{\nu=1}^{4} \bar\psi_\mu \alpha_{\mu\nu}^{(k)} \psi_\nu, \qquad (3.283)$$

where the kth component of $\alpha_{\mu\nu}$ has been denoted by $\alpha_{\mu\nu}^{(k)}$, so that

$$\mathbf{j} = -ec\psi^* \boldsymbol{\alpha} \psi. \qquad (3.284)$$

It can be readily verified that ρ and $\mathbf{j}$ satisfy the continuity equation (3.220), either by noting that the Lagrangian density (3.273) is invariant under the transformation (3.219) or by direct analysis following an analogous method to that used for the Klein–Gordon equation.

FIELD EQUATIONS

For the case of a Dirac electron in the presence of an electromagnetic field with vector potential **A** and scalar potential V, the Lagrangian density takes the modified form

$$\mathscr{L} = \psi^*\left\{i\hbar\left(\frac{\partial}{\partial t} + \frac{ie}{\hbar}V\right) - i\hbar c\boldsymbol{\alpha}\cdot\left(\nabla - \frac{ie}{\hbar c}\mathbf{A}\right) + m_0 c^2\beta\right\}\psi \quad (3.285)$$

in order that the Euler–Lagrange equation (3.274) should provide the appropriate equation

$$\left\{i\hbar\left(\frac{\partial}{\partial t} + \frac{ie}{\hbar}V\right) - i\hbar c\boldsymbol{\alpha}\cdot\left(\nabla - \frac{ie}{\hbar c}\mathbf{A}\right) + m_0 c^2\beta\right\}\psi = 0. \quad (3.286)$$

As for the case of an electron *in vacuo*, the charge and current densities are given by formulae (3.282) and (3.284) and satisfy the continuity equation (3.220). Once again the canonical momenta $\bar{\pi}_\mu$ vanish and the π_μ are given by (3.280), so that the Hamiltonian density becomes

$$\mathscr{H} = \sum_{\mu=1}^{4} \pi_\mu \left\{ c \sum_{\nu=1}^{4} \boldsymbol{\alpha}_{\mu\nu}\cdot\left(\nabla - \frac{ie}{\hbar c}\mathbf{A}\right)\psi_\nu \right. \\ \left. + \frac{im_0 c^2}{\hbar}\sum_{\nu=1}^{4}\beta_{\mu\nu}\psi_\nu - \frac{ie}{\hbar}V\psi_\mu \right\} \quad (3.287)$$

and Hamilton's equations give us equations (3.286) and its conjugate equation.

3.16 Linear equations

Our primary purpose in the present chapter has been to demonstrate how the various equations of mathematical physics may be derived by making the assumption that the integral

$$\int \mathscr{L} \, d\tau$$

of an appropriate Lagrangian density $\mathscr{L}$, is stationary with respect to infinitesimal variations of the field components. All the equations with which we have been concerned are linear in the field components and their derivatives. In the case of a linear field having a single real component ψ, the Lagrangian density must depend quadratically on ψ and its derivatives with respect to the coordinates x_μ ($\mu = 1, \ldots, m$). We can therefore achieve a unification of many of the topics treated in this chapter by inspecting a Lagrangian density which fulfills this requirement:

$$\mathscr{L} = \tfrac{1}{2} \sum_{\mu=1}^{m} \sum_{\nu=1}^{m} g_{\mu\nu} \frac{\partial \psi}{\partial x_\mu} \frac{\partial \psi}{\partial x_\nu} + \tfrac{1}{2}\lambda\psi^2 + f\psi \qquad (3.288)$$

where λ, f and all the $g_{\mu\nu}$ are functions of the coordinates $x_1, \ldots, x_m$. The resulting Euler–Lagrange equation is accordingly

$$\sum_{\mu=1}^{m} \sum_{\nu=1}^{m} \frac{\partial}{\partial x_\mu} \left(g_{\mu\nu} \frac{\partial \psi}{\partial x_\nu} \right) = \lambda\psi + f. \qquad (3.289)$$

If we now take $g_{\mu\nu} = \delta_{\mu\nu}$, this equation reduces to the simpler form

$$\sum_{\mu=1}^{m} \frac{\partial^2 \psi}{\partial x_\mu^2} = \lambda\psi + f \qquad (3.290)$$

which includes, as special cases, many of the equations we have considered in previous sections, viz.,

(i) Laplace equation: $\quad m = 3, \lambda = 0, f = 0$;
(ii) Poisson equation: $\quad m = 3, \lambda = 0$;
(iii) Helmholtz equation: $\quad m = 3, \lambda = -k^2, f = 0$;
(iv) wave equation: $\quad m = 4, \lambda = 0, f = 0$;
(v) Klein–Gordon equation: $\quad m = 4, \lambda = (m_0 c/\hbar)^2, f = 0$.

Returning to the general equation (3.289) we see that it also covers cases like the Maxwell field equation

$$\text{div } \mathbf{D} = 4\pi\rho \qquad (3.291)$$

for an *inhomogeneous anisotropic dielectric* which we have not had occasion to discuss until now. For such a medium the displacement vector $\mathbf{D}$ has components given by

$$\begin{aligned} D_x &= K_{11}E_x + K_{12}E_y + K_{13}E_z, \\ D_y &= K_{21}E_x + K_{22}E_y + K_{23}E_z, \\ D_z &= K_{31}E_x + K_{32}E_y + K_{33}E_z, \end{aligned} \qquad (3.292)$$

where E_x, E_y, E_z are the components of the electric field vector **E** and the K_{jk} are the elements of a 3 × 3 symmetric tensor characterizing the dielectric properties of the medium. Since the electric field may be expressed in terms of a scalar potential V by

$$\mathbf{E} = -\nabla V,$$

we may rewrite the Maxwell equation (3.291) as

$$\sum_{j=1}^{3} \sum_{k=1}^{3} \frac{\partial}{\partial x_j} \left(K_{jk} \frac{\partial V}{\partial x_k} \right) = -4\pi\rho, \qquad (3.293)$$

where x_1, x_2, x_3 denote the Cartesian coordinates x, y, z respectively. This has the form of the general field equation (3.289) from which it can be obtained at once by putting $g_{jk} = K_{jk}$, $f = -4\pi\rho$ and $\lambda = 0$.

3.17 Quantum field equations

We conclude this chapter by establishing the Schwinger dynamical principle[3] in the quantum theory of fields which will lead us directly to the quantum field equations in Lagrangian form. The procedure we shall adopt to obtain this variational principle is closely analogous to that described in section 2.8 where the derivation of Hamilton's principle in quantum mechanics is given.

Consider a quantized field which is characterized by operator functions $\psi_\sigma(x_\mu)$ ($\sigma = 1, \ldots, n$) of the coordinates x_μ ($\mu = 1, 2, 3, 4$) specifying position in space and time. We confine the discussion to localizable fields for which the operators x_μ, associated with the position coordinates, commute with the field operators ψ_σ as well as with each other. Next we choose a space-like surface S_1, composed of a continuous set of points which cannot be linked by any light signals and are therefore physically independent of each other, and construct a set of commuting Hermitian operators χ_1 out of the field quantities over the surface S_1. We now direct our attention to a unitary transformation, generated by a unitary operator U_{21}, from the set of commuting operators χ_1 attached to the space-like surface S_1 to a second set of commuting operators

χ_2 attached to another space-like surface S_2. Because the transformation is unitary, we know from the theory given on p. 72 that the two sets of eigenvalues χ_1 and χ_2 are the same and that the transformation function $\langle \chi_2', S_2 \mid \chi_1, S_1 \rangle$ is given by

$$\langle \chi_2', S_2 \mid \chi_1, S_1 \rangle = \langle \chi_1', S_1 \mid U_{21}^{-1} \mid \chi_1, S_1 \rangle. \qquad (3.294)$$

Following the treatment of Hamilton's principle in quantum mechanics given in section 2.8, we express an infinitesimal change in the transformation function as

$$\delta \langle \chi_2', S_2 \mid \chi_1, S_1 \rangle = \frac{i}{\hbar} \langle \chi_2', S_2 \mid \delta W_{21} \mid \chi_1, S_1 \rangle \qquad (3.295)$$

where δW_{21} is an infinitesimal Hermitian operator, given by the formula

$$\delta U_{21}^{-1} = \frac{i}{\hbar} U_{21}^{-1} \, \delta W_{21},$$

which satisfies

$$\delta W_{31} = \delta W_{32} + \delta W_{21}.$$

In conformity with this additive law we now write

$$W_{21} = \int_{S_1}^{S_2} \mathscr{L}\left(\psi_\sigma, \frac{\partial \psi_\sigma}{\partial x_\mu}\right) d\tau \qquad (3.296)$$

where the integration is performed over the volume contained between the two surfaces S_1 and S_2, and $\mathscr{L}(\psi_\sigma, \partial \psi_\sigma/\partial x_\mu)$ can be regarded as a Lagrangian density which is a relativistically invariant Hermitian function of the field operators ψ_σ and their derivatives with respect to the coordinates x_μ. Inserting (3.296) into (3.295), we immediately arrive at the Schwinger principle

$$\delta \langle \chi_2', S_2 \mid \chi_1, S_1 \rangle = \frac{i}{\hbar} \langle \chi_2', S_2 \mid \delta \int_{S_1}^{S_2} \mathscr{L} \, d\tau \mid \chi_1, S_1 \rangle. \qquad (3.297)$$

If the variation in the transformation function is due to infinitesimal changes in χ_1, S_1 and χ_2, S_2 arising from unitary transformations generated by infinitesimal Hermitian operators $F(S_1)$ and $F(S_2)$, we find that

$$\delta W_{21} = F(S_2) - F(S_1) \qquad (3.298)$$

FIELD EQUATIONS

by employing a similar method to that leading up to equation (2.103). Let us now suppose that the variation leaves the surfaces S_1 and S_2 unaltered, and that at each point of space-time, the field components are subjected to the variation

$$\psi_\sigma(x_\mu) \to \psi_\sigma(x_\mu) + \delta\psi_\sigma(x_\mu). \tag{3.299}$$

Then the ensuing change in the integral (3.296) assumes the form

$$\delta W_{21} = \int_{S_1}^{S_2} \sum_{\sigma=1}^{n} \delta\psi_\sigma \left\{ \frac{\partial \mathcal{L}}{\partial \psi_\sigma} - \sum_{\mu=1}^{4} \frac{\partial}{\partial x_\mu} \frac{\partial \mathcal{L}}{\partial(\partial\psi_\sigma/\partial x_\mu)} \right\} d\tau + [F(S)]_{S_1}^{S_2} \tag{3.300}$$

where

$$F(S) = \int_S \sum_{\sigma=1}^{n} \delta\psi_\sigma \sum_{\mu=1}^{4} \frac{\partial \mathcal{L}}{\partial(\partial\psi_\sigma/\partial x_\mu)} l_\mu \, dS \quad (S = S_1, S_2), \tag{3.301}$$

the l_μ being the direction cosines of the normal vector to the surface S at each point. This derivation of expression (3.300) for the variation δW_{21} depends upon the supposition that the commutation properties of the $\delta\psi_\sigma$ and the structure of the Lagrangian density $\mathcal{L}$ allow the $\delta\psi_\sigma$ to be placed outside the bracket without altering δW_{21}. Since the arbitrary variations $\delta\psi_\sigma$ are independent, it now follows from equations (3.298) and (3.300) that

$$\frac{\partial \mathcal{L}}{\partial \psi_\sigma} - \sum_{\mu=1}^{4} \frac{\partial}{\partial x_\mu} \frac{\partial \mathcal{L}}{\partial(\partial\psi_\sigma/\partial x_\mu)} = 0 \tag{3.302}$$

which are just the Lagrange equations of motion in quantum field theory.

References

1. Marshak, R. E., *Phys. Rev.*, **71**, 688 (1947).
2. Courant, R. and Hilbert, D., *Methods of Mathematical Physics*, Vol. I, Interscience, New York, 1953, p. 262.
3. Schwinger, J., *Phys. Rev.*, **82**, 914 (1951).

CHAPTER 4

Eigenvalue Problems

We now turn our attention from the expression of the equations of mathematical physics in the form of variational principles to another aspect of variational principles: their application to the determination of discrete eigenvalues, deferring until Chapter 5 any discussion of continuous eigenvalue problems. It is convenient to begin by examining the problem of the evaluation of discrete eigenvalues in classical dynamics. This will then lead us to Rayleigh's principle and to the Ritz variational procedure which will form the basis of our enquiry into the methods used to determine discrete eigenvalues in quantum mechanics.

4.1 Small oscillations of a dynamical system

We introduce the problem of the evaluation of discrete eigenvalues by investigating the oscillations of a conservative dynamical system with invariable constraints having n degrees of freedom. Suppose that $q_1, q_2, \ldots, q_n$ are chosen as the generalized coordinates specifying the system completely. The system will be in *equilibrium* at $q_1 = q_1^0, q_2 = q_2^0, \ldots, q_n = q_n^0$ if the generalized forces vanish for this configuration so that

$$\left(\frac{\partial V}{\partial q_r}\right)_0 = 0 \quad (r = 1, \ldots, n), \tag{4.1}$$

where $V(q_1, \ldots, q_n)$ is the potential energy of the system.

Let us suppose further that the equilibrium configuration is *stable*. Introducing a new set of coordinates ξ_r such that

$$q_r = q_r^0 + \xi_r \quad (r = 1, \ldots, n), \tag{4.2}$$

EIGENVALUE PROBLEMS

this supposition implies that for any small perturbation of the system from the equilibrium configuration, the coordinates ξ_r must remain small.

Using Taylor's theorem we have

$$V(q_1, \ldots, q_n) = V(q_1^0, \ldots, q_n^0) + \tfrac{1}{2} \sum_{r=1}^{n} \sum_{s=1}^{n} \left(\frac{\partial^2 V}{\partial q_r \, \partial q_s} \right)_0 \xi_r \xi_s + \cdots \tag{4.3}$$

since all the terms of the first order in the ξ_r vanish due to the condition (4.1). Hence, to the second order in the coordinates ξ_r, we get

$$V = \tfrac{1}{2} \sum_{r=1}^{n} \sum_{s=1}^{n} b_{rs} \xi_r \xi_s \tag{4.4}$$

choosing the potential to vanish at the equilibrium configuration given by $\xi_r = 0$ $(r = 1, \ldots, n)$ and putting

$$b_{rs} = \left(\frac{\partial^2 V}{\partial q_r \, \partial q_s} \right)_0 = b_{sr}. \tag{4.5}$$

Since the dynamical system is subject to invariable constraints, its kinetic energy may be expressed in the form

$$T = \tfrac{1}{2} \sum_{r=1}^{n} \sum_{s=1}^{n} A_{rs}(q_1, \ldots, q_n) \dot{q}_r \dot{q}_s, \tag{4.6}$$

where $A_{rs} = A_{sr}$, and so to the second order of small quantities we have

$$T = \tfrac{1}{2} \sum_{r=1}^{n} \sum_{s=1}^{n} a_{rs} \dot{\xi}_r \dot{\xi}_s, \tag{4.7}$$

a_{rs} being the value of A_{rs} at the equilibrium configuration so that

$$a_{rs} = A_{rs}(q_1^0, \ldots, q_n^0) = a_{sr}. \tag{4.8}$$

The Lagrangian function for the dynamical system is therefore given by

$$L = \tfrac{1}{2} \sum_{r=1}^{n} \sum_{s=1}^{n} (a_{rs} \dot{\xi}_r \dot{\xi}_s - b_{rs} \xi_r \xi_s) \tag{4.9}$$

and Lagrange's equations

$$\frac{d}{dt}\frac{\partial L}{\partial \dot{\xi}_r} - \frac{\partial L}{\partial \xi_r} = 0 \quad (r = 1, \ldots, n)$$

yield

$$\sum_{s=1}^{n} (a_{rs}\ddot{\xi}_s + b_{rs}\xi_s) = 0 \quad (r = 1, \ldots, n). \tag{4.10}$$

To solve this system of equations we put

$$\xi_s = c_s \sin(\omega t + \epsilon_s) \quad (s = 1, \ldots, n) \tag{4.11}$$

which represents an oscillation with *amplitude* c_s and *angular frequency* ω. Substitution into (4.10) provides the set of n homogeneous linear equations in the amplitudes $c_1, \ldots, c_n$ given by

$$\sum_{s=1}^{n} (b_{rs} - \omega^2 a_{rs})c_s = 0 \quad (r = 1, \ldots, n). \tag{4.12}$$

These equations are consistent if and only if the determinant of the coefficients of $c_1, \ldots, c_n$ vanishes so that

$$\det(b_{rs} - \lambda a_{rs}) = 0 \tag{4.13}$$

where $\lambda = \omega^2$. The n values of λ which satisfy this nth order *secular equation* furnish the angular frequencies ω for which (4.11) form a proper solution of the equations (4.10). These values of λ are called *proper values* or *eigenvalues*. For each eigenvalue λ_t we may solve the set of linear equations (4.12) for the amplitudes c_s ($s = 1, \ldots, n$).

If we denote the value of c_s corresponding to the eigenvalue λ_t by c_{st}, we may write (4.12) in the form

$$\sum_{s=1}^{n} b_{rs}c_{st} = \lambda_t \sum_{s=1}^{n} a_{rs}c_{st}. \tag{4.14}$$

Now the conjugate equation of (4.14) associated with the eigenvalue $\lambda_{t'}$ may be written

$$\sum_{r=1}^{n} \bar{c}_{rt'}b_{rs} = \bar{\lambda}_{t'} \sum_{r=1}^{n} \bar{c}_{rt'}a_{rs} \tag{4.15}$$

EIGENVALUE PROBLEMS

due to the symmetry properties $a_{rs} = a_{sr}$ and $b_{rs} = b_{sr}$, and so

$$(\lambda_t - \bar{\lambda}_{t'}) \sum_{r=1}^{n} \sum_{s=1}^{n} \bar{c}_{rt'} c_{st} a_{rs} = 0 \qquad (4.16)$$

which for $t' = t$ becomes

$$(\lambda_t - \bar{\lambda}_t) \sum_{r=1}^{n} \sum_{s=1}^{n} \bar{c}_{rt} c_{st} a_{rs} = 0. \qquad (4.17)$$

Putting

$$c_{st} = \gamma_{st} + i\gamma'_{st}$$

where γ_{st} and γ'_{st} are real quantities, we see that

$$\sum_{r=1}^{n} \sum_{s=1}^{n} \bar{c}_{rt} c_{st} a_{rs} = \sum_{r=1}^{n} \sum_{s=1}^{n} (\gamma_{rt}\gamma_{st} + \gamma'_{rt}\gamma'_{st}) a_{rs}$$
$$+ i \sum_{r=1}^{n} \sum_{s=1}^{n} (\gamma_{rt}\gamma'_{st} - \gamma_{st}\gamma'_{rt}) a_{rs}.$$

The imaginary term on the right-hand side vanishes because $a_{rs} = a_{sr}$, and the real term is non-vanishing due to the positive definite nature of the kinetic energy which ensures that

$$\sum_{r=1}^{n} \sum_{s=1}^{n} \gamma_{rt}\gamma_{st} a_{rs} \quad \text{and} \quad \sum_{r=1}^{n} \sum_{s=1}^{n} \gamma'_{rt}\gamma'_{st} a_{rs}$$

are positive. Hence the coefficient of $\lambda_t - \bar{\lambda}_t$ in (4.17) is real and non-zero and so $\lambda_t = \bar{\lambda}_t$, from which it follows that the eigenvalues are all real. In addition from (4.14) we see that all the elements c_{st} must be real, apart from a possible common phase factor which we may choose to be unity.

Multiplying (4.14) on both sides by c_{rt} and summing over r we get

$$\lambda_t = \frac{\sum\limits_{r=1}^{n} \sum\limits_{s=1}^{n} b_{rs} c_{rt} c_{st}}{\sum\limits_{r=1}^{n} \sum\limits_{s=1}^{n} a_{rs} c_{rt} c_{st}}. \qquad (4.18)$$

The denominator is just twice the kinetic energy evaluated for generalized velocities having the values $\dot{\xi}_r = c_{rt}$ and so is positive definite. The numerator is twice the potential energy evaluated for generalized coordinates $\xi_r = c_{rt}$. Since the equilibrium configuration is assumed to be stable, the angular frequencies ω_t must be real and so $\lambda_t = \omega_t^2$ is necessarily positive. It now follows from equation (4.18) that V must be entirely positive in the neighbourhood of the equilibrium point, and because the potential energy vanishes at the equilibrium point where $\xi_1 = \xi_2 = \cdots = \xi_n = 0$ we see that it necessarily attains a minimum value there.

Normal coordinates

Since λ_t and c_{st} are real we may write (4.16) as

$$(\lambda_t - \lambda_{t'}) \sum_{r=1}^{n} \sum_{s=1}^{n} c_{rt'} c_{st} a_{rs} = 0. \tag{4.19}$$

Then if all the roots of the secular equation (4.13) are different we see at once that

$$\sum_{r=1}^{n} \sum_{s=1}^{n} c_{rt'} c_{st} a_{rs} = 0 \qquad (t' \neq t). \tag{4.20}$$

When one or more of the roots of the secular equation are multiple, we can no longer directly infer the result (4.20) by virtue of (4.19). However, we can treat this case of degeneracy by first separating the repeated roots by making slight modifications in the coefficients a_{rs} and then subsequently proceeding to the required limiting case where the coefficients attain their true values.

Now the values of c_{rt} are not uniquely determined by equations (4.12) since any solution of these equations is arbitrary to a common factor. It is convenient to normalize the c_{rt} so that

$$\sum_{r=1}^{n} \sum_{s=1}^{n} c_{rt} c_{st} a_{rs} = 1 \qquad (t = 1, \ldots, n) \tag{4.21}$$

in which case we have

$$\sum_{r=1}^{n} \sum_{s=1}^{n} c_{rt'} c_{st} a_{rs} = \delta_{t't}. \tag{4.22}$$

EIGENVALUE PROBLEMS 139

This may be expressed in the matrix form

$$\mathbf{C'AC} = \mathbf{I} \tag{4.23}$$

where $\mathbf{A}$ is the matrix with elements a_{rs}, $\mathbf{C}$ is the matrix with elements c_{st}, $\mathbf{C'}$ denotes the transpose matrix of $\mathbf{C}$, and $\mathbf{I}$ is the unit matrix.

Introducing the diagonal matrix $\boldsymbol{\lambda}$ with elements $\lambda_{t't} = \lambda_t \delta_{t't}$ we see that (4.14), when rewritten in the form

$$\sum_{s=1}^{n} b_{rs} c_{st} = \sum_{s=1}^{n} \sum_{t'=1}^{n} a_{rs} c_{st'} \lambda_{t't}, \tag{4.24}$$

may be expressed as the matrix equation

$$\mathbf{BC} = \mathbf{AC}\boldsymbol{\lambda}, \tag{4.25}$$

where $\mathbf{B}$ is the matrix with elements b_{rs}. Hence

$$\mathbf{C'BC} = \mathbf{C'AC}\boldsymbol{\lambda} = \boldsymbol{\lambda} \tag{4.26}$$

using (4.23), and thus the matrix $\mathbf{C}$ simultaneously diagonalizes $\mathbf{A}$ and $\mathbf{B}$.

We now introduce a new set of coordinates η_s such that

$$\xi_r = \sum_{s=1}^{n} c_{rs} \eta_s. \tag{4.27}$$

Representing the coordinates ξ_r and η_s by the column matrices $\boldsymbol{\xi}$ and $\boldsymbol{\eta}$ respectively, we have

$$\boldsymbol{\xi} = \mathbf{C}\boldsymbol{\eta} \tag{4.28}$$

and

$$\boldsymbol{\xi'} = \boldsymbol{\eta'}\mathbf{C'} \tag{4.29}$$

where $\boldsymbol{\xi'}$ and $\boldsymbol{\eta'}$ are the corresponding row matrices.

Expressing V and T in the matrix forms

$$V = \tfrac{1}{2}\boldsymbol{\xi'}\mathbf{B}\boldsymbol{\xi} \tag{4.30}$$

and

$$T = \tfrac{1}{2}\dot{\boldsymbol{\xi}}'\mathbf{A}\dot{\boldsymbol{\xi}}, \tag{4.31}$$

where $\dot{\boldsymbol{\xi}}$ is a column matrix composed of the velocities $\dot{\xi}_r$ and $\dot{\boldsymbol{\xi}}'$ is the associated row matrix, we see that

$$V = \tfrac{1}{2}\boldsymbol{\eta}'\mathbf{C'BC}\boldsymbol{\eta} = \tfrac{1}{2}\boldsymbol{\eta}'\boldsymbol{\lambda}\boldsymbol{\eta} = \tfrac{1}{2}\sum_{s=1}^{n}\lambda_s\eta_s^2 \qquad (4.32)$$

and

$$T = \tfrac{1}{2}\dot{\boldsymbol{\eta}}'\mathbf{C'AC}\dot{\boldsymbol{\eta}} = \tfrac{1}{2}\dot{\boldsymbol{\eta}}'\dot{\boldsymbol{\eta}} = \tfrac{1}{2}\sum_{s=1}^{n}\dot{\eta}_s^2. \qquad (4.33)$$

Accordingly, in terms of the new generalized coordinates η_s and the generalized velocities $\dot{\eta}_s$, the Lagrangian function takes the form

$$L = \tfrac{1}{2}\sum_{s=1}^{n}(\dot{\eta}_s^2 - \omega_s^2\eta_s^2) \qquad (4.34)$$

which yields

$$\ddot{\eta}_s + \omega_s^2\eta_s = 0 \qquad (s = 1,\ldots,n) \qquad (4.35)$$

on employing Lagrange's equations.

Hence

$$\eta_s = c_s \sin(\omega_s t + \epsilon_s) \qquad (s = 1,\ldots,n) \qquad (4.36)$$

where the ω_s ($s = 1,\ldots,n$) are the natural angular frequencies of the dynamical system. The coordinates η_s are called *normal coordinates* since they each correspond to the oscillation of the dynamical system with a single frequency. Such an oscillation is referred to as a *normal mode*. The general motion of the system is composed of a combination of the normal modes of vibration.

4.2 Stationary property of angular frequencies

We now suppose that the dynamical system is constrained in such a manner that it has a single degree of freedom so that we may put

$$\eta_1 = \mu_1\eta, \qquad \eta_2 = \mu_2\eta,\ldots, \qquad \eta_n = \mu_n\eta \qquad (4.37)$$

where η is a variable parameter and $\mu_1, \mu_2,\ldots,\mu_n$ are certain constants. Then the kinetic energy T and the potential energy V

EIGENVALUE PROBLEMS

of the constrained system become

$$T = \tfrac{1}{2} \sum_{s=1}^{n} \mu_s^2 \dot{\eta}^2 \tag{4.38}$$

and

$$V = \tfrac{1}{2} \sum_{s=1}^{n} \omega_s^2 \mu_s^2 \eta^2. \tag{4.39}$$

Hence the Lagrange equation for the coordinate η is

$$\sum_{s=1}^{n} \mu_s^2 \ddot{\eta} + \sum_{s=1}^{n} \omega_s^2 \mu_s^2 \eta = 0 \tag{4.40}$$

and so

$$\eta = c \sin(\omega t + \epsilon), \tag{4.41}$$

where the angular frequency ω of the constrained system is given by

$$\omega^2 = \frac{\sum_{s=1}^{n} \omega_s^2 \mu_s^2}{\sum_{s=1}^{n} \mu_s^2}. \tag{4.42}$$

Since the mean values of the kinetic energy and the potential energy obtained by averaging over a long period of time are

$$\bar{T} = \tfrac{1}{4}\omega^2 c^2 \sum_{s=1}^{n} \mu_s^2 \tag{4.43}$$

and

$$\bar{V} = \tfrac{1}{4} c^2 \sum_{s=1}^{n} \omega_s^2 \mu_s^2, \tag{4.44}$$

it is plain that our value of ω^2 is given by setting

$$\bar{T} = \bar{V}, \tag{4.45}$$

or, equivalently, by choosing the mean value of the Lagrangian function to vanish.

If now $\delta\omega$ is the small change in ω resulting from small variations $\delta\mu_s$ in the μ_s, it follows from equation (4.42) that to the first order

$$\sum_{s=1}^{n} \mu_s^2 \omega \, \delta\omega = \sum_{s=1}^{n} (\omega_s^2 - \omega^2) \mu_s \, \delta\mu_s. \tag{4.46}$$

Choosing all the μ_s to vanish except for $s = s_1$ we get

$$\omega^2 = \omega_{s_1}^2 \tag{4.47}$$

which leads us to the result

$$\delta\omega = 0, \tag{4.48}$$

that is, the angular frequency ω has a stationary value when it coincides with any one of the angular frequencies ω_s of the unconstrained system. This is known as *Lagrange's theorem*.

4.3 Rayleigh's principle

If we let ω_1 and ω_n be respectively the least and greatest of the natural angular frequencies of the dynamical system under consideration, we see that

$$\omega^2 - \omega_1^2 = \frac{\sum_{s=1}^{n} (\omega_s^2 - \omega_1^2)\mu_s^2}{\sum_{s=1}^{n} \mu_s^2} \geqslant 0 \tag{4.49}$$

and

$$\omega_n^2 - \omega^2 = \frac{\sum_{s=1}^{n} (\omega_n^2 - \omega_s^2)\mu_s^2}{\sum_{s=1}^{n} \mu_s^2} \geqslant 0. \tag{4.50}$$

Hence

$$\omega_1^2 \leqslant \omega^2 \leqslant \omega_n^2 \tag{4.51}$$

from which it is clear that the value of ω given by equation (4.42) provides an upper bound to the least natural angular frequency ω_1 and a lower bound to the greatest natural angular frequency ω_n.

If the dynamical system is oscillating with natural angular frequency ω_1, all the μ_s vanish except μ_1 and ω attains its minimum value ω_1. Suppose now that values of $\mu_1, \ldots, \mu_n$ are chosen which differ slightly from the values they have in the case of the natural motion with angular frequency ω_1. Then the resulting value of the angular frequency ω of the constrained system, obtained by equating the mean value of the kinetic energy to the mean value of the potential energy, will be slightly greater than the fundamental angular frequency ω_1. This is *Rayleigh's principle*. Because of the stationary property of ω proved in the previous section and known as Lagrange's theorem, the error in the value of ω obtained in this way is of the second order of smallness and thus will be considerably less than the errors in the values of $\mu_1, \ldots, \mu_n$. A useful approximation to ω_1 should therefore be provided by ω.

By following an analogous procedure to the above, an approximation ω to ω_n may be derived such that $\omega \leqslant \omega_n$.

A simple illustration of the use of Rayleigh's principle is furnished by the vibrations of a light string to which a number of heavy particles have been attached. As an example we consider the case of a string of length $4a$, held stationary at its ends and stretched at tension P, to which three particles of equal masses m have been affixed so as to divide the string into four equal portions of length a. If y_1, y_2, y_3 are the displacements of the particles from their equilibrium positions, in order along the length of the string, the kinetic energy of the system is

$$T = \tfrac{1}{2}m(\dot{y}_1^2 + \dot{y}_2^2 + \dot{y}_3^2)$$

while the potential energy is

$$V = \frac{1}{2}\frac{P}{a}\{y_1^2 + (y_2 - y_1)^2 + (y_3 - y_2)^2 + y_3^2\}$$

using the formulae (3.86) and (3.85) respectively. Now if the string is constrained to vibrate with a single angular frequency ω in such a manner that

$$y_1 = \mu_1 y, \qquad y_2 = \mu_2 y, \qquad y_3 = \mu_3 y$$

where $y = c \sin(\omega t + \epsilon)$, then the mean kinetic and potential energies are given by

$$\bar{T} = \tfrac{1}{4}mc^2\omega^2(\mu_1^2 + \mu_2^2 + \mu_3^2)$$

and

$$\bar{V} = \frac{1}{4}\frac{Pc^2}{a}\{\mu_1^2 + (\mu_2 - \mu_1)^2 + (\mu_3 - \mu_2)^2 + \mu_3^2\}$$

so that, on equating $\bar{T}$ to $\bar{V}$, we get
$$\omega^2 = \frac{P}{ma} \frac{\mu_1^2 + (\mu_2 - \mu_1)^2 + (\mu_3 - \mu_2)^2 + \mu_3^2}{\mu_1^2 + \mu_2^2 + \mu_3^2}.$$
Approximating the shape of the string by a parabola so that
$$\mu_1 = \mu_3 = \tfrac{3}{4}\mu_2$$
yields
$$\omega^2 = \frac{10}{17} \frac{P}{ma} = 0.5882 \frac{P}{ma}$$
which is very close to the actual value ω_1 of the lowest angular frequency given by
$$\omega_1^2 = (2 - \sqrt{2}) \frac{P}{ma} = 0.5858 \frac{P}{ma}.$$

4.4 Vibrating string

As an example of the vibrations of a continuous system we now turn our attention to a non-uniform string stretched at tension P. If the displacement of the string from its equilibrium configuration is $y(x, t)$ at time t and at the point with coordinate x, then
$$\frac{\partial^2 y}{\partial x^2} = \frac{\rho(x)}{P} \frac{\partial^2 y}{\partial t^2} \tag{4.52}$$
where $\rho(x)$ is the mass per unit length at the point x.

If the string is vibrating with angular frequency ω we have
$$y(x, t) = u(x) \sin(\omega t + \epsilon) \tag{4.53}$$
where the amplitude $u(x)$ of the vibration at the point x satisfies the equation
$$\frac{d^2 u}{dx^2} + \frac{\omega^2 \rho}{P} u = 0. \tag{4.54}$$
For a string held fixed at the end points $x = 0, l$, the boundary conditions
$$u(0) = u(l) = 0 \tag{4.55}$$
must be obeyed by $u(x)$.

Equation (4.54) has solutions which satisfy the terminal conditions (4.55) only for particular values of ω called *eigenvalues*, the corresponding solutions $u(x)$ being called *eigenfunctions*.

EIGENVALUE PROBLEMS   145

We now suppose that $v(x)$ is an arbitrary function satisfying the boundary conditions $v(0) = v(l) = 0$. If the string is constrained to vibrate with amplitude $v(x)$ and angular frequency ω, the mean kinetic and potential energies are given by

$$\bar{T} = \tfrac{1}{4}\omega^2 \int_0^l \rho(x)\{v(x)\}^2 \, dx \qquad (4.56)$$

and

$$\bar{V} = \tfrac{1}{4}P \int_0^l \left(\frac{dv}{dx}\right)^2 dx \qquad (4.57)$$

respectively, as can be readily verified by referring to p. 93. Then the value of ω^2 obtained by equating the mean kinetic and potential energies is

$$\omega^2 = \frac{P \int_0^l \left(\frac{dv}{dx}\right)^2 dx}{\int_0^l \rho(x)\{v(x)\}^2 \, dx}. \qquad (4.58)$$

If we let $u_1(x)$ denote the solution of (4.54) corresponding to the least eigenvalue ω_1, then we have

$$\frac{\omega_1^2}{P}\rho(x) = -\frac{\dfrac{d^2 u_1}{dx^2}}{u_1}$$

and so

$$\left(\frac{dv}{dx}\right)^2 - \frac{\omega_1^2}{P}\rho v^2 = \left(\frac{dv}{dx}\right)^2 + \frac{\dfrac{d^2 u_1}{dx^2}}{u_1} v^2$$

$$= \frac{d}{dx}\left(\frac{\dfrac{du_1}{dx} v^2}{u_1}\right) + \left(\frac{dv}{dx}\right)^2 - \frac{2\dfrac{du_1}{dx}\dfrac{dv}{dx} v}{u_1} + \frac{\left(\dfrac{du_1}{dx}\right)^2 v^2}{u_1^2}$$

$$= \frac{d}{dx}\left(\frac{\dfrac{du_1}{dx} v^2}{u_1}\right) + \left(\frac{\dfrac{du_1}{dx} v - u_1 \dfrac{dv}{dx}}{u_1}\right)^2. \qquad (4.59)$$

Hence it follows that

$$\int_0^l \left(\frac{dv}{dx}\right)^2 dx - \frac{\omega_1^2}{P} \int_0^l \rho(x)\{v(x)\}^2 \, dx \geq \left[\frac{\frac{du_1}{dx} v^2}{u_1}\right]_0^l \quad (4.60)$$

and because of the boundary conditions imposed on v we see that the right-hand side of this inequality vanishes yielding

$$\int_0^l \left(\frac{dv}{dx}\right)^2 dx \geq \frac{\omega_1^2}{P} \int_0^l \rho(x)\{v(x)\}^2 \, dx, \quad (4.61)$$

that is

$$\omega^2 \geq \omega_1^2. \quad (4.62)$$

Thus the value of the angular frequency ω obtained by equating the mean values of the kinetic energy and the potential energy for an arbitrary amplitude function $v(x)$ satisfying the appropriate boundary conditions must be an upper bound to the lowest eigenvalue ω_1. This is *Rayleigh's principle for a continuous system.*

To exemplify the use of this principle we consider a string having uniform line density ρ. Taking the form

$$v(x) = \begin{cases} a\left[1 - \left(1 - \frac{2x}{l}\right)^n\right] & 0 \leq x \leq \frac{l}{2} \\ a\left[1 - \left(\frac{2x}{l} - 1\right)^n\right] & \frac{l}{2} \leq x \leq l \end{cases} \quad (4.63)$$

as an approximation to the amplitude of the string, we get

$$\overline{T} = \frac{\omega^2 \rho l a^2 n^2}{2(n+1)(2n+1)}$$

and

$$\overline{V} = \frac{P a^2 n^2}{l(2n-1)},$$

which gives

$$\omega^2 = \frac{2(n+1)(2n+1)}{2n-1} \frac{P}{\rho l^2}$$

on equating $\overline{T}$ and $\overline{V}$.

If we choose $n = 2$ the string is constrained to have the parabolic shape
$$v(x) = \frac{4a}{l^2} x(l - x)$$
and then
$$\omega^2 = \frac{10P}{\rho l^2}$$
which yields a value of ω slightly greater than the exact value of the fundamental frequency ω_1 given by
$$\omega_1^2 = \frac{\pi^2 P}{\rho l^2} = 9.870 \frac{P}{\rho l^2}.$$

In order to obtain the least upper bound to ω_1 provided by the form (4.63) we determine the value of n for which $d\omega/dn = 0$. This yields
$$n = \tfrac{1}{2}(1 + \sqrt{6})$$
giving
$$\omega^2 = (5 + 2\sqrt{6}) \frac{P}{\rho l^2} = 9.899 \frac{P}{\rho l^2}$$
which is only slightly greater than ω_1^2.

4.5 Sturm–Liouville eigenvalue problem

We now leave the special case of the vibrating string and consider the more general problem of an oscillating system described by the Sturm–Liouville equation

$$\frac{d}{dx}\left\{p(x) \frac{du}{dx}\right\} + \{q(x) + \lambda r(x)\}u(x) = 0 \qquad (4.64)$$

where λ is a constant. We have already noted in section 3.8 that the function $p(x)$ must have no zeros within the range of definition $a \leqslant x \leqslant b$, except at the end points $x = a, b$. Hence it cannot change sign and consequently it is usual to assume that it remains positive. In addition, for all the cases with which one normally deals, $r(x)$ does not change sign and so we may suppose that $r(x)$ also remains positive within the range of definition.

The Sturm–Liouville problem is concerned with the determination of the values of the constant λ for which there exist nontrivial solutions of the Sturm–Liouville equation satisfying prescribed boundary conditions. As in the previous discussion, these values of λ are called eigenvalues and the solutions associated with them are called eigenfunctions.

Now consider the two solutions $u_r(x)$ and $u_s(x)$ of equation (4.64) corresponding to the eigenvalues λ_r and λ_s respectively. Multiplying the equation associated with eigenvalue λ_s by $u_r(x)$ and multiplying the equation associated with eigenvalue λ_r by $u_s(x)$, subtracting and integrating from $x = a$ to $x = b$ we obtain

$$(\lambda_r - \lambda_s) \int_a^b r(x) u_r(x) u_s(x)\, dx = \int_a^b \frac{d}{dx}\left[p(x)\left\{u_r \frac{du_s}{dx} - \frac{du_r}{dx} u_s\right\}\right] dx$$

$$= \left[p(x)\left\{u_r \frac{du_s}{dx} - \frac{du_r}{dx} u_s\right\}\right]_a^b. \quad (4.65)$$

If $p(x)$ vanishes at $x = a$ and b, which are therefore singular points of the Sturm–Liouville equation, and if u_r and u_s are finite at the terminal points, we have at once that

$$(\lambda_r - \lambda_s) \int_a^b r(x) u_r(x) u_s(x)\, dx = 0. \quad (4.66)$$

Alternatively if the terminal points are not singular points but the solutions of equation (4.64) satisfy one of the boundary conditions:

(i) $u(a) = 0$ and $u(b) = 0$,
(ii) $(du/dx)_a = 0$ and $(du/dx)_b = 0$, $\quad$ (4.67)
(iii) $u(a) = u(b)$ and $p(a)(du/dx)_a = p(b)(du/dx)_b$,

we see that

$$\left[p(x)\left\{u_r \frac{du_s}{dx} - \frac{du_r}{dx} u_s\right\}\right]_a^b = 0$$

and consequently (4.66) is again true. In either case it follows that if $\lambda_r \neq \lambda_s$,

$$\int_a^b r(x) u_r(x) u_s(x)\, dx = 0 \quad (4.68)$$

and so the functions $u_r(x)$ form an *orthogonal set*. These functions are also said to be *normalized* if

$$\int_a^b r(x)\{u_r(x)\}^2\, dx = 1 \tag{4.69}$$

and then they compose an *orthonormal set of functions*. We may combine the two conditions (4.68) and (4.69) in the form

$$\int_a^b r(x)u_r(x)u_s(x)\, dx = \delta_{rs}. \tag{4.70}$$

The reality of the eigenvalues λ_r now follows because the functions $p(x)$, $q(x)$ and $r(x)$ are real. Thus if $u_r(x)$ is an eigenfunction associated with the eigenvalue λ_r we also have that $\bar{u}_r(x)$ must be an eigenfunction corresponding to the eigenvalue $\bar\lambda_r$. If we now suppose that λ_r is complex, we see that λ_r cannot be equal to $\bar\lambda_r$ and so

$$\int_a^b r(x)u_r(x)\bar{u}_r(x)\, dx = 0$$

which implies that $u_r(x)$ must vanish at all points in the range $a \leqslant x \leqslant b$. Since this contradicts the assumption that the eigenfunctions are non-trivial solutions of the Sturm–Liouville equation we conclude that $\lambda_r = \bar\lambda_r$ and so λ_r must, in fact, be real.

From the discussion given in section 3.8 we know that the Sturm–Liouville equation (4.64) can be derived by requiring the integral

$$I[v] = \int_a^b \left\{ p(x)\left(\frac{dv}{dx}\right)^2 - q(x)v^2 \right\} dx \tag{4.71}$$

to be stationary subject to the subsidiary condition

$$\int_a^b r(x)\{v(x)\}^2\, dx = 1 \tag{4.72}$$

being satisfied. Provided we restrict $v(x)$ to belong to the class of functions for which

$$\left| \int_a^b q(x)v^2\, dx \right| < \infty, \tag{4.73}$$

$I[v]$ will have a lower bound. We now vary the function $v(x)$ until we obtain a function $u_1(x)$ for which $I[v]$ attains its least possible value. Clearly the function $u_1(x)$ must be an eigenfunction of the Sturm–Liouville equation. Let us denote its associated eigenvalue by λ_1. Then we have

$$\begin{aligned} I[u_1] &= \int_a^b \left\{ p(x)\left(\frac{du_1}{dx}\right)^2 - q(x)u_1^2 \right\} dx \\ &= \left[p(x)u_1 \frac{du_1}{dx} \right]_a^b - \int_a^b u_1 \left[\frac{d}{dx}\left\{ p(x)\frac{du_1}{dx} \right\} + q(x)u_1 \right] dx \\ &= \lambda_1 \int_a^b r(x)u_1^2(x)\, dx \\ &= \lambda_1 \end{aligned} \qquad (4.74)$$

using the boundary conditions (4.67) and the subsidiary condition (4.72); and so we see that $u_1(x)$ must be the eigenfunction associated with the lowest eigenvalue of the Sturm–Liouville equation.

It is instructive to make a slight digression at this stage. An insidious difficulty arising in the calculus of variations stems from the fact that there are certain variational problems which do not have well-behaved solutions. As an illustration we consider the minimization of the functional

$$I[v] = \int_{-1}^{1} x^2 \left(\frac{dv}{dx}\right)^2 dx$$

where $v(x)$ satisfies the end conditions $v(-1) = -1$, $v(1) = 1$. This integral can be made arbitrarily small by choosing

$$v(x) = \begin{cases} -1 & (-1 < x < -\epsilon) \\ x/\epsilon & |x| < \epsilon \\ 1 & (\epsilon < x < 1) \end{cases}$$

and letting $\epsilon \to 0$. On the other hand it is clear that there does not exist any continuous function $v(x)$ for which $I[v]$ actually attains its minimum value zero. However, it can be shown that no difficulty of this nature arises in the Sturm–Liouville problem with the boundary conditions (4.67) and so we may now return to our main line of inquiry.

EIGENVALUE PROBLEMS

If we take a normalized function $v(x)$ satisfying one of the three prescribed boundary conditions (4.67) which is orthogonal to the eigenfunction $u_1(x)$ and vary $v(x)$ until we arrive at a function $u_2(x)$ for which $I[v]$ attains its least value, then $u_2(x)$ will also be an eigenfunction of the Sturm–Liouville equation associated with an eigenvalue $\lambda_2 = I[u_2]$ which is necessarily greater than λ_1. More generally if a normalized function $v(x)$ is chosen which satisfies one of the boundary conditions (4.67) and is orthogonal to all the eigenfunctions $u_1(x), u_2(x), \ldots, u_{n-1}(x)$ associated with eigenvalues $\lambda_1, \lambda_2, \ldots, \lambda_{n-1}$ respectively, where $\lambda_1 < \lambda_2 < \cdots < \lambda_{n-1}$, the minimum value of $I[v]$ will be reached for a function $u_n(x)$, which is an eigenfunction of equation (4.64) corresponding to eigenvalue $\lambda_n = I[u_n]$ with $\lambda_{n-1} < \lambda_n$. Thus, by employing the above procedure, we may in principle determine the eigenfunctions and associated eigenvalues of the Sturm–Liouville equation.

To conclude this section we consider the expansion of an arbitrary function $f(x)$ in terms of the orthonormal set of eigenfunctions $u_r(x)$. We set

$$f_n(x) = \sum_{r=1}^{n} c_r u_r(x) \tag{4.75}$$

and then define the *mean square integral*

$$\epsilon_n^2 = \int_a^b \{f(x) - f_n(x)\}^2 r(x)\, dx. \tag{4.76}$$

Since this may be rewritten in the form

$$\epsilon_n^2 = \int_a^b \{f(x)\}^2 r(x)\, dx - 2 \sum_{r=1}^{n} c_r \int_a^b f(x) u_r(x) r(x)\, dx + \sum_{r=1}^{n} c_r^2,$$

we see that ϵ_n^2 attains its minimum value

$$\int_a^b \{f(x)\}^2 r(x)\, dx - \sum_{r=1}^{n} c_r^2 \tag{4.77}$$

for the choice of coefficients

$$c_r = \int_a^b f(x) u_r(x) r(x)\, dx \tag{4.78}$$

which henceforth we shall suppose to have been made.

We now introduce the function

$$\phi_n(x) = \frac{1}{\epsilon_n}\{f(x) - f_n(x)\}. \quad (4.79)$$

Since it is orthogonal to all the eigenfunctions $u_1(x), u_2(x), \ldots,$ $u_n(x)$ as a consequence of the fact that the coefficients c_r have been chosen to have the values (4.78) which ensure that ϵ_n^2 is a minimum, we are led to the result

$$I[\phi_n] \geq \lambda_{n+1}. \quad (4.80)$$

Integrating by parts and assuming that $f(x)$ satisfies the same boundary conditions as the eigenfunctions $u_r(x)$ we obtain

$$I[\phi_n] = \frac{1}{\epsilon_n^2}\left\{I[f] - \sum_{r=1}^{n} \lambda_r c_r^2\right\}$$

and so

$$I[f] - \sum_{r=1}^{n} \lambda_r c_r^2 \geq \epsilon_n^2 \lambda_{n+1}.$$

Since we may always modify $q(x)$ so that the least eigenvalue λ_1 is zero, it follows that we can assume without loss of generality that all the eigenvalues $\lambda_2, \lambda_3, \ldots, \lambda_r, \ldots$ are positive. Then

$$\sum_{r=1}^{n} \lambda_r c_r^2$$

must likewise be positive and therefore

$$I[f] \geq \epsilon_n^2 \lambda_{n+1}. \quad (4.81)$$

Now it can be shown that the eigenvalues λ_n of the Sturm–Liouville equation tend to infinity as n becomes large. Hence, by virtue of the inequality (4.81), we see that $\epsilon_n \to 0$ as $n \to \infty$ and so

$$\sum_{r=1}^{n} c_r u_r(x)$$

converges in the mean to the function $f(x)$. Any set of functions $u_r(x)$ having this convergence property is said to form a *complete system*.

EIGENVALUE PROBLEMS

4.6 Ritz variational method

Our considerations now proceed to the famous method devised by Ritz for simultaneously obtaining upper bounds to the n lowest eigenvalues $\lambda_1, \lambda_2, \ldots, \lambda_n$ of the Sturm–Liouville equation.

Suppose that $v_1(x), v_2(x), \ldots, v_n(x)$ form a set of n linearly independent functions satisfying one of the boundary conditions (4.67) and let

$$v(x) = \sum_{r=1}^{n} c_r v_r(x), \tag{4.82}$$

where $c_1, c_2, \ldots, c_n$ are n arbitrary parameters. Then we know from the work of the previous section that the value of μ given by the vanishing of the integral

$$J[v] = \int_a^b \left[p(x)\left(\frac{dv}{dx}\right)^2 - \{q(x) + \mu r(x)\}v^2 \right] dx \tag{4.83}$$

satisfies the inequality

$$\mu \geq \lambda_1 \tag{4.84}$$

and so μ provides an upper bound to the least eigenvalue λ_1.

Putting

$$a_{rs} = \int_a^b r(x) v_r v_s \, dx \tag{4.85}$$

and

$$b_{rs} = \int_a^b \left\{ p(x) \frac{dv_r}{dx} \frac{dv_s}{dx} - q(x) v_r v_s \right\} dx, \tag{4.86}$$

we have

$$J[v] = \sum_{r=1}^{n} \sum_{s=1}^{n} (b_{rs} - \mu a_{rs}) c_r c_s \tag{4.87}$$

which is stationary if

$$\frac{\partial J}{\partial c_r} = 0 \quad (r = 1, \ldots, n), \tag{4.88}$$

that is, if

$$\sum_{s=1}^{n} (b_{rs} - \mu a_{rs}) c_s = 0 \quad (r = 1, \ldots, n). \tag{4.89}$$

This set of n linear algebraic equations are consistent if and only if
$$\det(b_{rs} - \mu a_{rs}) = 0. \tag{4.90}$$

Since $a_{rs} = a_{sr}$ and $b_{rs} = b_{sr}$, the n roots $\mu_1, \mu_2, \ldots, \mu_n$ of this secular equation are all real. Now from equations (4.89) it follows that J vanishes for the set of constants $c_1, c_2, \ldots, c_n$ associated with any one of the roots μ_t and hence all the roots are greater than the least eigenvalue λ_1 of the Sturm–Liouville equation. We order the roots of the secular equation (4.90) so that

$$\lambda_1 \leqslant \mu_1 \leqslant \mu_2 \leqslant \cdots \leqslant \mu_n. \tag{4.91}$$

Further if c_{st} is the value of c_s corresponding to the eigenvalue μ_t we can show that

$$\sum_{r=1}^{n} \sum_{s=1}^{n} c_{rt'} c_{st} a_{rs} = \delta_{t't} \tag{4.92}$$

and

$$\sum_{r=1}^{n} \sum_{s=1}^{n} c_{rt'} c_{st} b_{rs} = \mu_t \delta_{t't} \tag{4.93}$$

by following an analogous procedure to that described in section 4.1 yielding the formulae (4.23) and (4.26). Hence introducing the new set of functions $w_t(x)$ defined by

$$w_t(x) = \sum_{s=1}^{n} c_{st} v_s(x) \tag{4.94}$$

we see that

$$\int_a^b r(x) w_{t'} w_t \, dx = \sum_{r=1}^{n} \sum_{s=1}^{n} c_{rt'} c_{st} a_{rs} = \delta_{t't} \tag{4.95}$$

and

$$\int_a^b \left\{ p(x) \frac{dw_{t'}}{dx} \frac{dw_t}{dx} - q(x) w_{t'} w_t \right\} dx = \sum_{r=1}^{n} \sum_{s=1}^{n} c_{rt'} c_{st} b_{rs} = \mu_t \delta_{t't}. \tag{4.96}$$

It follows that if we put

$$v(x) = \sum_{t=1}^{n} d_t w_t(x), \tag{4.97}$$

EIGENVALUE PROBLEMS

where $d_1, d_2, \ldots, d_n$ are arbitrary constants, we may write the integral J in the simple form

$$J = \sum_{t=1}^{n} (\mu_t - \mu) d_t^2. \tag{4.98}$$

We now investigate the effect of including an additional term $c_{n+1}v_{n+1}(x)$ in the expansion of $v(x)$ and put

$$\alpha_t(\mu) = \int_a^b \left[p(x) \frac{dw_t}{dx} \frac{dv_{n+1}}{dx} - \{q(x) + \mu r(x)\} w_t v_{n+1} \right] dx, \tag{4.99}$$

$$\beta_{n+1}(\mu) = \int_a^b \left[p(x)\left(\frac{dv_{n+1}}{dx}\right)^2 - \{q(x) + \mu r(x)\} v_{n+1}^2 \right] dx, \tag{4.100}$$

in which case we have

$$J = \sum_{t=1}^{n} (\mu_t - \mu) d_t^2 + 2 \sum_{t=1}^{n} \alpha_t(\mu) d_t c_{n+1} + \beta_{n+1}(\mu) c_{n+1}^2. \tag{4.101}$$

This is stationary if

$$\frac{\partial J}{\partial d_t} = 2(\mu_t - \mu) d_t + 2\alpha_t(\mu) c_{n+1} = 0 \quad (t = 1, \ldots, n)$$

and

$$\frac{\partial J}{\partial c_{n+1}} = 2 \sum_{t=1}^{n} \alpha_t(\mu) d_t + 2\beta_{n+1}(\mu) c_{n+1} = 0.$$

Then the secular equation becomes

$$\Delta(\mu) \equiv \begin{vmatrix} \mu_1 - \mu & 0 & \cdots & 0 & \alpha_1 \\ 0 & \mu_2 - \mu & \cdots & 0 & \alpha_2 \\ \vdots & \vdots & \ddots & \vdots & \vdots \\ 0 & 0 & \cdots & \mu_n - \mu & \alpha_n \\ \alpha_1 & \alpha_2 & \cdots & \alpha_n & \beta_{n+1} \end{vmatrix} = 0. \tag{4.102}$$

As $\mu \to \infty$, $\Delta(\mu)$ takes the sign $(-1)^{n+1}$ while as $\mu \to -\infty$, $\Delta(\mu)$ becomes positive because the sign of $\Delta(\mu)$ for large values of $|\mu|$ is determined by its leading diagonal as a consequence of the inequality

$$\int_a^b r(x)v_{n+1}^2(x)\,dx - \sum_{t=1}^n \left\{\int_a^b r(x)w_t(x)v_{n+1}(x)\,dx\right\}^2$$
$$= \int_a^b r(x)\left\{v_{n+1}(x) - \sum_{t=1}^n w_t(x)\right.$$
$$\left. \times \int_a^b r(x')w_t(x')v_{n+1}(x')\,dx'\right\}^2 dx \geq 0.$$

Also

$$\Delta(\mu_1) = -\{\alpha_1(\mu_1)\}^2(\mu_2 - \mu_1)(\mu_3 - \mu_1)\ldots(\mu_n - \mu_1)$$
$$\Delta(\mu_2) = -\{\alpha_2(\mu_2)\}^2(\mu_1 - \mu_2)(\mu_3 - \mu_2)\ldots(\mu_n - \mu_2) \quad (4.103)$$
$$\vdots$$
$$\Delta(\mu_n) = -\{\alpha_n(\mu_n)\}^2(\mu_1 - \mu_n)(\mu_2 - \mu_n)\ldots(\mu_{n-1} - \mu_n),$$

and so we see that $\Delta(\mu_1) \leq 0, \Delta(\mu_2) \geq 0, \ldots$, and $\Delta(\mu_n)$ has the sign $(-1)^n$. Hence there is a root of the secular equation (4.102) lying below μ_1, a root in each interval (μ_t, μ_{t+1}), and a root lying above μ_n. We are therefore led to the conclusion that as additional terms are introduced into the expansion of $v(x)$, a given root μ_t of the secular equation is necessarily decreased and so it must provide an upper bound to the corresponding exact eigenvalue λ_t.

If $v_1(x), v_2(x), \ldots, v_r(x), \ldots$ form a complete set of functions and $\mu_t^{(n)}$ is a root of the nth order secular equation, the monotonically decreasing sequence of roots $\{\mu_t^{(n)}\}$ must converge from above to the eigenvalue λ_t of the Sturm–Liouville equation. On the other hand if the functions do not form a complete system, the sequence of roots will converge to a limiting value $\mu_t^{(\infty)}$ such that $\mu_t^{(\infty)} > \lambda_t$.

4.7 Wave motion

We now depart from the one dimensional case discussed hitherto and consider a three dimensional wave motion having angular frequency ω. Substituting

$$\Psi(\mathbf{r}, t) = \psi(\mathbf{r}) \sin(\omega t + \epsilon) \quad (4.104)$$

EIGENVALUE PROBLEMS

into the wave equation

$$\nabla^2 \Psi = \frac{1}{c^2} \frac{\partial^2 \Psi}{\partial t^2}, \tag{4.105}$$

we arrive at the Helmholtz equation

$$\nabla^2 \psi + \frac{\omega^2}{c^2} \psi = 0. \tag{4.106}$$

Multiplying across by ψ and integrating over the spacial coordinates we obtain

$$\frac{\omega^2}{c^2} = -\frac{\int \psi \nabla^2 \psi \, d\mathbf{r}}{\int \psi^2 \, d\mathbf{r}}. \tag{4.107}$$

Now suppose that at the boundary surface S of the volume of integration we have either

$$\text{(i) } \psi = 0 \quad \text{or} \quad \text{(ii) } \frac{\partial \psi}{\partial n} = 0, \tag{4.108}$$

where $\mathbf{n}$ is a unit vector in the direction of the outdrawn normal to S. Then by Green's theorem we see that

$$\int \psi \nabla^2 \psi \, d\mathbf{r} + \int (\nabla \psi)^2 \, d\mathbf{r} = \int_S \psi \frac{\partial \psi}{\partial n} \, dS = 0$$

and so it follows that

$$\frac{\omega^2}{c^2} = \frac{\int (\nabla \psi)^2 \, d\mathbf{r}}{\int \psi^2 \, d\mathbf{r}}. \tag{4.109}$$

Defining the functional

$$I[\chi] = \frac{\int (\nabla \chi)^2 \, d\mathbf{r}}{\int \chi^2 \, d\mathbf{r}}, \tag{4.110}$$

we may express equation (4.109) in the form

$$I[\psi] = \frac{\omega^2}{c^2} \tag{4.111}$$

where ψ is a solution of the Helmholtz equation (4.106) satisfying one of the boundary conditions (4.108).

Next consider the function $\psi + \delta\psi$ differing infinitesimally from ψ but satisfying the same boundary condition as ψ at the surface S. Then, to the first order of small quantities, the change in I due to the change $\delta\psi$ in ψ is given by

$$\delta I[\psi] \int \psi^2 \, d\mathbf{r} + 2I[\psi] \int (\delta\psi)\psi \, d\mathbf{r} = 2 \int (\nabla\delta\psi).(\nabla\psi) \, d\mathbf{r}.$$

But by Green's theorem we have

$$\int (\nabla\delta\psi).(\nabla\psi) \, d\mathbf{r} + \int \delta\psi \, \nabla^2\psi \, d\mathbf{r} = \int_S \delta\psi \frac{\partial \psi}{\partial n} \, dS$$

and since the right-hand side vanishes because either $\delta\psi = 0$ or $\partial\psi/\partial n = 0$ over the boundary surface S, it follows that

$$\delta I[\psi] = -\frac{2 \int \delta\psi \left(\nabla^2\psi + \frac{\omega^2}{c^2} \psi \right) d\mathbf{r}}{\int \psi^2 \, d\mathbf{r}}. \qquad (4.112)$$

Hence

$$\delta I[\psi] = 0 \qquad (4.113)$$

since ψ satisfies the Helmholtz equation. Conversely if $I[\psi]$ is stationary then ψ must be a solution of the Helmholtz equation.

Also it follows at once from (4.110) that

$$I[\chi] \geqslant 0 \qquad (4.114)$$

for all functions χ and so the function which makes $I[\chi]$ attain its least possible value and satisfies one of the prescribed boundary conditions (4.108), is an exact solution of (4.106). Denoting this function by ψ_1 and putting

$$I[\psi_1] = \frac{\omega_1^2}{c^2}, \qquad (4.115)$$

we see that ψ_1 satisfies the Helmholtz equation (4.106) with $\omega = \omega_1$ where ω_1 is the least eigenvalue.

Vibrating membrane

A simple illustrative example of the application of the Rayleigh–Ritz variational method is provided by the oscillations of a circular membrane kept fixed at its circumference. If the radius of the membrane is a, the function $\chi(r)$ representing the displacement of the membrane from its equilibrium configuration at a point whose radial distance from the centre is r, must satisfy $\chi(a) = 0$. For the mode of vibration associated with the lowest frequency, the displacement is independent of the angular coordinate θ. Taking

$$\chi(r) = (a^2 - r^2)^s \tag{4.116}$$

where the index s is to be regarded as a variable parameter, we have

$$I[\chi] = \frac{\int_0^a (d\chi/dr)^2 r\, dr}{\int_0^a \chi^2 r\, dr} = \frac{2s(2s+1)}{a^2(2s-1)}.$$

In the present instance the volume integrals arising in the definition (4.110) of I become replaced by integrals over the surface of the membrane which then reduce to single integrations over the radial distance r because χ is independent of θ.

The least value of I occurs when $\partial I/\partial s = 0$ which gives

$$s = \tfrac{1}{2}(1 + \sqrt{2})$$

yielding

$$\frac{\omega_1^2}{c^2} \leqslant \frac{3 + 2\sqrt{2}}{a^2} = \frac{5.828}{a^2},$$

where ω_1 is the angular frequency of the lowest oscillation. The exact value of $\omega_1^2 a^2/c^2$ is 5.7832 which is about 1% less than the least upper bound obtained above.

To obtain upper bounds to the first n angular frequencies for states of vibration having circular symmetry we may employ the function

$$\chi(r) = \sum_{s=1}^{n} c_s (a^2 - r^2)^s,$$

where the coefficients c_s form a set of n arbitrary parameters. If upper bounds to just the first two angular frequencies are required we may put

$$\chi(r) = c_1(a^2 - r^2) + c_2(a^2 - r^2)^2 \qquad (4.117)$$

which gives

$$a^2 I[\chi] = \frac{c_1^2 + \tfrac{4}{3}a^2 c_1 c_2 + \tfrac{2}{3}a^4 c_2^2}{\tfrac{1}{6}c_1^2 + \tfrac{1}{4}a^2 c_1 c_2 + \tfrac{1}{10}a^4 c_2^2}.$$

This expression is stationary when

$$\frac{\partial I}{\partial c_1} = 0, \qquad \frac{\partial I}{\partial c_2} = 0$$

which lead to the equations

$$\left(1 - \frac{\lambda}{6}\right)c_1 + \left(\frac{2}{3} - \frac{\lambda}{8}\right)a^2 c_2 = 0,$$

$$\left(\frac{2}{3} - \frac{\lambda}{8}\right)c_1 + \left(\frac{2}{3} - \frac{\lambda}{10}\right)a^2 c_2 = 0,$$

where λ is the stationary value of $a^2 I$. For these equations to be consistent, λ must satisfy the secular equation

$$\begin{vmatrix} 1 - \dfrac{\lambda}{6} & \dfrac{2}{3} - \dfrac{\lambda}{8} \\ \dfrac{2}{3} - \dfrac{\lambda}{8} & \dfrac{2}{3} - \dfrac{\lambda}{10} \end{vmatrix} = 0$$

which may be rewritten in the form of the quadratic equation

$$3\lambda^2 - 128\lambda + 640 = 0$$

whose solutions are

$$\lambda = \tfrac{8}{3}(8 \pm \sqrt{34}) \quad \text{or} \quad \lambda = 5.784 \text{ and } 36.9.$$

Thus we have

$$\frac{\omega_1^2 a^2}{c^2} \leqslant 5.784$$

and

$$\frac{\omega_2^2 a^2}{c^2} \leqslant 36.9,$$

EIGENVALUE PROBLEMS 161

where ω_2 is the angular frequency of the next highest vibrational state above the fundamental mode having circular symmetry.

It is seen that by using the trial function (4.117) we have obtained an upper bound to $\omega_1^2 a^2/c^2$ which is exceedingly close to the exact value of this quantity. On the other hand the exact value of $\omega_2^2 a^2/c^2$ is 30.5 which is considerably smaller than the upper bound 36.9 obtained above. It is apparent that whereas a good approximation to the lowest angular frequency can be obtained fairly readily, it is more difficult to derive a satisfactory approximation to a higher angular frequency.

4.8 Schrödinger equation for a particle

We can now leave the application of variational principles to eigenvalue problems in classical mechanics and investigate their role in quantum theory. The Schrödinger equation for a single particle of mass m and total energy E moving under the influence of a potential $V(\mathbf{r})$ has the form

$$-\frac{\hbar^2}{2m}\nabla^2\psi + V\psi = E\psi. \tag{4.118}$$

Multiplying across by $\bar{\psi}$ and integrating with respect to the spacial coordinates yields for the energy

$$E = \frac{\int \bar{\psi}\{-(\hbar^2/2m)\nabla^2 + V\}\psi \, d\mathbf{r}}{\int \bar{\psi}\psi \, d\mathbf{r}}. \tag{4.119}$$

Now introducing the functional

$$I[\chi] = \frac{\int \bar{\chi}\{-(\hbar^2/2m)\nabla^2 + V\}\chi \, d\mathbf{r}}{\int \bar{\chi}\chi \, d\mathbf{r}}, \tag{4.120}$$

we see that

$$I[\psi] = E \tag{4.121}$$

where ψ is the solution of the Schrödinger equation (4.118) with eigenenergy E.

We now let $\delta I[\psi]$ denote the change in $I[\psi]$ resulting from an infinitesimal change $\delta\psi$ in the solution ψ of the Schrödinger equation. Then, neglecting quantities of the second order of smallness, we obtain

$$\delta I[\psi] \int \bar{\psi}\psi \, d\mathbf{r} + I[\psi] \int \{(\delta\bar{\psi})\psi + (\delta\psi)\bar{\psi}\} \, d\mathbf{r}$$
$$= \int \delta\bar{\psi}\left(-\frac{\hbar^2}{2m}\nabla^2 + V\right)\psi \, d\mathbf{r} + \int \bar{\psi}\left(-\frac{\hbar^2}{2m}\nabla^2 + V\right)\delta\psi \, d\mathbf{r}. \tag{4.122}$$

This formula may be transformed into a more convenient form by the application of Green's theorem

$$\int (\bar{\psi}\nabla^2 \delta\psi - \delta\psi \, \nabla^2\bar{\psi}) \, d\mathbf{r} = \int \left(\bar{\psi}\frac{\partial}{\partial n}\delta\psi - \delta\psi\frac{\partial}{\partial n}\bar{\psi}\right) dS,$$

where the integration on the right-hand side is over the boundary surface of the volume of integration occurring in equation (4.122). At this stage we need to know the boundary conditions imposed upon the wave function ψ and the infinitesimal change $\delta\psi$ in the wave function. If we suppose that ψ and $\delta\psi$ decay more rapidly than r^{-1} for large distance r from the origin and if the volume integral is extended to the whole of space, the surface integral must vanish and then we have

$$\delta I[\psi] = $$
$$\frac{\int \delta\bar{\psi}\{-\left(\frac{\hbar^2}{2m}\right)\nabla^2 + V - E\}\psi \, d\mathbf{r} + \int \delta\psi\{-\left(\frac{\hbar^2}{2m}\right)\nabla^2 + V - E\}\bar{\psi} \, d\mathbf{r}}{\int \bar{\psi}\psi \, d\mathbf{r}}.$$
$$\tag{4.123}$$

Hence
$$\delta I[\psi] = 0 \tag{4.124}$$

since ψ satisfies the Schrödinger equation. Conversely if $I[\psi]$ is stationary then ψ is a solution of the Schrödinger equation.

Now Green's theorem may be written in the alternative form

$$\int \bar{\chi}\nabla^2\chi \, d\mathbf{r} + \int \nabla\bar{\chi}\cdot\nabla\chi \, d\mathbf{r} = \int \bar{\chi}\frac{\partial \chi}{\partial n} \, dS.$$

If χ decays more rapidly than r^{-1} for large r, the surface integral on the right-hand side must vanish and then we may rewrite (4.120) as

$$I[\chi] = \frac{\int \{(\hbar^2/2m)|\nabla\chi|^2 + V|\chi|^2\}\,d\mathbf{r}}{\int |\chi|^2\,d\mathbf{r}}. \qquad (4.125)$$

Hence if we choose only functions which satisfy the condition

$$\left|\int V|\chi|^2\,d\mathbf{r}\right| < \infty, \qquad (4.126)$$

the functional $I[\chi]$ will necessarily have a lower bound. It follows that the function $\chi = \psi_1$ for which $I[\chi]$ attains its minimum value, is an eigenfunction of the Schrödinger equation (4.118) with eigenenergy

$$E_1 = I[\psi_1]. \qquad (4.127)$$

Since E_1 is the least eigenenergy, the particle is said to be in its *ground state*.

4.9 Eigenenergies of a quantum mechanical system

We now generalize to the case of a quantum mechanical system having Hamiltonian operator H. Then the eigenenergies E_r and eigenfunctions ψ_r for this system satisfy the Schrödinger equation

$$(H - E)\psi = 0. \qquad (4.128)$$

Defining the functional $I[\chi]$ by means of the formula

$$I[\chi] = \frac{\int \bar\chi H\chi\,d\tau}{\int \bar\chi\chi\,d\tau} \qquad (4.129)$$

we get

$$I[\psi] = E, \qquad (4.130)$$

ψ being the solution of the Schrödinger equation (4.128) associated with eigenenergy E.

If an infinitesimal change $\delta\psi$ in ψ produces a change $\delta I[\psi]$ in $I[\psi]$, we see that

$$\delta I[\psi] \int \bar\psi\psi\, d\tau + I[\psi] \int \{(\delta\bar\psi)\psi + (\delta\psi)\bar\psi\}\, d\tau$$
$$= \int \delta\bar\psi\, H\psi\, d\tau + \int \bar\psi H\, \delta\psi\, d\tau. \qquad (4.131)$$

Since H is a real Hermitian operator we may put

$$\int \bar\psi H\, \delta\psi\, d\tau = \int \delta\psi\, H\bar\psi\, d\tau$$

provided the variation $\delta\psi$ is chosen to satisfy suitable boundary conditions, and then

$$\delta I[\psi] = \frac{\int \delta\bar\psi(H - E)\psi\, d\tau + \int \delta\psi(H - E)\bar\psi\, d\tau}{\int \bar\psi\psi\, d\tau}. \qquad (4.132)$$

Hence

$$\delta I[\psi] = 0 \qquad (4.133)$$

since ψ is a solution of the Schrödinger equation. Thus $I[\chi]$ is stationary for $\chi = \psi$ just as for the single particle case treated in the last section.

Upper bounds to eigenenergies

Making the assumption that the eigenfunctions ψ_r form a complete orthonormal set permits us to expand an arbitrary, bounded, quadratically integrable function χ in the form

$$\chi = \underset{r}{\mathsf{S}}\, a_r\psi_r \qquad (4.134)$$

where the a_r are constants and S denotes a summation over the discrete states and an integration over the continuous states of the quantum mechanical system. Then the expectation value of the Hamiltonian operator H for the function χ is given by

$$\int \bar\chi H\chi\, d\tau = \underset{r}{\mathsf{S}}\, E_r|a_r|^2.$$

Let E_1 denote the lowest eigenenergy, that is, the energy of the ground state of the quantum mechanical system. Then $E_1 \leq E_r$ for all r and so

$$\int \bar{\chi} H \chi \, d\tau \geq E_1 \sum_r |a_r|^2.$$

Since

$$\int \bar{\chi} \chi \, d\tau = \sum_r |a_r|^2$$

it follows that

$$I[\chi] \geq E_1 \tag{4.135}$$

where the functional $I[\chi]$ is given by (4.129). Thus an upper bound to the ground state eigenenergy E_1 can be obtained by substituting a suitable function ψ_T, which we shall call a *trial function*, into $I[\chi]$. If the trial function is chosen to depend upon a number of variable parameters c_i ($i = 1, \ldots, p$), a least upper bound to E_1 may be obtained by solving the p equations

$$\frac{\partial I}{\partial c_i} = 0 \quad (i = 1, \ldots, p) \tag{4.136}$$

and employing the derived set of values of the parameters c_i to calculate I.

Upper bounds to the higher eigenvalues can also be obtained. If we arrange the eigenenergies so that $E_r > E_s$ for $r > s$, then the function

$$\chi = \phi - \sum_{r=1}^{n-1} \psi_r \int \bar{\psi}_r \phi \, d\tau \tag{4.137}$$

where ϕ is an arbitrary quadratically integrable function, is orthogonal to all the eigenfunctions of the states below the nth state. Again expanding χ in the form (4.134), we see that $a_r = 0$ for $r = 1, \ldots, n - 1$ and so now we have

$$I[\chi] \geq E_n. \tag{4.138}$$

This means that if a trial function is chosen so that it is orthogonal to all the exact eigenfunctions of the states below the nth state, substitution into the functional $I[\chi]$ will provide an upper bound to E_n.

Hylleraas–Undheim method

Although, in principle, the procedure described in the previous section provides an upper bound to the eigenenergy of any excited state of a quantum mechanical system, it depends upon a knowledge of the exact wave functions of all the lower lying states which are not usually available. The Ritz variational method, which overcomes this difficulty, may be readily generalized from the case of the Sturm–Liouville equation discussed in section 4.6 to the case of the Schrödinger equation in wave mechanics. It is then referred to as the *Hylleraas–Undheim method* since it was first used by them to calculate eigenenergies of excited states of atoms.[1]

Let $\chi_1, \chi_2, \ldots, \chi_n$ form a set of n linearly independent functions. Introducing the function

$$\chi = \sum_{r=1}^{n} c_r \chi_r, \qquad (4.139)$$

where $c_1, c_2, \ldots, c_n$ are n variable parameters, we see that the functional $I[\chi]$ given by (4.129) is stationary if

$$\frac{\partial I}{\partial \bar{c}_r} = 0 \qquad (r = 1, \ldots, n)$$

which yields the set of n linear equations

$$\sum_{s=1}^{n} (H_{rs} - E S_{rs}) c_s = 0 \qquad (r = 1, \ldots, n), \qquad (4.140)$$

where

$$H_{rs} = \int \bar{\chi}_r H \chi_s \, d\tau, \qquad (4.141)$$

$$S_{rs} = \int \bar{\chi}_r \chi_s \, d\tau \qquad (4.142)$$

and the permissible values of E are the stationary values of I. These values are the n roots $E_t^{(n)}$ ($t = 1, \ldots, n$) of the secular equation

$$\det (H_{rs} - E S_{rs}) = 0. \qquad (4.143)$$

EIGENVALUE PROBLEMS

Following an analogous procedure to that given in section 4.6 we order the roots of the nth order determinantal equation (4.143) so that

$$E_1 \leq E_1^{(n)} \leq E_2^{(n)} \leq \cdots \leq E_n^{(n)} \tag{4.144}$$

where E_1 is the ground state eigenenergy.

We now construct a set of n orthonormal functions $\phi_1, \ldots, \phi_n$ out of the original set of functions $\chi_1, \ldots, \chi_n$ in such a way that the new matrix H_{rs} is diagonal, and then introduce an additional term $c_{n+1}\chi_{n+1}$ in the expansion of χ. Putting

$$A_t = \int \phi_t(H - E)\chi_{n+1}\, d\tau \tag{4.145}$$

and

$$B_{n+1} = \int \bar{\chi}_{n+1}(H - E)\chi_{n+1}\, d\tau, \tag{4.146}$$

the secular equation becomes

$$\Delta^{(n+1)}(E) \equiv \begin{vmatrix} E_1^{(n)} - E & 0 & \cdots & 0 & A_1 \\ 0 & E_2^{(n)} - E & \cdots & 0 & A_2 \\ \vdots & \vdots & & \vdots & \vdots \\ 0 & 0 & \cdots & E_n^{(n)} - E & A_n \\ \bar{A}_1 & \bar{A}_2 & \cdots & \bar{A}_n & B_{n+1} \end{vmatrix} = 0. \tag{4.147}$$

As $E \to \infty$, $\Delta^{(n+1)}(E)$ takes the sign $(-1)^{n+1}$ and as $E \to -\infty$, $\Delta^{(n+1)}(E)$ becomes positive. Also $\Delta^{(n+1)}(E_1^{(n)}) \leq 0$, $\Delta^{(n+1)}(E_2^{(n)}) \geq 0, \ldots$, and $\Delta^{(n+1)}(E_n^{(n)})$ has the sign $(-1)^n$. Hence the tth root $E_t^{(n+1)}$ of the secular equation (4.147) satisfies

$$E_t^{(n+1)} \leq E_t^{(n)} \quad (t = 1, \ldots, n) \tag{4.148}$$

and

$$E_{n+1}^{(n+1)} \geq E_n^{(n)}. \tag{4.149}$$

It follows that as extra terms are added to the expansion of χ, a given root $E_t^{(n)}$ of the secular equation is decreased and so

$$E_t \leq E_t^{(n)} \tag{4.150}$$

where E_t is the exact eigenenergy of the tth state of the system.

If there exist just N bound states of the system we see that

$$0 \leqslant E_t^{(n)} \tag{4.151}$$

for $t > N$. In particular

$$0 \leqslant E_{N+1}^{(N+1)}. \tag{4.152}$$

We now choose a set of functions $\chi_1, \ldots, \chi_N$ in such a way that $E_t^{(N)} < 0$ for $t = 1, \ldots, N$. Since $\Delta^{(N+1)}(E_N^{(N)})$ has the sign $(-1)^N$ and $\Delta^{(N+1)}(E_{N+1}^{(N+1)}) = 0$, it follows that $\Delta^{(N+1)}(E = 0)$ must have the sign $(-1)^N$. Now the product

$$\prod_{t=1}^{N} E_t^{(N)}$$

also has the sign $(-1)^N$ and so

$$H_{N+1,N+1} - \sum_{t=1}^{N} \frac{1}{E_t^{(N)}} |H_{t,N+1}|^2 \geqslant 0, \tag{4.153}$$

where

$$H_{t,N+1} = \int \bar{\phi}_t H \chi_{N+1} \, d\tau. \tag{4.154}$$

For the special case $N = 1$, corresponding to the existence of just a single bound state of the system, this result takes the simple form

$$H_{11} H_{22} - |H_{12}|^2 \leqslant 0 \tag{4.155}$$

since $H_{11} = E_1^{(1)}$ is negative.

Lower bounds to the ground state eigenenergy

Expressions which provide lower bounds to the ground state energy E_1 of a system have been derived by Temple[2] and Weinstein.[3] To establish Temple's formula we introduce the integral

$$J[\chi] = \int \bar{\chi}(H - E_1)(H - E_2)\chi \, d\tau, \tag{4.156}$$

where E_2 is the energy of the first excited state of the system and χ is an arbitrary bounded, quadratically integrable, normalized

EIGENVALUE PROBLEMS

function. Expanding χ in the form (4.134) we obtain

$$J = \sum_r (E_r - E_1)(E_r - E_2)|a_r|^2 \qquad (4.157)$$

and so

$$J \geqslant 0. \qquad (4.158)$$

Defining the integral

$$I_m = \int \bar{\chi} H^m \chi \, d\tau, \qquad (4.159)$$

we see that we may express J in the form

$$J = I_2 - E_2 I_1 + E_1(E_2 - I_1) \qquad (4.160)$$

and hence as long as $E_2 > I_1$ it follows from (4.158) that

$$E_1 \geqslant \frac{E_2 I_1 - I_2}{E_2 - I_1} = I_1 - \frac{I_2 - I_1^2}{E_2 - I_1} \qquad (4.161)$$

which is *Temple's formula*. Together with the upper bound

$$I_1 \geqslant E_1 \qquad (4.162)$$

obtained on p. 165, the lowest eigenenergy E_1 is now bounded both from above and below. Unfortunately, however, Temple's formula (4.161) suffers from the fact that it presupposes a knowledge of the exact eigenenergy E_2 of the first excited state which may not be available.

Weinstein's formula may be derived by considering the integral

$$K[\chi] = \int \bar{\chi}(H - I_1)^2 \chi \, d\tau. \qquad (4.163)$$

Using the expansion (4.134) once again we see that

$$K = \sum_r (E_r - I_1)^2 |a_r|^2. \qquad (4.164)$$

Rewriting this in the form

$$K = (E_1 - I_1)^2 + \sum_r \{(E_r - I_1)^2 - (E_1 - I_1)^2\}|a_r|^2 \qquad (4.165)$$

shows that

$$K \geqslant (E_1 - I_1)^2 \qquad (4.166)$$

provided I_1 is closer to the ground state energy E_1 than to the energy of any other state, a condition which is readily fulfilled. Since

$$K = I_2 - I_1^2 \qquad (4.167)$$

and I_1 satisfies the inequality (4.162), it follows that

$$E_1 \geqslant I_1 - \sqrt{(I_2 - I_1^2)}. \qquad (4.168)$$

This is *Weinstein's formula*.

If χ is identical to the ground state eigenfunction ψ_1, it is evident that $I_2 = I_1^2$ and then both Temple's and Weinstein's formulae become $E_1 \geqslant I_1$ which, together with (4.162), implies the obvious result that $I_1 = E_1$.

Weinstein's formula is a special case of a more general result due to Stevenson[4] which can be readily established in an analogous fashion to that given above by considering the functional

$$\int \bar{\chi}(H - \alpha)^2 \chi \, d\tau = I_2 - 2\alpha I_1 + \alpha^2$$

rather than (4.163). Stevenson shows that

$$E_1 \geqslant E_L(\alpha)$$

where

$$E_L(\alpha) = \alpha - \sqrt{(I_2 - 2\alpha I_1 + \alpha^2)},$$

provided the parameter α satisfies

$$E_1 \leqslant \alpha \leqslant \tfrac{1}{2}(E_1 + E_2).$$

If we choose $\alpha = I_1$ we regain Weinstein's formula (4.168) but the best value of α is evidently $\tfrac{1}{2}(E_1 + E_2)$ since $E_L(\alpha)$ is a monotonically increasing function of α.

4.10 Atomic eigenenergies

One of the most fruitful applications of the variational method has been to the calculation of the eigenenergies of simple atomic systems. We shall begin this section by considering the single electron system case for which the eigenvalue problem can be solved without approximation.

EIGENVALUE PROBLEMS

Hydrogenic atom

The Hamiltonian operator for a hydrogenic atom consisting of a single electron of mass m and charge e moving in the Coulomb field of a nucleus of charge Ze, is given by

$$H = -\frac{\hbar^2}{2m} \nabla^2 - \frac{Ze^2}{r} \qquad (4.169)$$

where r is the distance of the electron from the nucleus.

It is reasonable to expect the function representing the electron in its lowest state to be spherically symmetrical and to decay exponentially with increasing radial distance r and so we choose the normalized trial function

$$\chi(r) = \sqrt{\left(\frac{\alpha^3}{\pi a_0^3}\right)} e^{-\alpha r/a_0} \qquad (4.170)$$

where α is an arbitrary parameter and $a_0 = \hbar^2/me^2$ is the *Bohr radius* of the innermost orbit of a hydrogen atom. Then it can be shown without difficulty by simple integration that

$$I_1[\chi] = \int \bar{\chi} H \chi \, d\mathbf{r} = \frac{e^2}{2a_0}(\alpha^2 - 2Z\alpha). \qquad (4.171)$$

The minimum value of I_1 is attained when $\partial I_1/\partial \alpha = 0$ which requires that $\alpha = Z$ and then

$$I_1 = -\frac{e^2 Z^2}{2a_0}. \qquad (4.172)$$

Hence by (4.162) we see that the ground state eigenenergy must satisfy

$$E_1 \leqslant -\frac{e^2 Z^2}{2a_0}. \qquad (4.173)$$

A lower bound to E_1 can be found by employing Temple's or Weinstein's formula. For $\alpha = Z$ it can be easily verified that $I_2 = I_1^2$ and so

$$E_1 \geqslant I_1 = -\frac{e^2 Z^2}{2a_0}. \qquad (4.174)$$

It follows that the upper bound $-e^2Z^2/2a_0$ which we have derived for the ground state eigenenergy E_1 is equal to the exact value of this eigenenergy. This occurs because the function which we chose to represent the ground state orbital is identical to the actual ground state eigenfunction for $\alpha = Z$.

Two electron systems

We now come to the consideration of helium like systems which do not permit of exact solution due to the correlation between the two electrons. In fact, for the case of two electron systems, the variational method provides the most accurate means of determining the eigenenergies.

The Hamiltonian operator for an atomic system consisting of two electrons moving in the Coulomb field of a nucleus of charge Ze has the form

$$H = -\frac{\hbar^2}{2m}(\nabla_1^2 + \nabla_2^2) - Ze^2\left(\frac{1}{r_1} + \frac{1}{r_2}\right) + \frac{e^2}{r_{12}}, \quad (4.175)$$

where r_1 and r_2 are the distances of the two electrons from the nucleus and r_{12} is the distance between the two electrons.

The simplest form of trial function having suitable characteristics which we may use to describe the two electrons in the ground state of the atomic system is given by

$$\chi(\mathbf{r}_1, \mathbf{r}_2) = \frac{\alpha^3}{\pi a_0^3} e^{-\alpha(r_1 + r_2)/a_0} \quad (4.176)$$

where α is an arbitrary variable parameter. The external factor has been picked so as to ensure that the function satisfies the normalization condition

$$\iint |\chi(\mathbf{r}_1, \mathbf{r}_2)|^2 \, d\mathbf{r}_1 \, d\mathbf{r}_2 = 1. \quad (4.177)$$

For $\alpha = Z$ the function (4.176) is just the product of the wave functions of two independent electrons, each moving in the field of a nucleus of charge Ze. The screening of the nucleus by the electrons may be allowed for by permitting the parameter α to differ from Z. To obtain the optimum value of α we again apply

EIGENVALUE PROBLEMS   173

the variational method. Substituting (4.176) into (4.129) we get

$$I = \frac{e^2}{a_0}(\alpha^2 - 2Z\alpha) + e^2\left(\frac{\alpha^3}{\pi a_0^3}\right)^2 \int\int \frac{1}{r_{12}} e^{-2\alpha(r_1+r_2)/a_0} \, d\mathbf{r}_1 \, d\mathbf{r}_2, \quad (4.178)$$

where the first pair of terms arise from the kinetic and potential energy terms in the Hamiltonian (4.175) and is just twice the expression (4.171) found for a hydrogenic atom of nuclear charge Ze, while the second term is due to the Coulomb repulsion between the two electrons.

Using the expansion of $1/r_{12}$ in Legendre polynomials

$$\frac{1}{r_{12}} = \sum_{l=0}^{\infty} \gamma_l(r_1, r_2) P_l(\cos \Theta), \quad (4.179)$$

where

$$\gamma_l(r_1, r_2) = \begin{cases} r_2^l/r_1^{l+1} & (r_1 > r_2) \\ r_1^l/r_2^{l+1} & (r_2 > r_1) \end{cases} \quad (4.180)$$

and Θ is the angle between the position vectors $\mathbf{r}_1$ and $\mathbf{r}_2$ of the electrons relative to the nucleus, and employing the theorem

$$P_l(\cos \Theta) = P_l(\cos \theta_1) P_l(\cos \theta_2)$$
$$+ 2 \sum_{m=1}^{l} \frac{(l-m)!}{(l+m)!} P_l^m(\cos \theta_1) P_l^m(\cos \theta_2) \cos m(\phi_1 - \phi_2), \quad (4.181)$$

we obtain

$$\int\int \frac{1}{r_{12}} e^{-2\alpha(r_1+r_2)/a_0} \, d\mathbf{r}_1 \, d\mathbf{r}_2$$
$$= (4\pi)^2 \int_0^{\infty} e^{-2\alpha r_1/a_0} \left[\frac{1}{r_1}\int_0^{r_1} e^{-2\alpha r_2/a_0} r_2^2 \, dr_2 \right.$$
$$\left. + \int_{r_1}^{\infty} e^{-2\alpha r_2/a_0} r_2 \, dr_2\right] r_1^2 \, dr_1$$
$$= \frac{5\pi^2 a_0^5}{8\alpha^5} \quad (4.182)$$

since only the $l = 0$ term in (4.179) gives a non-zero contribution to the required integral.

Hence

$$I = \frac{e^2}{a_0}\{\alpha^2 - (2Z - \tfrac{5}{8})\alpha\} \quad (4.183)$$

which attains its minimum value when $\alpha = Z - \tfrac{5}{16}$ so that the least upper bound to the ground state eigenenergy obtainable with the function (4.176) is

$$-(Z - \tfrac{5}{16})^2 \frac{e^2}{a_0}. \qquad (4.184)$$

If we put $Z = 2, 3, 4$ in (4.184), we obtain $-2\cdot 848$, $-7\cdot 223$, $-13\cdot 598$ e^2/a_0 for the least upper bounds to the eigenenergies of the ground states of He, Li$^+$, Be^{2+} respectively. A sensitive test of the accuracy of the function (4.176) is provided by the energy required to remove an electron from each of these systems, that is, their *ionization energies*. Since the ground state eigenenergies of He$^+$, Li^{2+}, Be^{3+} are -2, $-4\cdot 5$, -8 e^2/a_0, we see that (4.184) yields $0\cdot 848$, $2\cdot 723$, $5\cdot 598$ e^2/a_0 for the ionization energies of He, Li$^+$, Be^{2+} respectively, in close agreement with the experimentally determined values $0\cdot 904$, $2\cdot 780$, $5\cdot 655$ e^2/a_0 for these quantities, the errors being $6\cdot 2$, $2\cdot 1$, $1\cdot 0\%$ respectively.

In the case of the negative ion of atomic hydrogen H$^-$, setting $Z = 1$ in (4.184) gives $-0\cdot 473$ e^2/a_0 for the least upper bound to the ground state eigenenergy, which yields $-0\cdot 027$ e^2/a_0 for the energy necessary to detach the extra electron from the negative ion. However, this quantity, the *electron affinity* of atomic hydrogen, must be positive since H$^-$ is known to be stable. Hence, whereas the function (4.176) is quite satisfactory for the atom and positive ion cases, it proves to be inadequate for the case of the negative ion. In view of the extreme simplicity of the form of the function which was employed in the previous discussion it is hardly surprising that we should have failed to predict the stability of the negative ion of atomic hydrogen.

An obvious way of improving the description of the two electron system is to use a more flexible function $u(\mathbf{r})$ to represent the individual electrons though continuing to employ the separable form

$$\chi(\mathbf{r}_1, \mathbf{r}_2) = u(\mathbf{r}_1)u(\mathbf{r}_2). \qquad (4.185)$$

Choosing

$$u(\mathbf{r}) = \mathcal{N}(e^{-\alpha r/a_0} + c\, e^{-k\alpha r/a_0}), \qquad (4.186)$$

α, k and c being arbitrary parameters and $\mathcal{N}$ being a normalization factor, yields the values for the least upper bound E_U to the ground state eigenenergies of H^-, He, Li^+, Be^{2+} displayed in Table 4.1, where they are compared with the values of E_U obtained using other forms of trial function. Although we see that the use of (4.185) together with (4.186) does not result in much improvement over the simple form of function (4.176), it closely approaches the best that can be obtained with the Hartree–Fock approximation, discussed in some detail in section 4.11, which separates the motions of the two electrons by employing the functional form (4.185).

One way of obtaining a positive electron affinity for the hydrogen atom is to use a function which allows the screening parameters of the two electrons to differ. Since the spins of the electrons in the ground state are opposed, the space wave function must be symmetric with respect to interchange of the coordinates of the two electrons in order to satisfy the condition imposed by the *Pauli exclusion principle* that the total wave function must be antisymmetric with respect to interchange of space and spin coordinates. Hence we put

$$\chi(\mathbf{r}_1, \mathbf{r}_2) = \mathcal{N}\{u(\alpha, \mathbf{r}_1)u(\beta, \mathbf{r}_2) + u(\beta, \mathbf{r}_1)u(\alpha, \mathbf{r}_2)\} \quad (4.187)$$

where

$$u(\lambda, \mathbf{r}) = e^{-\lambda r/a_0}, \quad (4.188)$$

α and β being arbitrary parameters and $\mathcal{N}$ a normalization constant. This gives the values of E_U shown in Table 4.1 which result in the positive electron affinity $0\cdot013\ e^2/a_0$ for atomic hydrogen and the ionization energies $0\cdot875$, $2\cdot749$, $5\cdot623\ e^2/a_0$ for He, Li^+, Be^{2+}. Thus a substantial gain has been obtained for H^- by employing the function (4.187) rather than (4.176) though the improvement is less for He, Li^+, Be^{2+}.

Another way of obtaining a positive electron affinity for atomic hydrogen is to allow for correlation by introducing an explicit dependence upon the distance r_{12}. A simple form of function which does this is

$$\chi(\mathbf{r}_1, \mathbf{r}_2) = \mathcal{N} e^{-\alpha(r_1 + r_2)/a_0}(1 + cr_{12}) \quad (4.189)$$

where α and c are arbitrary parameters. The values obtained with this function are also given in Table 4.1. For He, Li$^+$, Be^{2+} they are somewhat better than the values obtained with (4.187) but for H$^-$ the upper bound is not quite so satisfactory.

Table 4.1. Values of the least upper bounds E_U to the ground state eigenenergies of the two electron systems H$^-$, He, Li$^+$, Be^{2+} obtained by using (i) function (4.176), (ii) function (4.185) with (4.186), (iii) function (4.187) with (4.188) and (iv) function (4.189)

	$-E_U$ (in units of e^2/a_0)				
	(i)	(ii)	(iii)	(iv)	Experiment
H$^-$	0·473	0·483	0·513	0·509	0·528
He	2·848	2·861	2·875	2·891	2·904
Li$^+$	7·223	7·236	7·249	7·268	7·280
Be^{2+}	13·598	13·611	13·623	13·644	13·656

To obtain more accurate values of E_U it is necessary to use more complicated functions. However, this requires us to consider the application of the variational method to the two electron system in greater detail. Since the ground state of a two electron system has zero angular momentum, the wave function describing the system in this state may be expressed entirely in terms of the coordinates r_1, r_2, r_{12}.[5] In order to employ these coordinates we need to write the double volume element $d\mathbf{r}_1 \, d\mathbf{r}_2$ in terms of them. Using the spherical polar coordinates r_1, θ_1, ϕ_1 referred to a coordinate system with origin at the atomic nucleus, we have

$$d\mathbf{r}_1 = r_1^2 \, dr_1 \sin \theta_1 \, d\theta_1 \, d\phi_1.$$

Now relative to a coordinate system with origin at the point with position vector $\mathbf{r}_1$ referred to the nucleus and polar axis in the direction of $-\mathbf{r}_1$, the point with position vector $\mathbf{r}_2$ has spherical polar coordinates $r_{12}, \theta_{12}, \phi_{12}$ and so we may express the volume element $d\mathbf{r}_2$ in the form

$$d\mathbf{r}_2 = r_{12}^2 \, dr_{12} \sin \theta_{12} \, d\theta_{12} \, d\phi_{12}.$$

Since
$$r_2^2 = r_1^2 + r_{12}^2 - 2r_1 r_{12} \cos \theta_{12}$$
it follows that
$$r_2 \, dr_2 = r_1 r_{12} \sin \theta_{12} \, d\theta_{12}$$
and hence
$$d\mathbf{r}_1 \, d\mathbf{r}_2 = r_1 \, dr_1 \, r_2 \, dr_2 \, r_{12} \, dr_{12} \sin \theta_1 \, d\theta_1 \, d\phi_1 \, d\phi_{12},$$
where
$$0 \leqslant r_1 < \infty, \quad 0 \leqslant r_2 < \infty, \quad |r_1 - r_2| \leqslant r_{12} \leqslant r_1 + r_2$$
and
$$0 \leqslant \theta_1 \leqslant \pi, \quad 0 \leqslant \phi_1 \leqslant 2\pi, \quad 0 \leqslant \phi_{12} \leqslant 2\pi.$$

Since the wave function χ is independent of the angles, integration over θ_1, ϕ_1 and ϕ_{12} yields an external factor of $8\pi^2$ and so the volume element becomes

$$d\tau = 8\pi^2 \, r_1 \, dr_1 \, r_2 \, dr_2 \, r_{12} \, dr_{12}. \tag{4.190}$$

Now putting
$$\chi(r_1, r_2, r_{12}) = \phi(kr_1, kr_2, kr_{12}) \tag{4.191}$$

where k is a scaling parameter, and substituting into (4.129) gives

$$I = \frac{k^2 \mathcal{M} - k \mathcal{L}}{\mathcal{N}} \tag{4.192}$$

where

$$\mathcal{L} = \int \left\{ Ze^2 \left(\frac{1}{r_1} + \frac{1}{r_2} \right) - \frac{e^2}{r_{12}} \right\} \phi^2(r_1, r_2, r_{12}) \, d\tau, \tag{4.193}$$

$$\mathcal{M} = -\frac{\hbar^2}{2m} \int \phi(r_1, r_2, r_{12}) (\nabla_1^2 + \nabla_2^2) \phi(r_1, r_2, r_{12}) \, d\tau$$
$$= \frac{\hbar^2}{2m} \int \{(\nabla_1 \phi)^2 + (\nabla_2 \phi)^2\} \, d\tau, \tag{4.194}$$

and

$$\mathcal{N} = \int \phi^2(r_1, r_2, r_{12}) \, d\tau. \tag{4.195}$$

The optimum value of I occurs when $\partial I / \partial k = 0$ which yields

$$k = \frac{\mathcal{L}}{2\mathcal{M}} \tag{4.196}$$

and

$$I = -\frac{\mathcal{L}^2}{4\mathcal{M}\mathcal{N}}. \tag{4.197}$$

The expectation values of the kinetic energy and potential energy are given by

$$\langle T \rangle = -\frac{\frac{\hbar^2}{2m}\int \chi(r_1, r_2, r_{12})(\nabla_1^2 + \nabla_2^2)\chi(r_1, r_2, r_{12})\, d\tau}{\int \chi^2(r_1, r_2, r_{12})\, d\tau} = \frac{k^2 \mathcal{M}}{\mathcal{N}} \tag{4.198}$$

and

$$\langle V \rangle = \frac{\int \left\{\frac{e^2}{r_{12}} - Ze^2\left(\frac{1}{r_1} + \frac{1}{r_2}\right)\right\}\chi^2(r_1, r_2, r_{12})\, d\tau}{\int \chi^2(r_1, r_2, r_{12})\, d\tau} = -\frac{k\mathcal{L}}{\mathcal{N}} \tag{4.199}$$

and so for $k = \mathcal{L}/2\mathcal{M}$, we see that

$$2\langle T \rangle = -\langle V \rangle = -2I = \frac{\mathcal{L}^2}{2\mathcal{M}\mathcal{N}} \tag{4.200}$$

which is a special instance of the virial theorem, proved for the case of classical mechanics in section 1.8 by also introducing a scaling parameter.

Instead of the coordinates r_1, r_2, r_{12} we may employ the coordinates

$$s = r_1 + r_2, \quad t = r_1 - r_2, \quad u = r_{12} \tag{4.201}$$

first introduced by Hylleraas, where $0 \leqslant u \leqslant s < \infty$ and $-u \leqslant t \leqslant u$. In terms of these coordinates it can be shown without undue difficulty that

$$d\tau = \pi^2 u(s^2 - t^2)\, ds\, dt\, du \tag{4.202}$$

and

$$\tfrac{1}{2}\{(\nabla_1 \phi)^2 + (\nabla_2 \phi)^2\} = \left(\frac{\partial \phi}{\partial s}\right)^2 + \left(\frac{\partial \phi}{\partial t}\right)^2 + \left(\frac{\partial \phi}{\partial u}\right)^2$$
$$+ \frac{2}{u(s^2 - t^2)}\left\{s(u^2 - t^2)\frac{\partial \phi}{\partial s}\frac{\partial \phi}{\partial u} + t(s^2 - u^2)\frac{\partial \phi}{\partial u}\frac{\partial \phi}{\partial t}\right\}.$$

EIGENVALUE PROBLEMS

Hence, since ϕ must be symmetric with respect to the transformation $t \to -t$, we obtain

$$\mathscr{L} = 2\pi^2 e^2 \int_0^\infty ds \int_0^s du \int_0^u dt \, (4Zsu - s^2 + t^2)\phi^2(s, t, u), \tag{4.203}$$

$$\mathscr{M} = \frac{2\pi^2 \hbar^2}{m} \int_0^\infty ds \int_0^s du \int_0^u dt \left[u(s^2 - t^2) \left\{ \left(\frac{\partial \phi}{\partial s}\right)^2 + \left(\frac{\partial \phi}{\partial t}\right)^2 + \left(\frac{\partial \phi}{\partial u}\right)^2 \right\} + 2s(u^2 - t^2) \frac{\partial \phi}{\partial s} \frac{\partial \phi}{\partial u} + 2t(s^2 - u^2) \frac{\partial \phi}{\partial u} \frac{\partial \phi}{\partial t} \right], \tag{4.204}$$

and

$$\mathscr{N} = 2\pi^2 \int_0^\infty ds \int_0^s du \int_0^u dt \, u(s^2 - t^2)\phi^2(s, t, u). \tag{4.205}$$

Listed in Table 4.2 are the values of E_U for the hydrogen negative ion and for the helium atom obtained by a number of investigators using trial functions of varying complexity based on the expansion

$$\phi(s, t, u) = e^{-s} \sum_{l,m,n=0}^\infty c(l, m, n) s^l t^m u^n \tag{4.206}$$

where the $c(l, m, n)$ are variable parameters. Also given in this table are values of the lower bound E_L to the ground state eigenenergy of helium obtained by using Temple's formula (4.161) with energy E_2 of the first excited state put equal to the experimentally determined value $-2 \cdot 146 \, e^2/a_0$.

We come finally to the procedure developed by Pekeris[6] for dealing with two electron systems. Because the coordinates r_1, r_2, r_{12} are not independent it is impossible to expand the function χ in terms of an orthonormal set of functions based on these coordinates or on the coordinates introduced by Hylleraas. Instead, we employ the perimetric coordinates

$$u = \epsilon(r_2 - r_1 + r_{12}), \quad v = \epsilon(r_1 - r_2 + r_{12}),$$
$$w = 2\epsilon(r_1 + r_2 - r_{12}), \tag{4.207}$$

which, by virtue of their independence, enable us to expand the function χ in the form

$$\chi(u, v, w) = e^{-\frac{1}{2}(u+v+w)} \sum_{l,m,n=0}^{\infty} A(l, m, n) L_l(u) L_m(v) L_n(w) \quad (4.208)$$

where $L_n(x)$ denotes a Laguerre polynomial of order n. Taking $\epsilon = \sqrt{(-2mE/\hbar^2)}$ and applying the Ritz variational method described on pp. 166 to 168 then leads to a set of linear equations in the parameters $A(l, m, n)$, the vanishing of the determinant of the coefficients resulting in an eigenvalue equation for E. Using a procedure essentially equivalent to this, Pekeris has calculated the ground state eigenenergies of the sequence of two electron systems to astonishing accuracy, obtaining for the ground state energy of H$^-$ the value $-0\cdot527751006\ e^2/a_0$ and for the ground

Table 4.2. Values of the upper bound E_U and lower bound E_L to the ground state eigenenergies of H$^-$ and He in units of e^2/a_0

H$^-$		He			
Number of parameters	$-E_U$	Number of parameters	$-E_U$	$-E_L$	$E_U - E_L$
3	0·5253	3	2·9024	2·965	0·063
6	0·5264	6	2·9032	2·9256	0·0223
11	0·52756	10	2·90363	2·90891	0·00529
20	0·52764	18	2·903715	2·904932	0·001217
24	0·52772	39	2·9037225	2·9038737	0·0001512

state energy of He the value $-2\cdot903724375\ e^2/a_0$ by solving secular equations based on determinants of order 444 and 1078 respectively[7]. The results of his calculations for the detachment energy of H$^-$ and the ionization energies of He, Li$^+$, Be^{2+} are given in Table 4.3. Extrapolating from the values of the ionization energy given by determinants of finite order to determinants of infinite order and making small corrections for the motion of the nucleus, relativistic and radiative effects gives complete agreement with experimental determinations to a remarkable number of significant figures.

Excited states. We have seen on p. 165 that $I[\chi]$ provides an upper bound to the eigenenergy of an excited state of a system if the function χ is orthogonal to the exact eigenfunctions of all the lower lying states. For definiteness we now consider the 2^1S, 2^1P, 2^3S and 2^3P excited states of a two electron system. The spin function of a singlet state is antisymmetric with respect to interchange of the spin coordinates of the two electrons. Hence the associated space wave function must be symmetric with respect to interchange of the space coordinates of the two electrons in order to ensure that the total wave function is antisymmetric with respect to interchange of all the coordinates of the two electrons as required by the Pauli exclusion principle. On the other hand the spin function of a triplet state is symmetric and hence must be associated with an antisymmetric space wave function. It follows that the space wave functions of a singlet state and a triplet state

Table 4.3. Values of the ionization energy calculated by Pekeris[6] using an expansion in Laguerre polynomials based on perimetric coordinates

	Ionization energy (cm^{-1})		
	(i)	(ii)	Experiment
H$^-$	6087·311	6083·08	
He	198317·342	198310·67	198310·82 ± 0·15
Li$^+$	610072·683	610079·61	610079 ± 25
Be^{2+}	1241177·63	1241259·4	1241225 ± 100

(i) Calculated using a determinant of order 203.
(ii) Calculated by extrapolating to a determinant of infinite order and correcting for mass polarization, relativity and the Lamb shift.

are necessarily orthogonal to each other due to their different symmetries and so the space wave functions of the 2^3S and 2^3P states are automatically orthogonal to the 2^1S and 2^1P excited states of helium as well as to the 1^1S ground state. In addition, the space wave function of a P state is necessarily orthogonal to that of an S state due to the differing spacial distributions of the two functions and so the space wave function of the 2^3P state is also orthogonal to that of the 2^3S state while the 2^1P state has a space wave function which is orthogonal to those of the 1^1S and 2^1S states. Thus we see that the space wave functions of all the states considered above are necessarily orthogonal to each other apart from the 1^1S and 2^1S states for which special provision has to be made.

Employing the simple form of trial functions

$$\chi(\mathbf{r}_1, \mathbf{r}_2) = \mathcal{N}[\{\exp(-\alpha r_1 - \beta r_2)/a_0\}(\beta r_2 - 1)$$
$$- \{\exp(-\alpha r_2 - \beta r_1)/a_0\}(\beta r_1 - 1)] \quad (4.209)$$

for the 2^3S state and

$$\chi(\mathbf{r}_1, \mathbf{r}_2) = \mathcal{N}[\{\exp(-\alpha r_1 - \beta r_2)/a_0\} r_2 \cos\theta_2$$
$$\pm \{\exp(-\alpha r_2 - \beta r_1)/a_0\} r_1 \cos\theta_1] \quad (4.210)$$

for the 2^1P and 2^3P states, all of which have the appropriate symmetries, the values $-2\cdot167$, $-2\cdot122$ and $-2\cdot131$ e^2/a_0 are obtained for the eigenenergies which are slightly in excess of the experimentally determined values $-2\cdot1752$, $-2\cdot1238$ and $-2\cdot1332$ e^2/a_0 respectively.

To obtain an upper bound to the eigenenergy of the 2^1S state of a two electron system it is necessary to use the Ritz variational method described on pp. 166 to 168 which simultaneously provides an upper bound to the ground state eigenenergy. This was first done by Hylleraas and Undheim[1] who obtained $-2\cdot1449$ e^2/a_0 as an upper bound to the eigenenergy of the 2^1S state of helium which may be compared with the experimentally determined value $-2\cdot1460$ e^2/a_0 for this energy.

Extremely accurate calculations of the eigenenergies of the 2^1S, 2^3S, 2^1P and 2^3P states of helium have been performed by Pekeris and his collaborators[7,8] who obtained the values $-2\cdot1459740$, $-2\cdot1752294$, $-2\cdot1238414$ and $-2\cdot1331633$ e^2/a_0 for these energies. The accuracy of their calculations can be assessed by referring to Table 4.4 where a comparison is made between the theoretical values of the ionization energies for the above states of helium and the experimentally determined values.

Table 4.4. Values of the ionization energies for excited states of helium

State	Ionization energy (cm^{-1})	
	Theory*	Experiment
2^1S	32033·21	32033·26 ± 0.03
2^3S	38454·72	38454·73 ± 0.05
2^1P	27176·64	27175·8
2^3P	29222·14	29223·86

* Calculated by Pekeris and associates[7,8].

4.11 Hartree–Fock equations

So far we have confined our attention to the determination of the energy levels of two electron systems. Unfortunately the equivalent problem for complex atoms cannot be treated with the same precision. A useful approach to the problem is to employ the *Hartree–Fock approximation* which assumes that the total eigenfunction of an atom consisting of N electrons bound to a nucleus of charge Ze may be expressed in the form

$$\chi(s_1\mathbf{r}_1, s_2\mathbf{r}_2, \ldots, s_N\mathbf{r}_N)$$
$$= \frac{1}{\sqrt{(N!)}} \begin{vmatrix} \phi_1(s_1\mathbf{r}_1) & \phi_1(s_2\mathbf{r}_2) & \ldots & \phi_1(s_N\mathbf{r}_N) \\ \phi_2(s_1\mathbf{r}_1) & \phi_2(s_2\mathbf{r}_2) & \ldots & \phi_2(s_N\mathbf{r}_N) \\ \vdots & \vdots & & \vdots \\ \phi_N(s_1\mathbf{r}_1) & \phi_N(s_2\mathbf{r}_2) & \ldots & \phi_N(s_N\mathbf{r}_N) \end{vmatrix} \quad (4.211)$$

or, more generally, as a linear combination of such determinants, where $\phi_i(s\mathbf{r})$ is a single electron orbital associated with a set of quantum numbers a_i, and s denotes the spin and $\mathbf{r}$ the space coordinates of the electron. The function (4.211) is clearly antisymmetric with respect to the interchange of the spin and space coordinates of any pair of electrons in the atom. Also if any two

184 VARIATIONAL PRINCIPLES

electrons have identical sets of quantum numbers, the function (4.211) clearly vanishes and thus the Pauli exclusion principle is satisfied. Further if the orbitals ϕ_i are chosen to be orthogonal and normalized so that

$$\sum_s \int \bar{\phi}_i(s\mathbf{r})\phi_j(s\mathbf{r})\, d\mathbf{r} = (\phi_i \mid \phi_j) = \delta_{ij} \qquad (4.212)$$

it follows that

$$\sum_{s_1,\ldots,s_N} \int |\chi(s_1\mathbf{r}_1,\ldots,s_N\mathbf{r}_N)|^2\, d\mathbf{r}_1 \ldots d\mathbf{r}_N = 1 \qquad (4.213)$$

and so the total eigenfunction χ is normalized.

Since the Hamiltonian of the atomic system may be written in the form

$$H = \sum_{i=1}^{N} H_0(\mathbf{r}_i) + \sum_{\substack{i,j=1 \\ i>j}}^{N} \frac{e^2}{r_{ij}} \qquad (4.214)$$

where

$$H_0(\mathbf{r}_i) = -\frac{\hbar^2}{2m}\nabla_i^2 - \frac{Ze^2}{r_i} \qquad (4.215)$$

is the Hamiltonian of a single electron in the Coulomb field of a nucleus of charge Ze, it can be seen that

$$I[\chi] = \sum_{s_1,\ldots,s_N} \int \bar{\chi} H \chi\, d\mathbf{r}_1 \ldots d\mathbf{r}_N$$

$$= \sum_{i=1}^{N} (\phi_i \mid H_0 \mid \phi_i) + \sum_{\substack{i,j=1 \\ i>j}}^{N} \left\{ \left(\phi_i\phi_j \left| \frac{e^2}{r_{12}} \right| \phi_i\phi_j\right) - \left(\phi_i\phi_j \left| \frac{e^2}{r_{12}} \right| \phi_j\phi_i\right) \right\} \qquad (4.216)$$

where

$$(\phi_i \mid H_0 \mid \phi_i) = \sum_s \int \bar{\phi}_i(s\mathbf{r}) H_0(\mathbf{r}) \phi_i(s\mathbf{r})\, d\mathbf{r} \qquad (4.217)$$

and

$$\left(\phi_i\phi_j \left| \frac{e^2}{r_{12}} \right| \phi_k\phi_l\right)$$
$$= \sum_{s_1,s_2} \iint \bar{\phi}_i(s_1\mathbf{r}_1)\bar{\phi}_j(s_2\mathbf{r}_2) \frac{e^2}{r_{12}} \phi_k(s_1\mathbf{r}_1)\phi_l(s_2\mathbf{r}_2)\, d\mathbf{r}_1\, d\mathbf{r}_2. \qquad (4.218)$$

To determine the equations satisfied by the orbitals ϕ_i which make $I[\chi]$ a least upper bound to the eigenenergy of the ground state of the atom we employ the stationary property of the integral (4.216) subject to the auxiliary conditions (4.212). If δI denotes the change in $I[\chi]$ due to the infinitesimal variations $\delta\phi_i$ in the ϕ_i we have

$$\delta I = \sum_{i=1}^{N} (\delta\phi_i \mid H_0 \mid \phi_i)$$

$$+ \sum_{\substack{i,j=1 \\ i>j}}^{N} \left\{ \left(\delta\phi_i \phi_j \left|\frac{e^2}{r_{12}}\right| \phi_i\phi_j\right) + \left(\phi_i \delta\phi_j \left|\frac{e^2}{r_{12}}\right| \phi_i\phi_j\right) \right.$$

$$\left. - \left(\delta\phi_i \phi_j \left|\frac{e^2}{r_{12}}\right| \phi_j\phi_i\right) - \left(\phi_i \delta\phi_j \left|\frac{e^2}{r_{12}}\right| \phi_j\phi_i\right) \right\} \quad (4.219)$$

since we may regard the ϕ_i and $\bar\phi_i$ as independent functions. Introducing Lagrange multipliers λ_{ij} with $\lambda_{ij} = \lambda_{ji}$ and writing the stationary condition in the form

$$\delta\left\{ I - \sum_{i,j=1}^{N} \lambda_{ij}(\phi_i \mid \phi_j) \right\} = 0 \quad (4.220)$$

and remembering that the variations $\delta\bar\phi_i$ are completely independent, we arrive at the set of equations

$$H_0(\mathbf{r}_1)\phi_i(s_1\mathbf{r}_1) + \sum_{\substack{j=1 \\ j\neq i}}^{N} \left\{ \phi_i(s_1\mathbf{r}_1)\left(\phi_j \left|\frac{e^2}{r_{12}}\right| \phi_j\right) - \phi_j(s_1\mathbf{r}_1)\left(\phi_j \left|\frac{e^2}{r_{12}}\right| \phi_i\right) \right\}$$

$$- \sum_{j=1}^{N} \lambda_{ij}\phi_j(s_1\mathbf{r}_1) = 0 \quad (4.221)$$

where

$$\left(\phi_j \left|\frac{e^2}{r_{12}}\right| \phi_i\right) = \sum_{s_2} \int \bar\phi_j(s_2\mathbf{r}_2) \frac{e^2}{r_{12}} \phi_i(s_2\mathbf{r}_2) \, d\mathbf{r}_2. \quad (4.222)$$

Equations (4.221) are known as the *Hartree–Fock equations*. They are usually solved by a numerical procedure involving successive approximation.

If we had used the *Hartree approximation* based on the simple product of single electron orbitals

$$\chi = \phi_1(s_1\mathbf{r}_1)\phi_2(s_2\mathbf{r}_2)\ldots\phi_N(s_N\mathbf{r}_N) \quad (4.223)$$

instead of the symmetrized form (4.211) we would have obtained the set of equations

$$\left\{ H_0(\mathbf{r}_1) + \sum_{\substack{j=1 \\ j \neq i}}^{N} \left(\phi_j \left| \frac{e^2}{r_{12}} \right| \phi_j \right) - \lambda_{ii} \right\} \phi_i(s_1\mathbf{r}_1) = 0 \quad (4.224)$$

where the λ_{ii} can be identified with the energies of the individual electrons to a rough approximation.

For the case of two electron systems the Hartree–Fock approximation to the total eigenfunction is

$$\chi = \frac{1}{\sqrt{2}} \{\phi_1(s_1\mathbf{r}_1)\phi_2(s_2\mathbf{r}_2) - \phi_1(s_2\mathbf{r}_2)\phi_2(s_1\mathbf{r}_1)\} \quad (4.225)$$

where

$$\phi_i(s\mathbf{r}) = \sigma_i(s)u_i(\mathbf{r}), \quad (4.226)$$

$\sigma_i(s)$ being the spin function and $u_i(\mathbf{r})$ the space function of an electron. In the ground state of the system the two electrons have the same space functions $u(\mathbf{r})$ but different spin functions $\sigma_1(s)$ and $\sigma_2(s)$ so that we have

$$\chi = \frac{1}{\sqrt{2}} \{\sigma_1(s_1)\sigma_2(s_2) - \sigma_1(s_2)\sigma_2(s_1)\} u(\mathbf{r}_1)u(\mathbf{r}_2). \quad (4.227)$$

Thus the total spin function is antisymmetric as required for a singlet state of the atomic system while the total space function has the symmetric form $u(\mathbf{r}_1)u(\mathbf{r}_2)$ used on pp. 172 to 174. The Hartree approximation leads to the total eigenfunction

$$\chi = \sigma_1(s_1)\sigma_2(s_2)u(\mathbf{r}_1)u(\mathbf{r}_2)$$

which has a properly symmetrized space function but an incorrect spin function.

Finally it should be noted here that for excited states it is often necessary to take linear combinations of determinants in order to obtain a properly symmetrized spin function. For example the

2^1S state of a two electron system is described in the Hartree-Fock approximation by the function

$$\chi = \tfrac{1}{2} \begin{vmatrix} \sigma_1(s_1)u_1(\mathbf{r}_1) & \sigma_1(s_2)u_1(\mathbf{r}_2) \\ \sigma_2(s_1)u_2(\mathbf{r}_1) & \sigma_2(s_2)u_2(\mathbf{r}_2) \end{vmatrix} - \tfrac{1}{2} \begin{vmatrix} \sigma_2(s_1)u_1(\mathbf{r}_1) & \sigma_2(s_2)u_1(\mathbf{r}_2) \\ \sigma_1(s_1)u_2(\mathbf{r}_1) & \sigma_1(s_2)u_2(\mathbf{r}_2) \end{vmatrix}$$
$$= \frac{1}{\sqrt{2}} \{\sigma_1(s_1)\sigma_2(s_2) - \sigma_1(s_2)\sigma_2(s_1)\} \frac{1}{\sqrt{2}} \{u_1(\mathbf{r}_1)u_2(\mathbf{r}_2) + u_1(\mathbf{r}_2)u_2(\mathbf{r}_1)\}$$
(4.228)

where u_1 and u_2 are the space function of electrons in the $1s$ and $2s$ orbitals.

4.12 Molecular energy curves

When a pair of atomic systems approach each other their mutual interaction leads to an alteration in the potential energy of the combined systems. In many instances the potential energy decreases to a minimum value at some internuclear separation and thus a stable molecule can be formed. The present section is devoted to the application of the variational method to the calculation of the potential energy of a diatomic molecular system as a function of the distance between the atomic nuclei.

Hydrogen molecular ion

The simplest example we can consider is the hydrogen molecular ion H_2^+ composed of a single electron bound to a pair of protons which we suppose to be situated at points A and B whose distance apart is R. In the neighbourhood of either nucleus the electron moves under the influence of an electric field which closely approximates to the Coulomb field of a single proton and so we take a linear combination of atomic hydrogen orbitals as the function representing the electron. This may be written in the form

$$\chi = a\chi_A + b\chi_B \qquad (4.229)$$

where a and b are variable parameters and χ_A and χ_B are ground state wave functions of hydrogen atoms with their nuclei situated at the points A and B respectively. Because the function χ is composed of a linear combination of atomic orbitals our approximation is often referred to as the *lcao approximation*.

If the distances of the electron from the nuclei at A and B are denoted by r_A and r_B respectively we have

$$\chi_A = \sqrt{\left(\frac{1}{\pi a_0^3}\right)} e^{-r_A/a_0}, \qquad \chi_B = \sqrt{\left(\frac{1}{\pi a_0^3}\right)} e^{-r_B/a_0} \qquad (4.230)$$

while the Hamiltonian operator of the single electron molecular system has the form

$$H = -\frac{\hbar^2}{2m} \nabla^2 - \frac{e^2}{r_A} - \frac{e^2}{r_B} + \frac{e^2}{R}. \qquad (4.231)$$

Thus we have for an upper bound to the energy of the molecular ion at internuclear separation R

$$I[\chi] = \int \chi H \chi \, d\mathbf{r} \bigg/ \int \chi^2 \, d\mathbf{r}$$
$$= \frac{a^2 H_{AA} + 2ab H_{AB} + b^2 H_{BB}}{a^2 S_{AA} + 2ab S_{AB} + b^2 S_{BB}} \qquad (4.232)$$

where

$$H_{AB} = \int \chi_A H \chi_B \, d\mathbf{r}, \qquad S_{AB} = \int \chi_A \chi_B \, d\mathbf{r}. \qquad (4.233)$$

I provides a least upper bound to the energy when

$$\frac{\partial I}{\partial a} = 0, \qquad \frac{\partial I}{\partial b} = 0 \qquad (4.234)$$

which lead to the linear equations

$$a(H_{AA} - ES_{AA}) + b(H_{AB} - ES_{AB}) = 0,$$
$$a(H_{AB} - ES_{AB}) + b(H_{BB} - ES_{BB}) = 0 \qquad (4.235)$$

where the permissible values of E are the stationary values of I. These two equations are consistent provided

$$\begin{vmatrix} H_{AA} - ES_{AA} & H_{AB} - ES_{AB} \\ H_{AB} - ES_{AB} & H_{BB} - ES_{BB} \end{vmatrix} = 0. \qquad (4.236)$$

Since χ_A and χ_B are both normalized and because the Hamiltonian is symmetric with respect to the two nuclei A and B we see that

$$S_{AA} = S_{BB} = 1, \qquad H_{AA} = H_{BB} \qquad (4.237)$$

and so it follows from the secular equation (4.236) that

$$(H_{AA} - E)^2 - (H_{AB} - ES_{AB})^2 = 0 \tag{4.238}$$

which has the two solutions

$$E_{\pm} = \frac{H_{AA} \pm H_{AB}}{1 \pm S_{AB}}. \tag{4.239}$$

Substituting into (4.235) now gives $a = \pm b$ which means that the function χ assumes the two alternative forms

$$\chi_{\pm} = \frac{1}{\sqrt{\{2(1 \pm S_{AB})\}}} (\chi_A \pm \chi_B) \tag{4.240}$$

where the outside factor has been chosen so as to normalize the wave function.

To proceed further we need to obtain expressions for S_{AB}, H_{AA} and H_{AB} in terms of the internuclear separation R. Remembering that

$$\left(-\frac{\hbar^2}{2m}\nabla^2 - \frac{e^2}{r_A}\right)\chi_A = -\frac{e^2}{2a_0}\chi_A$$

with a similar equation for χ_B, we see that

$$H_{AA} = \frac{e^2}{R} - \frac{e^2}{2a_0} - e^2 \int \frac{1}{r_B}\chi_A^2 \, d\mathbf{r}$$

and

$$H_{AB} = \left(\frac{e^2}{R} - \frac{e^2}{2a_0}\right) S_{AB} - e^2 \int \frac{1}{r_B} \chi_A \chi_B \, d\mathbf{r}.$$

Introducing the confocal elliptic coordinates

$$\xi = \frac{r_A + r_B}{R}, \quad \eta = \frac{r_A - r_B}{R} \tag{4.241}$$

and the azimuthal angle ϕ, where $1 \leq \xi < \infty$, $-1 \leq \eta \leq 1$, $0 \leq \phi \leq 2\pi$, employing the expression for the volume element

$$d\mathbf{r} = \frac{R^3}{8}(\xi^2 - \eta^2) \, d\xi \, d\eta \, d\phi \tag{4.242}$$

and performing the elementary integrations over ξ, η, ϕ we arrive at the formulae

$$S_{AB} = \left(1 + \frac{R}{a_0} + \frac{R^2}{3a_0^2}\right) e^{-R/a_0}, \tag{4.243}$$

$$H_{AA} = -\frac{e^2}{2a_0} + \frac{e^2}{R}\left(1 + \frac{R}{a_0}\right) e^{-2R/a_0}, \tag{4.244}$$

$$H_{AB} = \left(\frac{e^2}{R} - \frac{e^2}{2a_0}\right) S_{AB} - \frac{e^2}{a_0}\left(1 + \frac{R}{a_0}\right) e^{-R/a_0}. \tag{4.245}$$

In the limit of large R we see that S_{AB} and H_{AB} both vanish exponentially while $H_{AA} \to -e^2/2a_0$ so that $E_\pm$ tend to $-e^2/2a_0$, the ground state energy of a hydrogen atom. For small R, the internuclear Coulomb repulsion dominates and $E_\pm$ behave as e^2/R. At intermediate values of R the energy E_- increases monotonically from $-e^2/2a_0$ to ∞ with decreasing R and thus corresponds to a repulsive state of the molecular ion, while the energy E_+ attains a minimum value for an internuclear separation close to $R = 2a_0$ and is therefore associated with an attractive state. At $R = 2a_0$ we find that $E_+ = -0.5538\ e^2/a_0$ and $E_- = -0.1609\ e^2/a_0$. These values must be upper bounds to the actual energies of the lowest states of positive and negative symmetry, the $1s\sigma$ ground state and $2p\sigma$ excited state of H_2^+, at $R = 2a_0$. In fact exact calculation yields -0.6026 and $-0.1675\ e^2/a_0$ respectively for these energies.

Hydrogen molecule

Another simple example of some considerable interest is the hydrogen molecule which consists of two electrons bound to a pair of protons. The first method we shall use to investigate this case was introduced by Heitler and London[9] and is analogous to that employed in the preceding subsection to consider the hydrogen molecular ion. When the two nuclei A and B are far apart the molecule is composed essentially of two hydrogen atoms and so we may represent the electrons by either

$$\chi_\mathrm{I}(r_{A1}, r_{B2}) = \chi_A(1)\chi_B(2) \quad \text{or} \quad \chi_\mathrm{II}(r_{A2}, r_{B1}) = \chi_A(2)\chi_B(1)$$

where the numbers 1 and 2 are used to distinguish between the electrons. Proceeding as in the previous subsection by taking a linear combination of these two functions, we find that the energy of the molecule becomes

$$E_{\pm} = \frac{H_{\mathrm{I\,I}} \pm H_{\mathrm{I\,II}}}{1 \pm S_{\mathrm{I\,II}}} \tag{4.246}$$

where the Hamiltonian operator for the hydrogen molecule has the form

$$H = -\frac{\hbar^2}{2m}(\nabla_1^2 + \nabla_2^2) - e^2\left(\frac{1}{r_{A1}} + \frac{1}{r_{B1}} + \frac{1}{r_{A2}} + \frac{1}{r_{B2}}\right) + \frac{e^2}{r_{12}} + \frac{e^2}{R} \tag{4.247}$$

and the matrix elements arising in (4.246) are given by

$$H_{\mathrm{I\,I}} = \iint \chi_{\mathrm{I}} H \chi_{\mathrm{I}} \, d\mathbf{r}_1 \, d\mathbf{r}_2, \tag{4.248}$$

$$H_{\mathrm{I\,II}} = \iint \chi_{\mathrm{I}} H \chi_{\mathrm{II}} \, d\mathbf{r}_1 \, d\mathbf{r}_2, \tag{4.249}$$

and

$$S_{\mathrm{I\,II}} = \iint \chi_{\mathrm{I}} \chi_{\mathrm{II}} \, d\mathbf{r}_1 \, d\mathbf{r}_2. \tag{4.250}$$

The wave function associated with the two expressions for the energy have the symmetric and antisymmetric forms

$$\chi_{\pm} = \frac{1}{\sqrt{\{2(1 \pm S_{\mathrm{I\,II}})\}}} \{\chi_{\mathrm{I}}(r_{A1}, r_{B2}) \pm \chi_{\mathrm{II}}(r_{A2}, r_{B1})\}. \tag{4.251}$$

Since χ_{I} and χ_{II} satisfy the equations

$$\left\{-\frac{\hbar^2}{2m}(\nabla_1^2 + \nabla_2^2) - e^2\left(\frac{1}{r_{A1}} + \frac{1}{r_{B2}}\right)\right\}\chi_{\mathrm{I}} = -\frac{e^2}{a_0}\chi_{\mathrm{I}}$$

and

$$\left\{-\frac{\hbar^2}{2m}(\nabla_1^2 + \nabla_2^2) - e^2\left(\frac{1}{r_{B1}} + \frac{1}{r_{A2}}\right)\right\}\chi_{\mathrm{II}} = -\frac{e^2}{a_0}\chi_{\mathrm{II}}$$

we see that the matrix elements (4.248), (4.249) and (4.250) may be expressed as

$$H_{\text{I I}} = \iint \chi_A(1)\chi_B(2)\left\{\frac{e^2}{R} - \frac{e^2}{a_0} + \frac{e^2}{r_{12}} - e^2\left(\frac{1}{r_{B1}} + \frac{1}{r_{A2}}\right)\right\}$$
$$\times \chi_A(1)\chi_B(2)\, d\mathbf{r}_1\, d\mathbf{r}_2,$$

$$H_{\text{I II}} = \iint \chi_A(1)\chi_B(2)\left\{\frac{e^2}{R} - \frac{e^2}{a_0} + \frac{e^2}{r_{12}} - e^2\left(\frac{1}{r_{A1}} + \frac{1}{r_{B2}}\right)\right\}$$
$$\times \chi_A(2)\chi_B(1)\, d\mathbf{r}_1\, d\mathbf{r}_2$$

and

$$S_{\text{I II}} = \left\{\int \chi_A \chi_B\, d\mathbf{r}\right\}^2.$$

Introducing the overlap integral

$$S = \int \chi_A \chi_B\, d\mathbf{r}, \qquad (4.252)$$

the Coulomb integral

$$J = e^2 \iint \chi_A(1)\chi_B(2)\left(\frac{1}{r_{12}} - \frac{1}{r_{B1}} - \frac{1}{r_{A2}}\right)\chi_A(1)\chi_B(2)\, d\mathbf{r}_1\, d\mathbf{r}_2$$
$$= e^2 \iint \frac{\{\chi_A(1)\}^2\{\chi_B(2)\}^2}{r_{12}}\, d\mathbf{r}_1\, d\mathbf{r}_2 - e^2 \int \frac{\{\chi_A(1)\}^2}{r_{B1}}\, d\mathbf{r}_1$$
$$- e^2 \int \frac{\{\chi_B(2)\}^2}{r_{A2}}\, d\mathbf{r}_2, \quad (4.253)$$

and the exchange integral

$$K = e^2 \iint \chi_A(1)\chi_B(2)\left(\frac{1}{r_{12}} - \frac{1}{r_{A1}} - \frac{1}{r_{B2}}\right)\chi_A(2)\chi_B(1)\, d\mathbf{r}_1\, d\mathbf{r}_2$$
$$= e^2 \iint \frac{\chi_A(1)\chi_B(1)\chi_A(2)\chi_B(2)}{r_{12}}\, d\mathbf{r}_1\, d\mathbf{r}_2$$
$$- Se^2 \int \frac{\chi_A(1)\chi_B(1)}{r_{A1}}\, d\mathbf{r}_1 - Se^2 \int \frac{\chi_A(2)\chi_B(2)}{r_{B2}}\, d\mathbf{r}_2, \qquad (4.254)$$

we may rewrite these matrix elements in the form

$$H_{\text{I I}} = \frac{e^2}{R} - \frac{e^2}{a_0} + J, \qquad (4.255)$$

$$H_{\text{I II}} = \left(\frac{e^2}{R} - \frac{e^2}{a_0}\right)S^2 + K, \qquad (4.256)$$

and

$$S_{\text{I II}} = S^2. \qquad (4.257)$$

The first term of J corresponds to the Coulomb interaction between two electron charge distributions $\{\chi_A(1)\}^2$ and $\{\chi_B(2)\}^2$ while the second and third terms (which are actually equal) correspond to the Coulomb interaction between one of the nuclei and the electron charge distribution about the other nucleus. K is called an exchange integral because it involves the exchange of electrons between the two nuclei.

If the integrations are performed and calculations carried out, it is found that E_- produces a repulsive energy curve while E_+ results in an attractive energy curve with a minimum value of $-1\cdot 116\ e^2/a_0$ occurring at the internuclear distance $R = 1\cdot 64 a_0$. The observed value of the equilibrium distance between the nuclei of the hydrogen molecule is $1\cdot 40 a_0$ while the energy required to dissociate H_2 into two ground state hydrogen atoms (ignoring the zero-point vibrational energy) is observed to be 4·75 eV compared to the value 3·16 eV obtained from the calculation described above.

Since the wave function χ_+ obtained for the ground electronic state of the hydrogen molecule is symmetric with respect to interchange of the two electrons, the associated spin function must be antisymmetric in order that the Pauli exclusion principle be satisfied and thus the lowest electronic state of H_2 is a singlet state as for the case of the helium atom. On the other hand the repulsive state described by the wave function must be a triplet state with a symmetric spin function.

A simple way of improving the function describing the electrons in a hydrogen molecule is to introduce a screening parameter α by taking

$$\chi_A = \sqrt{\left(\frac{\alpha^3}{\pi a_0^3}\right)}\, e^{-\alpha r_A/a_0}, \qquad \chi_B = \sqrt{\left(\frac{\alpha^3}{\pi a_0^3}\right)}\, e^{-\alpha r_B/a_0}. \qquad (4.258)$$

This modification was made by Wang[10] and gives a minimum value of $-1\cdot 139\ e^2/a_0$ for the energy E_+ of the lowest electronic state occurring at the internuclear separation $R = 1\cdot 40 a_0$, the associated optimum value of α being 1·166.

The second approach to the problem of the H_2 molecule which we consider in this section is that based upon molecular orbitals.

For simplicity we represent each electron by an orbital which is a linear combination of atomic orbitals. The obvious choice is the molecular orbital

$$\frac{1}{\sqrt{\{2(1+S)\}}}(\chi_A + \chi_B)$$

which was successfully used to describe the single electron in the H_2^+ molecular ion. Thus we take for the wave function of the lowest state of H_2

$$\chi = \frac{1}{2(1+S)}\{\chi_A(1) + \chi_B(1)\}\{\chi_A(2) + \chi_B(2)\} \quad (4.259)$$

which yields as an upper bound to the energy of the two electrons

$$\int \chi H \chi \, d\mathbf{r}_1 \, d\mathbf{r}_2$$
$$= \frac{e^2}{R} + \frac{e^2}{4(1+S)^2} \iint \frac{\{\chi_A(1) + \chi_B(1)\}^2 \{\chi_A(2) + \chi_B(2)\}^2}{r_{12}} d\mathbf{r}_1 \, d\mathbf{r}_2$$
$$- \frac{1}{1+S} \int (\chi_A + \chi_B)\left(\frac{\hbar^2}{2m}\nabla^2 + \frac{e^2}{r_A} + \frac{e^2}{r_B}\right)(\chi_A + \chi_B) \, d\mathbf{r}.$$
$$(4.260)$$

This gives $-1\cdot099 \, e^2/a_0$ for the minimum value of the energy for the lowest electronic state. If we now introduce a screening parameter α into the atomic orbitals as in (4.258) we obtain the minimum energy $-1\cdot128 \, e^2/a_0$ occurring at the internuclear separation $R = 1\cdot38 a_0$ with $\alpha = 1\cdot193$.[11] Evidently the molecular orbital approach is inferior to the Heitler–London treatment.

More elaborate trial functions have been used by many investigators. For example James and Coolidge[12] chose the function having the form

$$\chi = e^{-k(\xi_1 + \xi_2)/a_0} \sum_{mnpqs} c(mnpqs)(\xi_1^m \xi_2^n \eta_1^p \eta_2^q + \xi_2^m \xi_1^n \eta_2^p \eta_1^q) u^s \quad (4.261)$$

where ξ_1, η_1 and ξ_2, η_2 are the confocal elliptic coordinates for the two electrons and $u = 2r_{12}/R$. This function is symmetric with

EIGENVALUE PROBLEMS   195

respect to the interchange of the coordinates of the electrons. To ensure that the function is also symmetric with respect to the nuclei A and B we must have $p + q$ even. Using 13 terms in the expansion (4.261), James and Coolidge obtained $-1{\cdot}1735\, e^2/a_0$ as the upper bound to the minimum value of the energy of the lowest electronic state of H_2 leading to a dissociation energy of 4.72 eV (again ignoring the zero-point vibrational energy) in very close agreement with observation.

4.13 Virial theorem

When we were considering the eigenenergy of the ground state of a two electron system on p. 178 we showed that the virial theorem is satisfied for the optimum value of the scaling parameter k. We shall now prove this result for the more general case of a molecular system consisting of N electrons and nuclei. The Hamiltonian operator for this system may be expressed in the form

$$H = T + V \qquad (4.262)$$

where

$$T = -\sum_{i=1}^{N} \frac{\hbar^2}{2m_i} \nabla_i^2, \qquad V = \sum_{\substack{i,j=1 \\ i>j}}^{N} \frac{Z_i Z_j e^2}{r_{ij}}, \qquad (4.263)$$

m_i and $Z_i e$ being the mass and charge respectively of the ith particle and r_{ij} being the distance between the ith and jth particles.

Introducing the integrals

$$\mathscr{L} = -\int \phi V \phi \, d\tau, \qquad \mathscr{M} = \int \phi T \phi \, d\tau, \qquad \mathscr{N} = \int \phi \phi \, d\tau \quad (4.264)$$

where ϕ is any quadratically integrable function of the position vectors $\mathbf{r}_1, \ldots, \mathbf{r}_N$ of the particles of the molecular system and $d\tau$ is a volume element in the $3N$ dimensional configuration space of $\mathbf{r}_1, \ldots, \mathbf{r}_N$, and setting

$$\phi(k\mathbf{r}_1, \ldots, k\mathbf{r}_N) = \chi(\mathbf{r}_1, \ldots, \mathbf{r}_N) \qquad (4.265)$$

where k is a variable scaling parameter, we see that

$$I[\chi] = \frac{\int \bar{\chi} H \chi \, d\tau}{\int \bar{\chi}\chi \, d\tau} = \frac{k^2 \mathscr{M} - k\mathscr{L}}{\mathscr{N}}. \tag{4.266}$$

The optimum value of k is obtained when $\partial I/\partial k = 0$ which yields

$$k = \frac{\mathscr{L}}{2\mathscr{M}}. \tag{4.267}$$

Since the expectation values of the kinetic energy and potential energy of the system are given by

$$\langle T \rangle = \frac{k^2 \mathscr{M}}{\mathscr{N}}, \quad \langle V \rangle = -\frac{k\mathscr{L}}{\mathscr{N}}$$

for the function $\chi(\mathbf{r}_1, \ldots, \mathbf{r}_N)$, we see that

$$2\langle T \rangle = -\langle V \rangle = -2\langle H \rangle \tag{4.268}$$

where $\langle H \rangle = I[\chi]$ is the expectation value of the total energy. This result is just the quantum mechanical virial theorem.

Putting $k = R^{-1}$ for a diatomic molecule with internuclear separation R we get

$$I[\chi] = \left(\frac{1}{R^2}\mathscr{M}' - \frac{1}{R}\mathscr{L}'\right) \Big/ \mathscr{N}' \tag{4.269}$$

where $\mathscr{L}', \mathscr{M}', \mathscr{N}'$ are defined in an analogous fashion to $\mathscr{L}, \mathscr{M}, \mathscr{N}$ except that the integrations are now performed over the coordinates of the electrons only and T is replaced by the kinetic energy operator T' which contains no terms involving the coordinates of the nuclei. We now find that

$$R \frac{\partial I}{\partial R} = -2\langle T' \rangle - \langle V \rangle$$

and hence at the equilibrium position of the nuclei, where I is a minimum so that $\partial I/\partial R = 0$, we have

$$2\langle T' \rangle = -\langle V \rangle. \tag{4.270}$$

EIGENVALUE PROBLEMS

4.14 Variational principle for a general eigenvalue equation

We next proceed to the generalization of the variational principle (4.133), associated with the eigenenergy equation of a quantum mechanical system, to the eigenvalue equation

$$L\psi = \lambda M\psi \qquad (4.271)$$

where L and M denote certain differential or integral operators. This can be achieved by introducing the adjoint equation

$$\tilde{L}\tilde{\psi} = \lambda \tilde{M}\tilde{\psi} \qquad (4.272)$$

where the adjoint operators $\tilde{L}$ and $\tilde{M}$ together with the adjoint solution $\tilde{\psi}$ are defined in such a way that the relations

$$\int \tilde{\psi} L\psi \, d\tau = \int \psi \tilde{L}\tilde{\psi} \, d\tau, \qquad \int \tilde{\psi} M\psi \, d\tau = \int \psi \tilde{M}\tilde{\psi} \, d\tau \qquad (4.273)$$

are satisfied, the integrations being performed over the configuration space of the system under consideration. We note that this definition of an adjoint operator $\tilde{L}$ is not the same as that for the Hermitian adjoint operator L^*, introduced in the section 2.7 on quantum mechanics, being related by the formula $\tilde{L} = \bar{L}^*$ and consequently are in accordance only if the operator L is real.

Next we multiply both sides of equation (4.271) by $\tilde{\psi}$ and integrate which yields

$$\lambda = I[\psi] \qquad (4.274)$$

where

$$I[\psi] = \frac{\int \tilde{\psi} L\psi \, d\tau}{\int \tilde{\psi} M\psi \, d\tau}. \qquad (4.275)$$

The change in $I[\psi]$ due to infinitesimal variations $\delta\psi$ and $\delta\tilde{\psi}$ in ψ and $\tilde{\psi}$ respectively is given by

$$\delta I[\psi] \int \tilde{\psi} M\psi \, d\tau + I[\psi]\left\{\int \delta\tilde{\psi} \, M\psi \, d\tau + \int \tilde{\psi} M \, \delta\psi \, d\tau\right\}$$
$$= \int \delta\tilde{\psi} \, L\psi \, d\tau + \int \tilde{\psi} L \, \delta\psi \, d\tau$$

neglecting quantities of the second order of smallness. Now, provided the variations $\delta\psi$ and $\delta\tilde{\psi}$ satisfy suitable boundary conditions, we may put

$$\int \tilde{\psi} L\, \delta\psi\, d\tau = \int \delta\psi\, \tilde{L}\tilde{\psi}\, d\tau, \qquad \int \tilde{\psi} M\, \delta\psi\, d\tau = \int \delta\psi\, \tilde{M}\tilde{\psi}\, d\tau$$

and so

$$\delta I[\psi] = \frac{\int \delta\tilde{\psi}(L\psi - \lambda M\psi)\, d\tau + \int \delta\psi(\tilde{L}\tilde{\psi} - \lambda\tilde{M}\tilde{\psi})\, d\tau}{\int \tilde{\psi} M\psi\, d\tau}. \qquad (4.276)$$

By virtue of the fact that ψ and $\tilde{\psi}$ are solutions of equation (4.271) and (4.272) respectively it is at once apparent that

$$\delta I[\psi] = 0 \qquad (4.277)$$

which means that $I[\psi]$ is stationary.

If $\tilde{L} = \bar{L}$ and $\tilde{M} = \bar{M}$ we see immediately that $\tilde{\psi} = \bar{\psi}$ in which case (4.275) becomes replaced by

$$I[\psi] = \frac{\int \bar{\psi} L\psi\, d\tau}{\int \bar{\psi} M\psi\, d\tau}. \qquad (4.278)$$

4.15 Perturbation theory

The relationship between the variational method and perturbation theory is of considerable interest and consequently we shall devote this section and the succeeding subsections to the investigation of this relationship and also to the use of trial functions based on perturbation theory in applications of the variational method.

We concern ourselves with the generalized eigenvalue equation discussed in the previous section but shall assume that L and M are Hermitian so that $\tilde{L} = \bar{L}$, $\tilde{M} = \bar{M}$ and $\tilde{\psi} = \bar{\psi}$. Let L_0 be the operator associated with the unperturbed system and let L' be the operator corresponding to the small perturbation which we

EIGENVALUE PROBLEMS

suppose to act upon the system. Then we take the function ψ representing the perturbed system in a state with eigenvalue λ to satisfy the equation

$$L\psi = \lambda M\psi \qquad (4.279)$$

where

$$L = L_0 + L'. \qquad (4.280)$$

We now suppose that ϕ_r is a member of the complete set of orthogonal and normalized eigenfunctions of the operator L_0 associated with the eigenvalue μ_r so that

$$L_0 \phi_r = \mu_r M \phi_r \qquad (4.281)$$

with

$$\int \phi_r M \phi_s \, d\tau = \delta_{rs}. \qquad (4.282)$$

To determine the effect of the perturbation to a given order we replace L' by $\eta L'$ where η is a parameter which we shall eventually let tend to unity. Expanding ψ and λ in the form

$$\psi = \sum_{n=0}^{\infty} \eta^n \psi_n, \qquad \lambda = \sum_{n=0}^{\infty} \eta^n \lambda_n, \qquad (4.283)$$

substituting into the eigenvalue equation (4.279) and equating the coefficients of equal powers of η we obtain

$$(L_0 - \lambda_0 M)\psi_0 = 0, \qquad (4.284)$$

$$(L_0 - \lambda_0 M)\psi_1 + (L' - \lambda_1 M)\psi_0 = 0, \qquad (4.285)$$

$$\vdots$$

$$(L_0 - \lambda_0 M)\psi_n + (L' - \lambda_1 M)\psi_{n-1} = \sum_{m=2}^{n} \lambda_m M \psi_{n-m} \quad (n \geq 2). \qquad (4.286)$$

It follows from (4.284) and (4.281) that ψ_0 must be one of the unperturbed functions ϕ_r. We make the choice:

$$\psi_0 = \phi_0, \qquad \lambda_0 = \mu_0. \qquad (4.287)$$

Expanding ψ_n in the form
$$\psi_n = \sum_r a_r^{(n)} \phi_r \tag{4.288}$$
we see that
$$\int \bar{\phi}_0 (L_0 - \lambda_0 M) \psi_n \, d\tau = \sum_r a_r^{(n)} \int \bar{\phi}_0 (\mu_r - \mu_0) M \phi_r \, d\tau = 0 \tag{4.289}$$
due to the orthogonality property of the functions ϕ_r and so, multiplying (4.285) and (4.286) across by $\bar{\phi}_0$ and integrating over all configuration space, we get
$$\lambda_1 = L'_{00} \tag{4.290}$$
and
$$\lambda_n = \int \bar{\phi}_0 (L' - \lambda_1 M) \psi_{n-1} \, d\tau - \sum_{m=1}^{n-2} \lambda_{m+1} \int \bar{\phi}_0 M \psi_{n-m-1} \, d\tau, \tag{4.291}$$
where
$$L'_{rs} = \int \bar{\phi}_r L' \phi_s \, d\tau. \tag{4.292}$$
Inserting (4.288) into (4.285) and (4.286), multiplying across by $\bar{\phi}_r$ and integrating gives for $r \neq 0$:
$$a_r^{(1)} = \frac{L'_{r0}}{\mu_0 - \mu_r} \tag{4.293}$$
and
$$a_r^{(n)} = \frac{1}{\mu_0 - \mu_r} \left\{ \sum_s a_s^{(n-1)} (L'_{rs} - \lambda_1 \delta_{rs}) - \sum_{m=1}^{n-1} \lambda_{m+1} a_r^{(n-m-1)} \right\}. \tag{4.294}$$

The coefficients $a_0^{(n)}$ may be determined by normalizing the function ψ to any required order in η.

We have already obtained the explicit expression (4.290) for λ_1. It follows from (4.291) that
$$\lambda_2 = \int \bar{\phi}_0 (L' - \lambda_1 M) \psi_1 \, d\tau = \sum_r a_r^{(1)} (L'_{0r} - \lambda_1 \delta_{0r})$$
$$= \sum_{r \neq 0} \frac{L'_{0r} L'_{r0}}{\mu_0 - \mu_r} \tag{4.295}$$

using expression (4.293) for $a_r^{(1)}$. Also, from (4.294) we see that for $r \neq 0$

$$a_r^{(2)} = \frac{1}{\mu_0 - \mu_r} \sum_s a_s^{(1)}(L'_{rs} - \lambda_1 \delta_{rs})$$

$$= \frac{1}{\mu_0 - \mu_r} \left\{ \sum_{s \neq 0} \frac{L'_{rs} L'_{s0}}{\mu_0 - \mu_s} - \frac{L'_{00} L'_{r0}}{\mu_0 - \mu_r} + a_0^{(1)} L'_{r0} \right\} \quad (4.296)$$

and so

$$\lambda_3 = \int \phi_0 (L' - \lambda_1 M) \psi_2 \, d\tau - \lambda_2 \int \phi_0 M \psi_1 \, d\tau$$

$$= \sum_r a_r^{(2)}(L'_{0r} - \lambda_1 \delta_{0r}) - \lambda_2 a_0^{(1)}$$

$$= \sum_{r \neq 0} \sum_{s \neq 0} \frac{L'_{0r} L'_{rs} L'_{s0}}{(\mu_0 - \mu_r)(\mu_0 - \mu_s)} - \sum_{r \neq 0} \frac{L'_{00} L'_{0r} L'_{r0}}{(\mu_0 - \mu_r)^2}. \quad (4.297)$$

By allowing $\eta \to 1$ we obtain the third order expression for the eigenvalue

$$\lambda = \sum_{n=0}^{3} \lambda_n = \mu_0 + L'_{00} + \sum_{r \neq 0} \frac{L'_{0r} L'_{r0}}{\mu_0 - \mu_r}$$

$$+ \sum_{r \neq 0} \sum_{s \neq 0} \frac{L'_{0r} L'_{rs} L'_{s0}}{(\mu_0 - \mu_r)(\mu_0 - \mu_s)} - \sum_{r \neq 0} \frac{L'_{00} L'_{0r} L'_{r0}}{(\mu_0 - \mu_r)^2}. \quad (4.298)$$

Higher order expressions for λ can be obtained by successive substitution.

The perturbation theory expressions for the eigenvalue λ can also be derived by using the variational principle (4.277) together with trial functions based upon perturbation theory. Taking the zero order eigenfunction ϕ_0 as trial function we obtain

$$I[\phi_0] = \frac{\int \phi_0 (L_0 + L') \phi_0 \, d\tau}{\int \phi_0 M \phi_0 \, d\tau} = \mu_0 + L'_{00} \quad (4.299)$$

which is the first order expression for λ. If the operator L is positive definite then we must have

$$\lambda \leq \mu_0 + L'_{00}. \quad (4.300)$$

We now take as trial function the first order eigenfunction

$$\phi_0 + \underset{r \neq 0}{S} \frac{L'_{r0}}{\mu_0 - \mu_r} \phi_r. \tag{4.301}$$

Then

$$I = \frac{\int \left(\phi_0 + \underset{r \neq 0}{S} \frac{L'_{0r}}{\mu_0 - \mu_r} \phi_r\right)(L_0 + L')\left(\phi_0 + \underset{s \neq 0}{S} \frac{L'_{s0}}{\mu_0 - \mu_s} \phi_s\right) d\tau}{\int \left(\phi_0 + \underset{r \neq 0}{S} \frac{L'_{0r}}{\mu_0 - \mu_r} \phi_r\right) M \left(\phi_0 + \underset{s \neq 0}{S} \frac{L'_{s0}}{\mu_0 - \mu_s} \phi_s\right) d\tau}$$

$$= \frac{\mu_0 + L'_{00} + 2\underset{r \neq 0}{S} \frac{L'_{0r}L'_{r0}}{\mu_0 - \mu_r} + \underset{r \neq 0}{S} \frac{\mu_r L'_{0r} L'_{r0}}{(\mu_0 - \mu_r)^2} + \underset{r \neq 0}{S}\underset{s \neq 0}{S} \frac{L'_{0r} L'_{rs} L'_{s0}}{(\mu_0 - \mu_r)(\mu_0 - \mu_s)}}{1 + \underset{r \neq 0}{S} \frac{L'_{0r} L'_{r0}}{(\mu_0 - \mu_r)^2}} \tag{4.302}$$

which is identical to (4.298) to the third order. Thus the variational principle provides an eigenvalue correct to the third order in the perturbation employing an eigenfunction to the first order only. More generally it can be readily verified that the variational principle must give the eigenvalue correct to the $(2n + 1)$th order using a trial function which is correct to the nth order.

Polarizability of an atom

As an application of the variational method using a trial function based on perturbation theory, we consider the effect of a constant electric field F acting on a hydrogen atom in its ground state. The Schrödinger equation for this system is

$$(H_0 + H')\psi = E\psi \tag{4.303}$$

where

$$H_0 = -\frac{\hbar^2}{2m} \nabla^2 - \frac{e^2}{r} \tag{4.304}$$

is the unperturbed Hamiltonian of the hydrogen atom and

$$H' = -eFr \cos \theta \tag{4.305}$$

is the perturbation due to the electric field, the axis of coordinates being taken along the direction of the field.

In order to obtain the eigenenergy E accurate to the third order in the perturbation we require a knowledge of the eigenfunction ψ of the system correct to the first order. To achieve this we put

$$\psi = \psi_0 + f\psi_0 \qquad (4.306)$$

where ψ_0 is the ground state eigenfunction of the unperturbed system associated with eigenenergy E_0 and f is a function to be determined which depends on the perturbation to the first order only. Substituting into the Schrödinger equation (4.303) and retaining only terms up to the first order in the perturbation we get

$$(H' - E_1)\psi_0 + (H_0 - E_0)f\psi_0 = 0 \qquad (4.307)$$

where

$$E_1 = \int \psi_0 H' \psi_0 \, d\mathbf{r}$$

is the first order correction to the eigenenergy. Using equation (4.307) it follows that

$$I[\psi] = \frac{\int \bar{\psi}(H_0 + H')\psi \, d\mathbf{r}}{\int \bar{\psi}\psi \, d\mathbf{r}}$$

$$= E_0 + \frac{E_1 + E_1 \int \psi_0 f \psi_0 \, d\mathbf{r} + \int \psi_0 H' f \psi_0 \, d\mathbf{r} + \int \psi_0 f H' f \psi_0 \, d\mathbf{r}}{1 + 2\int \psi_0 f \psi_0 \, d\mathbf{r} + \int \psi_0 f^2 \psi_0 \, d\mathbf{r}}.$$

$$(4.308)$$

If we now assume that ψ is normalized to the first order so that

$$\int \psi_0 f \psi_0 \, d\mathbf{r} = 0, \qquad (4.309)$$

we find to the third order in the perturbation

$$I[\psi] = E_0 + E_1 + \int \psi_0 H' f \psi_0 \, d\mathbf{r} - E_1 \int \psi_0 f^2 \psi_0 \, d\mathbf{r}$$
$$+ \int \psi_0 f H' f \psi_0 \, d\mathbf{r}. \quad (4.310)$$

To proceed further we need to determine the function $f(\mathbf{r})$. Since

$$H_0 \psi_0 = E_0 \psi_0$$

we obtain

$$\psi_0 \nabla^2 f + 2 \nabla \psi_0 \cdot \nabla f = \frac{2m}{\hbar^2} (H' - E_1) \psi_0 \quad (4.311)$$

using equation (4.307) for f.

So far our analysis has been quite general making no assumption regarding the form of the perturbation. Taking $H' = -eFr \cos \theta$, we see immediately that $E_1 = 0$ and that we may express f in the form

$$f(\mathbf{r}) = R(r) \cos \theta. \quad (4.312)$$

Substituting into (4.311) gives

$$\frac{d^2 R}{dr^2} + 2\left(\frac{1}{r} - \frac{1}{a_0}\right) \frac{dR}{dr} - \frac{2}{r^2} R = -\frac{2r}{ea_0} F \quad (4.313)$$

which has the particular solution

$$R(r) = \frac{a_0^2 F}{e} \left(\frac{r}{a_0} + \frac{r^2}{2a_0^2}\right), \quad (4.314)$$

the solution of the homogeneous equation being excluded since it has the asymptotic behaviour e^{2r/a_0} for large r. Because H' and f are both proportional to $\cos \theta$ we see that

$$\int \psi_0 f H' f \psi_0 \, d\mathbf{r} = 0 \quad (4.315)$$

and so we have now

$$I[\psi] = E_0 + \int \psi_0 H' f \psi_0 \, d\mathbf{r}$$
$$= E_0 - \frac{F^2}{\pi a_0} \int e^{-2r/a_0} r \left(\frac{r}{a_0} + \frac{r^2}{2a_0^2}\right) \cos^2 \theta \, d\mathbf{r}$$
$$= E_0 - \frac{9}{4} F^2 a_0^3 \quad (4.316)$$

on carrying out the elementary integrations. Thus the effect of the perturbing electric field F is to lower the ground state energy of the hydrogen atom by the amount $\frac{1}{2}\alpha F^2$ where $\alpha = (9/2)a_0^3$ is called the *polarizability* of the atom. This result is exact to the third order in the perturbation, the first and third order correction terms vanishing for this particular case.

Van der Waals force

In the example of the polarization of a hydrogen atom by an electric field it was possible to obtain an exact expression for the first order perturbation theory eigenfunction and hence to obtain the energy to the third order in the perturbation without making any further approximation. However, in the case of the interaction between two neutral hydrogen atoms, known as the van der Waals force, it is not possible to obtain an exact formula for the first order eigenfunction. We must therefore apply the variational principle directly using a suitable trial function.

We regard the two hydrogen atoms when situated at an infinite distance apart as the unperturbed system having Hamiltonian operator

$$H_0 = -\frac{\hbar^2}{2m}(\nabla_1^2 + \nabla_2^2) - \frac{e^2}{r_{A1}} - \frac{e^2}{r_{B2}} \qquad (4.317)$$

where r_{A1} and r_{B2} are the distances of electrons 1 and 2 from their parent nuclei A and B respectively. If the atoms are in their ground states with eigenfunctions $\chi_A(1)$ and $\chi_B(2)$, the unperturbed function describing the total system takes the form

$$\chi_0(r_{A1}, r_{B2}) = \chi_A(1)\chi_B(2) \qquad (4.318)$$

with eigenenergy $E_0 = -e^2/a_0$. As the atoms approach each other they will interact, the perturbing Hamiltonian being given by

$$H' = \frac{e^2}{R} + \frac{e^2}{r_{12}} - \frac{e^2}{r_{B1}} - \frac{e^2}{r_{A2}}, \qquad (4.319)$$

where R is the internuclear separation and r_{12} is the distance between the two electrons. For large R we may expand H' in powers of $1/R$. The leading term of the expansion is the electric dipole–dipole interaction

$$H' = \frac{e^2}{R^3}(x_{A1}x_{B2} + y_{A1}y_{B2} - 2z_{A1}z_{B2}) \qquad (4.320)$$

where x_{A1}, y_{A1}, z_{A1} and x_{B2}, y_{B2}, z_{B2} are the Cartesian coordinates of the two electrons referred to reference frames with their origins at A and B and their z-axes along the internuclear line.

We now take as trial function

$$\chi(\mathbf{r}_1, \mathbf{r}_2) = \chi_0(r_{A1}, r_{B2})(1 + cH') \qquad (4.321)$$

where c is a variable parameter. Substituting into the functional $I[\chi]$ given by (4.129) yields

$$I[\chi] = \frac{\iint \chi_0(1 + cH')(H_0 + H')\chi_0(1 + cH')\, d\mathbf{r}_1\, d\mathbf{r}_2}{\iint \{\chi_0(1 + cH')\}^2\, d\mathbf{r}_1\, d\mathbf{r}_2}. \qquad (4.322)$$

It can be readily verified that

$$\int \chi_0 H' \chi_0\, d\mathbf{r}_1\, d\mathbf{r}_2, \quad \int \chi_0 H'^3 \chi_0\, d\mathbf{r}_1\, d\mathbf{r}_2$$

and

$$\int \chi_0 H' H_0 H' \chi_0\, d\mathbf{r}_1\, d\mathbf{r}_2$$

all vanish so that

$$I[\chi] = \frac{E_0 + 2c \int \chi_0 H'^2 \chi_0\, d\mathbf{r}_1\, d\mathbf{r}_2}{1 + c^2 \int \chi_0 H'^2 \chi_0\, d\mathbf{r}_1\, d\mathbf{r}_2} \qquad (4.323)$$

which gives

$$I[\chi] = E_0 + (2c - E_0 c^2) \int \chi_0 H'^2 \chi_0\, d\mathbf{r}_1\, d\mathbf{r}_2 \qquad (4.324)$$

retaining only terms up to the second order in the perturbation.

The least value of $I[\chi]$ occurs when $c = 1/E_0$ and then

$$I[\chi] = E_0 + \frac{1}{E_0} \int \chi_0 H'^2 \chi_0 \, d\mathbf{r}_1 \, d\mathbf{r}_2. \tag{4.325}$$

But

$$\int \chi_0 H'^2 \chi_0 \, d\mathbf{r}_1 \, d\mathbf{r}_2 = \frac{e^4}{R^6} \int \chi_0 (x_{A1}^2 x_{B2}^2 + y_{A1}^2 y_{B2}^2 + 4z_{A1}^2 z_{B2}^2) \chi_0 \, d\mathbf{r}_1 \, d\mathbf{r}_2$$

$$= \frac{2e^4}{3R^6} \int \chi_0 r_{A1}^2 r_{B2}^2 \chi_0 \, d\mathbf{r}_1 \, d\mathbf{r}_2$$

since the cross terms integrate to zero. Because

$$\int \chi_A(1) r_{A1}^2 \chi_A(1) \, d\mathbf{r}_1 = 3a_0^2$$

it follows that

$$\int \chi_0 H'^2 \chi_0 \, d\mathbf{r}_1 \, d\mathbf{r}_2 = \frac{6e^4 a_0^4}{R^6} \tag{4.326}$$

and so

$$I[\chi] = E_0 - \frac{6e^2 a_0^5}{R^6}. \tag{4.327}$$

Hence $-6e^2 a_0^5/R^6$ provides an upper bound to the interaction energy between two hydrogen atoms when they are at a large distance apart. More detailed calculations by Pauling and Beach[13] in which the parameter c is replaced by a function of $r_{A1} r_{B2}$ shows that the interaction energy actually behaves as $-6 \cdot 499 \, e^2 a_0^5/R^6$ for large R. Thus, considering the simplicity of the trial function we have used in this section, the result obtained is quite accurate.

4.16 Variational principle for an arbitrary operator

So far our considerations have been entirely concerned with variational principles for the eigenvalues of eigenvalue equations having the general form (4.271) of which a particularly important instance is the eigenenergy problem in wave mechanics. To conclude this chapter we shall establish the variational principle due to Delves[14] for the expectation value

$$\langle W \rangle = \int \bar{\psi}_0 W \psi_0 \, d\tau \tag{4.328}$$

of an arbitrary Hermitian operator W with respect to a normalized eigenfunction ψ_0 satisfying the Schrödinger equation

$$H\psi_0 = E_0\psi_0, \quad (4.329)$$

E_0 being the associated eigenenergy and H the Hamiltonian operator.

This can be achieved by introducing the equation

$$(H + \alpha W)\psi = E\psi \quad (4.330)$$

where the eigenenergy E and the eigenfunction ψ depend upon the value of the parameter α. Differentiating with respect to α and then letting α tend to zero we obtain

$$(H - E_0)\psi_1 = (E_1 - W)\psi_0 \quad (4.331)$$

where

$$\psi_1 = \left(\frac{d\psi}{d\alpha}\right)_{\alpha=0}, \quad E_1 = \left(\frac{dE}{d\alpha}\right)_{\alpha=0}. \quad (4.332)$$

Multiplying equation (4.331) across by $\bar{\psi}_0$ and integrating over the spacial coordinates of the system gives

$$\int \bar{\psi}_0(H - E_0)\psi_1 \, d\tau = \int \bar{\psi}_0(E_1 - W)\psi_0 \, d\tau.$$

But the integral on the left-hand side of this equation vanishes because H is an Hermitian operator and ψ_0 satisfies (4.329), and so it follows that

$$E_1 = \int \bar{\psi}_0 W \psi_0 \, d\tau = \langle W \rangle \quad (4.333)$$

which is analogous to the perturbation theory result (4.290).

We now define the functional

$$J[\chi_0, \chi_1] = \frac{\int \bar{\chi}_0 W \chi_0 \, d\tau + \int \bar{\chi}_0(H - E_0)\chi_1 \, d\tau + \int \chi_0(H - E_0)\bar{\chi}_1 \, d\tau}{\int \bar{\chi}_0 \chi_0 \, d\tau}. \quad (4.334)$$

Then we have

$$J[\psi_0, \psi_1] = \langle W \rangle. \quad (4.335)$$

EIGENVALUE PROBLEMS

Further, to the first order of small quantities, the variation in J due to small arbitrary changes $\delta\chi_0$ and $\delta\chi_1$ in the functions χ_0 and χ_1 is given by

$$\delta J[\chi_0, \chi_1] \int \bar{\chi}_0 \chi_0 \, d\tau + J[\chi_0, \chi_1] \left\{ \int (\delta\bar{\chi}_0)\chi_0 \, d\tau + \int (\delta\chi_0)\bar{\chi}_0 \, d\tau \right\}$$

$$= \int \delta\bar{\chi}_0 W \chi_0 \, d\tau + \int \bar{\chi}_0 W \, \delta\chi_0 \, d\tau$$

$$+ \int \delta\bar{\chi}_0 (H - E_0)\chi_1 \, d\tau + \int \delta\chi_0 (H - E_0)\bar{\chi}_1 \, d\tau$$

$$+ \int \bar{\chi}_0 (H - E_0) \, \delta\chi_1 \, d\tau + \int \chi_0 (H - E_0) \, \delta\bar{\chi}_1 \, d\tau$$

and so

$$\delta J[\psi_0, \psi_1] = \int \delta\bar{\psi}_0 \{(H - E_0)\psi_1 - (E_1 - W)\psi_0\} \, d\tau$$

$$+ \int \delta\psi_0 \{(H - E_0)\bar{\psi}_1 - (E_1 - W)\bar{\psi}_0\} \, d\tau$$

$$+ \int \delta\bar{\psi}_1 (H - E_0)\psi_0 \, d\tau + \int \delta\psi_1 (H - E_0)\bar{\psi}_0 \, d\tau$$

using the Hermitian property of H and W. Hence we have

$$\delta J[\psi_0, \psi_1] = 0 \qquad (4.336)$$

as a consequence of the fact that ψ_0 and ψ_1 satisfy equations (4.329) and (4.331) respectively. Thus J is stationary for the exact solutions ψ_0 and ψ_1.

In order to apply the variational principle (4.336) derived above, we employ trial functions ψ_{0T} and ψ_{1T} depending upon a set of variable parameters which we determine by making $J[\psi_{0T}, \psi_{1T}]$ stationary. Our variational approximation to $\langle W \rangle$ then takes the form

$$\langle W \rangle = \int \bar{\psi}_{0T} W \psi_{0T} \, d\tau + \int \bar{\psi}_{0T}(H - E_0)\psi_{1T} \, d\tau$$

$$+ \int \psi_{0T}(H - E_0)\bar{\psi}_{1T} \, d\tau, \qquad (4.337)$$

where we have chosen ψ_{0T} to be normalized to unity.

A method of utilizing this result which is of considerable value is to take trial functions ψ_{0T} and ψ_{1T} derived from perturbation theory. Suppose that H_0 denotes the unperturbed Hamiltonian and that the zero order solution ψ_0^0 satisfies

$$H_0\psi_0^0 = E_0^0\psi_0^0. \tag{4.338}$$

Writing

$$E_1^0 = \int \bar{\psi}_0^0 W \psi_0^0 \, d\tau \tag{4.339}$$

and introducing the function ψ_1^0 satisfying the equation

$$(H_0 - E_0^0)\psi_1^0 = (E_1^0 - W)\psi_0^0, \tag{4.340}$$

we may take ψ_0^0 and ψ_1^0 as our trial functions ψ_{0T} and ψ_{1T}. In this case, to the first order in the perturbation H', the variational approximation to the expectation value of W becomes

$$\langle W \rangle = \int \bar{\psi}_0^0 W \psi_0^0 \, d\tau + \int \bar{\psi}_0^0 H' \psi_1^0 \, d\tau + \int \psi_0^0 H' \bar{\psi}_1^0 \, d\tau$$
$$- E_0^1 \left\{ \int \bar{\psi}_0^0 \psi_1^0 \, d\tau + \int \psi_0^0 \bar{\psi}_1^0 \, d\tau \right\} \tag{4.341}$$

where

$$E_0^1 = \int \bar{\psi}_0^0 H' \psi_0^0 \, d\tau, \tag{4.342}$$

a result first established by Dalgarno and Stewart[15] and employed by them and co-workers in numerous applications in atomic physics.

A simple example of the application of this result is provided by the determination of diamagnetic susceptibilities. In the case of the helium isoelectronic sequence, the molar diamagnetic susceptibility in $cm^3/(gram\ ion)$ units is given by

$$-7 \cdot 93 \times 10^{-7} \int \bar{\psi}_0 W \psi_0 \, d\mathbf{r}_1 \, d\mathbf{r}_2 \tag{4.343}$$

where

$$W = r_1^2 + r_2^2 \tag{4.344}$$

and $\psi_0(\mathbf{r}_1, \mathbf{r}_2)$ is the ground state wave function of the particular two electron system under consideration, the unit of length being the Bohr radius a_0.

For the zero order unperturbed function we choose

$$\psi_0^0(\mathbf{r}_1, \mathbf{r}_2) = \frac{Z^3}{\pi} e^{-Z(r_1+r_2)} \tag{4.345}$$

where Z is the nuclear charge of the atomic system. Then, in Rydberg units, the zero and first order energies are

$$E_0^0 = -2Z^2, \quad E_0^1 = \frac{5Z}{4} \tag{4.346}$$

and the zero order expression for the expectation value of W is given by

$$\int \bar{\psi}_0^0 W \psi_0^0 \, d\mathbf{r}_1 \, d\mathbf{r}_2 = \frac{6}{Z^2}. \tag{4.347}$$

Now the solution of equation (4.340) which vanishes exponentially for large r_1 and r_2 is

$$\psi_1^0(\mathbf{r}_1, \mathbf{r}_2) = -\left(\frac{r_1^3 + r_2^3}{6Z} + \frac{r_1^2 + r_2^2}{2Z^2}\right)\psi_0^0(\mathbf{r}_1, \mathbf{r}_2) \tag{4.348}$$

and so, using (4.341), we get to the first order

$$\langle W \rangle = \frac{6}{Z^2} + \frac{153}{32}\frac{1}{Z^3}. \tag{4.349}$$

This gives $-6 \cdot 7 \times 10^{-7}$ cm^3/(gram ion) for the molar diamagnetic susceptibility of Li$^+$ (corresponding to $Z = 3$) compared with $-7 \cdot 06 \times 10^{-7}$ obtained by using a Hartree wave function to evaluate $\langle W \rangle$.

References

1. Hylleraas, E. A. and Undheim, B., *Z. Physik*, **65**, 759 (1930).
2. Temple, G., *Proc. Roy. Soc.*, **A119**, 276 (1928).
3. Weinstein, D. H., *Proc. Nat. Acad. Sci. U.S.*, **20**, 529 (1934).
4. Stevenson, A. F., *Phys. Rev.*, **53**, 199 (1938).
5. Hylleraas, E. A., *Z. Physik*, **48**, 469 (1928); **54**, 347 (1929).
6. Pekeris, C. L., *Phys. Rev.*, **112**, 1649 (1958).
7. Pekeris, C. L., *Phys. Rev.* **115**, 1216 (1959); **126**, 1470 (1962).
8. Pekeris, C. L., Schiff, B. and Lifson, H., *Phys. Rev.* **126**, 1057 (1962).

9. Heitler, W. and London F., *Z. Physik*, **44**, 455 (1927).
10. Wang, S. C., *Phys. Rev.*, **31**, 579 (1928).
11. Pauling, L. and Wilson, E. B., *Introduction to Quantum Mechanics*, McGraw-Hill, New York, 1935.
12. James, H. M. and Coolidge, A. S., *J. Chem. Phys.*, **1**, 825 (1933).
13. Pauling, L. and Beach, J. Y., *Phys. Rev.*, **47**, 686 (1935).
14. Delves, L. M., *Nucl. Phys.*, **41**, 497 (1963).
15. Dalgarno, A. and Stewart, A. L., *Proc. Roy. Soc.*, **A247**, 245 (1958).

CHAPTER 5

Scattering Theory

Many of the most important recent developments in the application of variational principles to physical problems have taken place in the theory of scattering, and in this final chapter we shall direct our attention to the derivation of variational principles for the scattering amplitude and for scattering phase shifts as well as for the scattering length, and show how they have been employed to investigate elastic collisions between electrons and hydrogen atoms and between neutrons and deuterons. The special case of zero energy incident particles will be examined in considerable detail since extremely valuable upper bounds to the scattering length can be established. However, to commence this chapter, we shall look at the instructive but relatively simple example of the scattering of waves by a one-dimensional potential barrier and then study the more general case of scattering by a perturbation located over a surface before proceeding to our main objective: the examination of the scattering of waves by a three-dimensional field of force.

5.1 One-dimensional potential barrier

Consider the scattering of waves by a one-dimensional potential barrier $U(x)$ which attains its maximum value close to the origin and which vanishes as $x \to \pm\infty$. Then the time-independent equation describing the wave motion takes the form

$$\frac{d^2u}{dx^2} + \{k^2 - U(x)\}u = 0 \tag{5.1}$$

where $k = 2\pi/\lambda$ is the *wave number* and λ is the wave length.

Now suppose that a wave of amplitude A is incident in the positive x direction upon the barrier. We may represent this incident wave by $A\,e^{ikx}$. The presence of the potential barrier will produce a reflected wave $B\,e^{-ikx}$ for large negative x and a transmitted wave $C\,e^{ikx}$ for large positive x so that the asymptotic behaviour of the appropriate solution of the wave equation (5.1) is

$$u(x) \sim A\,e^{ikx} + B\,e^{-ikx} \qquad (x \to -\infty) \qquad (5.2)$$

and

$$u(x) \sim C\,e^{ikx} \qquad (x \to +\infty). \qquad (5.3)$$

The quantity $f = C/A$ is called the *transmission amplitude*, and the transmitted intensity per unit incident intensity given by $T = |f|^2$ is known as the transmission coefficient. The reflection coefficient $R = |B/A|^2$ is just the reflected intensity per unit incident intensity and, in the absence of absorption, we must have $T + R = 1$.

We now note that the solution of the equation

$$\frac{d^2u}{dx^2} + k^2 u = F(x) \qquad (5.4)$$

satisfying the boundary conditions (5.2) and (5.3) can be expressed in the integral equation form

$$u(x) = A\,e^{ikx} + \int_{-\infty}^{\infty} G(x, x') F(x')\,dx' \qquad (5.5)$$

where the Green's function $G(x, x')$ satisfies

$$\frac{d^2 G(x, x')}{dx^2} + k^2 G(x, x') = \delta(x - x'), \qquad (5.6)$$

$\delta(x)$ being the one-dimensional Dirac function which vanishes for $x \neq 0$ but integrates to unity:

$$\int_{-\infty}^{\infty} \delta(x)\,dx = 1. \qquad (5.7)$$

Since we require the second term on the right-hand side of (5.5) to behave as e^{ikx} for large positive x and as e^{-ikx} for large negative x, we choose

$$G(x, x') = \frac{1}{2ik} e^{ik|x - x'|} \qquad (5.8)$$

which one can readily verify is a particular solution of equation (5.6) having the correct asymptotic behaviour. Hence we may express the appropriate solution of the wave equation (5.1) in the form

$$u(x) = A\,e^{ikx} + \frac{1}{2ik}\int_{-\infty}^{\infty} e^{ik|x-x'|}U(x')u(x')\,dx'. \quad (5.9)$$

Rewriting this equation as

$$u(x) = A\,e^{ikx} + \frac{1}{2ik}e^{ikx}\int_{-\infty}^{x} e^{-ikx'}U(x')u(x')\,dx'$$
$$+ \frac{1}{2ik}e^{-ikx}\int_{x}^{\infty} e^{ikx'}U(x')u(x')\,dx', \quad (5.10)$$

we see that the amplitude of the reflected wave is given by

$$B = \frac{1}{2ik}\int_{-\infty}^{\infty} e^{ikx'}U(x')u(x')\,dx' \quad (5.11)$$

while the amplitude of the transmitted wave takes the form

$$C = A + \frac{1}{2ik}\int_{-\infty}^{\infty} e^{-ikx'}U(x')u(x')\,dx'. \quad (5.12)$$

Solving for A in terms of the transmission amplitude $f = C/A$, we see that we may rewrite (5.9) as

$$u(x) = \frac{1}{2ik}\frac{1}{f-1}e^{ikx}\int_{-\infty}^{\infty} e^{-ikx'}U(x')u(x')\,dx'$$
$$+ \frac{1}{2ik}\int_{-\infty}^{\infty} e^{ik|x-x'|}U(x')u(x')\,dx'. \quad (5.13)$$

Next we introduce the *adjoint solution* $\tilde{u}(x)$ of the wave equation (5.1) corresponding to the case in which the incident wave approaches the barrier in the negative x direction. It is given by

$$\tilde{u}(x) = \frac{1}{2ik}\frac{1}{f-1}e^{-ikx}\int_{-\infty}^{\infty} \tilde{u}(x')U(x')\,e^{ikx'}\,dx'$$
$$+ \frac{1}{2ik}\int_{-\infty}^{\infty} \tilde{u}(x')U(x')\,e^{ik|x-x'|}\,dx' \quad (5.14)$$

where the transmission amplitude f is clearly the same for waves incident in the positive x and in the negative x directions. Multiplying each term of equation (5.13) on the left by $\tilde{u}(x)U(x)$ and then integrating over the coordinate x, we arrive at the result

$$\frac{1}{f-1} = \frac{2ik \int_{-\infty}^{\infty} \tilde{u}(x)U(x)u(x)\,dx - \int_{-\infty}^{\infty}\int_{-\infty}^{\infty} \tilde{u}(x)U(x)e^{ik|x-x'|}U(x')u(x')\,dx\,dx'}{\int_{-\infty}^{\infty} \tilde{u}(x)U(x)\,e^{ikx}\,dx \int_{-\infty}^{\infty} e^{-ikx'}U(x')u(x')\,dx'}.$$
(5.15)

It can be readily verified that this expression is stationary with respect to infinitesimal variations of $u(x)$ and $\tilde{u}(x)$ provided these functions satisfy the integral equations (5.13) and (5.14) respectively. Furthermore it has the valuable property of being homogeneous in $u(x)$ and $\tilde{u}(x)$ so that if we multiply u or $\tilde{u}$ by a constant factor, the expression remains unaltered. The stationary property of the homogeneous expression (5.15) was first noted by Schwinger and is usually referred to as the *Schwinger variational principle*.

An approximate formula for the transmission amplitude can be obtained by inserting suitable trial functions into (5.15). If we take the unperturbed incident waves $u_T(x) = A\,e^{ikx}$ and $\tilde{u}_T(x) = A\,e^{-ikx}$ as trial functions, we get

$$f = 1 + \frac{\left[\int_{-\infty}^{\infty} U(x)\,dx\right]^2}{2ik\int_{-\infty}^{\infty} U(x)\,dx - \int_{-\infty}^{\infty}\int_{-\infty}^{\infty} e^{-ikx}U(x)\,e^{ik|x-x'|}U(x')\,e^{ikx'}\,dx\,dx'}.$$
(5.16)

5.2 Scattering at a surface

We now turn to the three-dimensional problem of the scattering of waves by a perturbation located over a surface S. Outside the surface we suppose that the wave motion is governed by the Helmholtz equation

$$\nabla^2 \psi + k^2 \psi = 0 \tag{5.17}$$

SCATTERING THEORY

and for the incident wave we take the plane wave

$$A\, e^{ikz}$$

propagated in the direction of the z axis of a frame of reference which we choose as the polar axis of a spherical polar system of coordinates r, θ, ϕ. Due to the presence of the surface perturbation, a scattered wave is produced which has the form of an outgoing spherical wave

$$A\frac{e^{ikr}}{r}f(\theta, \phi)$$

for large radial distances r from the origin, where $f(\theta, \phi)$ is called the *scattering amplitude*. Then for large r, the function ψ describing the wave motion in the presence of the surface perturbation has the asymptotic form

$$\psi \sim A\left\{e^{ikz} + \frac{1}{r}e^{ikr}f(\theta, \phi)\right\}. \tag{5.18}$$

Suppose now that the perturbation requires the function ψ to satisfy either the boundary condition

(i) $$\frac{\partial \psi}{\partial n} = 0 \tag{5.19}$$

over the surface S, the derivative being in the direction of the normal to the surface, or the boundary condition

(ii) $$\psi = 0 \tag{5.20}$$

over the surface S.

To obtain an integral equation for ψ, we introduce the Green's function $G(\mathbf{r}, \mathbf{r}')$ satisfying the equation

$$(\nabla'^2 + k^2)G(\mathbf{r}, \mathbf{r}') = \delta(\mathbf{r} - \mathbf{r}') \tag{5.21}$$

where $\delta(\mathbf{r})$ is the three-dimensional Dirac function which vanishes everywhere except at the origin $r = 0$ and integrates to unity:

$$\int \delta(\mathbf{r})\, d\mathbf{r} = 1. \tag{5.22}$$

Then we see that

$$\psi(\mathbf{r}) = \int [\psi(\mathbf{r}')\nabla'^2 G(\mathbf{r},\mathbf{r}') - G(\mathbf{r},\mathbf{r}')\nabla'^2 \psi(\mathbf{r}')]\, d\mathbf{r}' \quad (5.23)$$

and so, on applying Green's theorem, we get

$$\psi(\mathbf{r}) = \oint_S \left[\psi(\mathbf{r}')\frac{\partial}{\partial n'} G(\mathbf{r},\mathbf{r}') - G(\mathbf{r},\mathbf{r}')\frac{\partial}{\partial n'} \psi(\mathbf{r}')\right] dS' \quad (5.24)$$

where the derivatives are in the direction of the outdrawn normal **n** to the surface S.

We suppose first that the boundary condition (i) applies. This yields

$$\psi(\mathbf{r}) = \oint_S \psi(\mathbf{r}')\frac{\partial}{\partial n'} G(\mathbf{r},\mathbf{r}')\, dS' \quad (5.25)$$

which behaves as an outgoing spherical wave for large values of r if we choose the Green's function to have the form

$$G(\mathbf{r},\mathbf{r}') = -\frac{1}{4\pi}\frac{e^{ik|\mathbf{r}-\mathbf{r}'|}}{|\mathbf{r}-\mathbf{r}'|} \quad (5.26)$$

readily verifiable as a particular solution of equation (5.21). Hence we may write the solution of the wave equation (5.17) satisfying the boundary conditions (5.18) and (5.19) as

$$\psi(\mathbf{r}) = A\, e^{i\mathbf{k}_i \cdot \mathbf{r}} - \frac{1}{4\pi}\oint_S \psi(\mathbf{r}')\frac{\partial}{\partial n'}\left(\frac{e^{ik|\mathbf{r}-\mathbf{r}'|}}{|\mathbf{r}-\mathbf{r}'|}\right) dS' \quad (5.27)$$

where $\mathbf{k}_i$ is a vector of magnitude k pointing in the direction of propagation of the incident wave. Since for large r

$$k|\mathbf{r}-\mathbf{r}'| \sim kr - \mathbf{k}_s \cdot \mathbf{r}' \quad (5.28)$$

where $\mathbf{k}_s$ is a vector of magnitude k pointing in the direction of scattering given by $\mathbf{r}$, we see that

$$\psi(\mathbf{r}) \sim A\, e^{i\mathbf{k}_i \cdot \mathbf{r}} + \frac{1}{4\pi}\frac{e^{ikr}}{r}\oint_S i\mathbf{n}' \cdot \mathbf{k}_s \psi(\mathbf{r}')\, e^{-i\mathbf{k}_s \cdot \mathbf{r}'}\, dS' \quad (5.29)$$

and so the scattering amplitude takes the form

$$f(\mathbf{k}_s, \mathbf{k}_i) = \frac{1}{4\pi A}\oint_S i\mathbf{n}' \cdot \mathbf{k}_s \psi(\mathbf{r}')\, e^{-i\mathbf{k}_s \cdot \mathbf{r}'}\, dS'. \quad (5.30)$$

SCATTERING THEORY

Solving for A in terms of f and substituting into (5.27), leads to the integral equation

$$\psi(\mathbf{r}) = \frac{e^{i\mathbf{k}_i \cdot \mathbf{r}}}{4\pi f} \oint_S i\mathbf{n}' \cdot \mathbf{k}_s \psi(\mathbf{r}') \, e^{-i\mathbf{k}_s \cdot \mathbf{r}'} \, dS'$$
$$- \frac{1}{4\pi} \oint_S \psi(\mathbf{r}') \frac{\partial}{\partial n'} \left(\frac{e^{ik|\mathbf{r}-\mathbf{r}'|}}{|\mathbf{r}-\mathbf{r}'|} \right) dS' \qquad (5.31)$$

which is homogeneous in the function ψ. We now take the normal derivative $\partial/\partial n$ over the surface S on both sides of this equation. Because $\partial\psi/\partial n$ vanishes over S, we get

$$\frac{\mathbf{n} \cdot \mathbf{k}_i \, e^{i\mathbf{k}_i \cdot \mathbf{r}}}{f} \oint_S \mathbf{n}' \cdot \mathbf{k}_s \psi(\mathbf{r}') \, e^{-i\mathbf{k}_s \cdot \mathbf{r}'} \, dS'$$
$$= -\oint_S \psi(\mathbf{r}') \frac{\partial^2}{\partial n \partial n'} \left(\frac{e^{ik|\mathbf{r}-\mathbf{r}'|}}{|\mathbf{r}-\mathbf{r}'|} \right) dS'.$$

Multiplying across by the adjoint solution

$$\tilde{\psi}(\mathbf{r}) = A \, e^{-i\mathbf{k}_s \cdot \mathbf{r}} - \frac{1}{4\pi} \oint_S \tilde{\psi}(\mathbf{r}') \frac{\partial}{\partial n'} \left(\frac{e^{ik|\mathbf{r}-\mathbf{r}'|}}{|\mathbf{r}-\mathbf{r}'|} \right) dS' \qquad (5.32)$$

corresponding to the situation in which the incident wave is propagated in the direction of $-\mathbf{k}_s$, and integrating with respect to $\mathbf{r}$ over the surface S, we obtain an expression for the scattering amplitude which is homogeneous in the functions ψ and $\tilde{\psi}$:

$$f(\mathbf{k}_s, \mathbf{k}_i) = -\frac{\oint_S \mathbf{n} \cdot \mathbf{k}_i \tilde{\psi}(\mathbf{r}) \, e^{i\mathbf{k}_i \cdot \mathbf{r}} \, dS \oint_S \mathbf{n}' \cdot \mathbf{k}_s \, e^{-i\mathbf{k}_s \cdot \mathbf{r}'} \psi(\mathbf{r}') \, dS'}{\oint_S \oint_S \tilde{\psi}(\mathbf{r}) \frac{\partial^2}{\partial n \partial n'} \left(\frac{e^{ik|\mathbf{r}-\mathbf{r}'|}}{|\mathbf{r}-\mathbf{r}'|} \right) \psi(\mathbf{r}') \, dS \, dS'}. \qquad (5.33)$$

We see at once that

$$f(\mathbf{k}_s, \mathbf{k}_i) = f(-\mathbf{k}_i, -\mathbf{k}_s),$$

a result known as the *reciprocity theorem*. Also it can be readily verified that expression (5.33) is stationary with respect to infinitesimal variations of ψ and $\tilde{\psi}$ provided these functions satisfy the integral equations (5.27) and (5.32) respectively.

If the boundary condition (ii) is satisfied by ψ over the surface S, we see that

$$\psi(\mathbf{r}) = A\,e^{i\mathbf{k}_i\cdot\mathbf{r}} + \frac{1}{4\pi}\oint_S \frac{e^{ik|\mathbf{r}-\mathbf{r}'|}}{|\mathbf{r}-\mathbf{r}'|}\frac{\partial}{\partial n'}\psi(\mathbf{r}')\,dS' \quad (5.34)$$

and so the scattering amplitude is given by

$$f(\mathbf{k}_s,\mathbf{k}_i) = \frac{1}{4\pi A}\oint_S e^{-i\mathbf{k}_s\cdot\mathbf{r}'}\frac{\partial}{\partial n'}\psi(\mathbf{r}')\,dS' \quad (5.35)$$

which leads to the homogeneous integral equation

$$\psi(\mathbf{r}) = \frac{e^{i\mathbf{k}_i\cdot\mathbf{r}}}{4\pi f}\oint_S e^{-i\mathbf{k}_s\cdot\mathbf{r}'}\frac{\partial}{\partial n'}\psi(\mathbf{r}')\,dS'$$
$$+ \frac{1}{4\pi}\oint_S \frac{e^{ik|\mathbf{r}-\mathbf{r}'|}}{|\mathbf{r}-\mathbf{r}'|}\frac{\partial}{\partial n'}\psi(\mathbf{r}')\,dS'. \quad (5.36)$$

Since ψ vanishes over S we get, on multiplying across by $(\partial/\partial n)\tilde{\psi}(\mathbf{r})$ and integrating with respect to $\mathbf{r}$ over the surface S, the following expression for the scattering amplitude:

$$f = -\frac{\oint_S \frac{\partial}{\partial n}\tilde{\psi}(\mathbf{r})\,e^{i\mathbf{k}_i\cdot\mathbf{r}}\,dS \oint_S e^{-i\mathbf{k}_s\cdot\mathbf{r}'}\frac{\partial}{\partial n'}\psi(\mathbf{r}')\,dS'}{\oint_S \oint_S \frac{\partial}{\partial n}\tilde{\psi}(\mathbf{r})\frac{e^{ik|\mathbf{r}-\mathbf{r}'|}}{|\mathbf{r}-\mathbf{r}'|}\frac{\partial}{\partial n'}\psi(\mathbf{r}')\,dS\,dS'} \quad (5.37)$$

which is also stationary for infinitesimal variations of the functions ψ and $\tilde{\psi}$.

The stationary expressions (5.33) and (5.37) are of the form originated by Schwinger.

Scattering at a spherical surface

In this subsection we consider the application of the variational principles derived previously to the scattering of waves at a spherical surface of radius a. This is a problem that can be solved exactly without undue difficulty and thus does not necessitate the use of the variational method for its solution. However it is instructive to apply the variational method to the scattering at a spherical surface since less tractable problems may be treated by using an analogous procedure and because in any case it serves as an interesting illustrative example.

SCATTERING THEORY

We approach the present problem by expanding the function $\psi(\mathbf{r})$ describing the wave motion in terms of *Legendre polynomials* $P_l(\mu)$ satisfying the equation

$$\frac{d}{d\mu}\left\{(1-\mu^2)\frac{dP_l}{d\mu}\right\} + l(l+1)P_l(\mu) = 0. \tag{5.38}$$

Choosing the centre of the sphere as origin and taking the direction of propagation of the incident wave as polar axis for a spherical polar system of coordinates r, ϑ, φ, we may write

$$\psi(\mathbf{r}) = \sum_{l=0}^{\infty} R_l(r)P_l(\cos\vartheta) \tag{5.39}$$

because of the cylindrical symmetry of the problem. Now substituting into the Helmholtz equation (5.17), we obtain

$$\frac{1}{r^2}\frac{d}{dr}\left(r^2\frac{dR_l}{dr}\right) + \left\{k^2 - \frac{l(l+1)}{r^2}\right\}R_l(r) = 0 \tag{5.40}$$

whose general solution may be expressed in the form

$$R_l(r) = a_l j_l(kr) + b_l n_l(kr) \tag{5.41}$$

where

$$j_l(\rho) = \sqrt{\left(\frac{\pi}{2\rho}\right)} J_{l+1/2}(\rho) \tag{5.42}$$

is called a *spherical Bessel function*, and

$$n_l(\rho) = -(-1)^l \sqrt{\left(\frac{\pi}{2\rho}\right)} J_{-l-1/2}(\rho) \tag{5.43}$$

is called a *spherical Neumann function*, the nth order Bessel function being denoted by $J_n(\rho)$. The asymptotic forms of these functions for large ρ are given by

$$\left.\begin{array}{l} j_l(\rho) \sim \dfrac{1}{\rho}\sin(\rho - \tfrac{1}{2}l\pi) \\[4pt] n_l(\rho) \sim -\dfrac{1}{\rho}\cos(\rho - \tfrac{1}{2}l\pi) \end{array}\right\} \tag{5.44}$$

and since we shall require a solution of (5.17) having the character of an outgoing spherical wave for large values of r, it is convenient to introduce also the *spherical Hankel function*

$$h_l(\rho) = j_l(\rho) + in_l(\rho) \tag{5.45}$$

whose asymptotic form is given by

$$h_l(\rho) \sim (-i)^{l+1} \frac{1}{\rho} e^{i\rho} \tag{5.46}$$

for large ρ.

Now it is a well-known result that[1]

$$e^{i\mathbf{k}_i \cdot \mathbf{r}} = \sum_{l=0}^{\infty} i^l (2l+1) j_l(kr) P_l(\cos \vartheta) \tag{5.47}$$

and so it immediately follows that the solutions of the Helmholtz equation (5.17) having the asymptotic behaviour for large r

$$\psi(\mathbf{r}) \sim e^{i\mathbf{k}_i \cdot \mathbf{r}} + \frac{e^{ikr}}{r} f(\vartheta, \varphi) \tag{5.48}$$

and satisfying the boundary conditions

(i) $\qquad \dfrac{\partial \psi}{\partial r} = 0 \quad$ at $r = a$ $\qquad$ (5.49)

and

(ii) $\qquad \psi = 0 \quad$ at $r = a$ $\qquad$ (5.50)

are respectively

(i) $\psi(\mathbf{r}) = \sum_{l=0}^{\infty} i^l (2l+1) \left\{ j_l(kr) - \dfrac{j_l'(ka)}{h_l'(ka)} h_l(kr) \right\} P_l(\cos \vartheta)$
$\tag{5.51}$

and

(ii) $\psi(\mathbf{r}) = \sum_{l=0}^{\infty} i^l (2l+1) \left\{ j_l(kr) - \dfrac{j_l(ka)}{h_l(ka)} h_l(kr) \right\} P_l(\cos \vartheta)$
$\tag{5.52}$

where the primes denote derivatives. Thus we have been able to obtain exact solutions to our problem by means of a direct analysis.

We now re-examine the same scattering problem using variational principles. For points on the surface of the sphere we may write

$$\psi(\mathbf{r}) = \sum_{l=0}^{\infty} C_l P_l(\cos \vartheta) \qquad (r = a) \tag{5.53}$$

SCATTERING THEORY

where the coefficients C_l form an infinite set of variable parameters. Then the adjoint function is given by

$$\tilde{\psi}(\mathbf{r}) = \sum_{l=0}^{\infty} C_l P_l[\cos(\pi - \Theta)] \qquad (r = a)$$

where Θ is the angle between $\mathbf{r}$ and $\mathbf{k}_s$. On using the formula (4.181), this may be rewritten as

$$\tilde{\psi}(\mathbf{r}) = \sum_{l=0}^{\infty} (-1)^l C_l \bigg\{ P_l(\cos\theta) P_l(\cos\vartheta) \\ + 2 \sum_{m=1}^{l} \frac{(l-m)!}{(l+m)!} P_l^m(\cos\theta) P_l^m(\cos\vartheta) \cos m(\phi - \varphi) \bigg\}, \tag{5.54}$$

θ and ϕ being the spherical polar angles of the wave vector $\mathbf{k}_s$.

Further, at the surface of the sphere

$$\mathbf{n} \cdot \mathbf{k}_i \, e^{i\mathbf{k}_i \cdot \mathbf{r}} = k \sum_{l=0}^{\infty} i^l (2l+1) j_l(ka) \cos\vartheta \, P_l(\cos\vartheta)$$

$$= k \sum_{l=0}^{\infty} i^l j_l(ka) \{l P_{l-1}(\cos\vartheta) + (l+1) P_{l+1}(\cos\vartheta)\}$$

since

$$(2l+1)\mu P_l(\mu) = l P_{l-1}(\mu) + (l+1) P_{l+1}(\mu),$$

while

$$\mathbf{n} \cdot \mathbf{k}_s \, e^{-i\mathbf{k}_s \cdot \mathbf{r}} = k \sum_{l=0}^{\infty} (-i)^l j_l(ka) \\ \times \bigg[l \bigg\{ P_{l-1}(\cos\theta) P_{l-1}(\cos\vartheta) \\ + 2 \sum_{m=1}^{l-1} \frac{(l-m-1)!}{(l+m-1)!} P_{l-1}^m(\cos\theta) \\ \times P_{l-1}^m(\cos\vartheta) \cos m(\phi - \varphi) \bigg\} \\ + (l+1) \bigg\{ P_{l+1}(\cos\theta) P_{l+1}(\cos\vartheta) \\ + 2 \sum_{m=1}^{l+1} \frac{(l-m+1)!}{(l+m+1)!} P_{l+1}^m(\cos\theta) \\ \times P_{l+1}^m(\cos\vartheta) \cos m(\phi - \varphi) \bigg\} \bigg].$$

Hence, making use of the orthonormality property of associated Legendre functions

$$\int_{-1}^{1} P_l^m(\mu) P_{l'}^m(\mu)\, d\mu = \frac{2}{2l+1} \frac{(l+m)!}{(l-m)!} \delta_{ll'},$$

we obtain

$$\oint_S \mathbf{n}\cdot\mathbf{k}_i \tilde{\psi}(\mathbf{r})\, e^{i\mathbf{k}_i\cdot\mathbf{r}}\, dS = -\oint_S \mathbf{n}\cdot\mathbf{k}_s\, e^{-i\mathbf{k}_s\cdot\mathbf{r}} \psi(\mathbf{r})\, dS$$

$$= 4\pi i k a^2 \sum_{l=0}^{\infty} \frac{(-i)^l C_l P_l(\cos\theta)}{2l+1}$$
$$\times \{(l+1)j_{l+1}(ka) - lj_{l-1}(ka)\}$$
$$= -4\pi i k a^2 \sum_{l=0}^{\infty} (-i)^l C_l j_l'(ka) P_l(\cos\theta) \quad (5.55)$$

since

$$(2l+1)j_l'(\rho) = lj_{l-1}(\rho) - (l+1)j_{l+1}(\rho).$$

Also it is known that

$$\frac{e^{ik|\mathbf{r}-\mathbf{r}'|}}{|\mathbf{r}-\mathbf{r}'|} = ik \sum_{l=0}^{\infty} (2l+1) j_l(kr_<) h_l(kr_>) P_l(\cos\Theta') \quad (5.56)$$

where

$$\cos\Theta' = \cos\vartheta\cos\vartheta' + \sin\vartheta\sin\vartheta'\cos(\varphi-\varphi'),$$

and $r_<$ and $r_>$ are the lesser and greater of r and r'. Hence

$$\oint\oint \tilde{\psi}(\mathbf{r}) \frac{\partial^2}{\partial r \partial r'}\left(\frac{e^{ik|\mathbf{r}-\mathbf{r}'|}}{|\mathbf{r}-\mathbf{r}'|}\right) \psi(\mathbf{r}')\, dS\, dS'$$

$$= 16\pi^2 i k^3 a^4 \sum_{l=0}^{\infty} \frac{(-1)^l C_l^2}{2l+1} j_l'(ka) h_l'(ka) P_l(\cos\theta) \quad (5.57)$$

and so, on applying the variational principle (5.33) in conjunction with formulae (5.55) and (5.57), we get

$$f(\theta) = -\frac{\left[\sum_{l=0}^{\infty} (-i)^l C_l j_l'(ka) P_l(\cos\theta)\right]^2}{ik \sum_{l=0}^{\infty} \frac{(-1)^l C_l^2}{2l+1} j_l'(ka) h_l'(ka) P_l(\cos\theta)} \quad (5.58)$$

SCATTERING THEORY   225

for the scattering amplitude in the case when $\partial \psi / \partial r$ vanishes over the surface of the sphere. Now using the stationary property of $f(\theta)$, we have $\partial f / \partial C_{l_0} = 0$ which yields

$$\frac{C_{l_0} h'_{l_0}(ka)}{2l_0 + 1} = \frac{i^{l_0} \sum_{l=0}^{\infty} \frac{(-1)^l C_l^2}{2l+1} j'_l(ka) h'_l(ka) P_l(\cos \theta)}{\sum_{l=0}^{\infty} (-i)^l C_l j'_l(ka) P_l(\cos \theta)}.$$

An inspection of this infinite set of equations shows that it is satisfied by the values

$$C_l = \frac{i^l(2l+1)}{h'_l(ka)} \tag{5.59}$$

giving for the scattering amplitude

$$f(\theta) = -\frac{1}{ik} \sum_{l=0}^{\infty} (2l+1) \frac{j'_l(ka)}{h'_l(ka)} P_l(\cos \theta). \tag{5.60}$$

This expression can also be seen to follow at once from the formula (5.51) for the function $\psi(\mathbf{r})$ derived previously.

The case for which ψ is chosen to vanish over the spherical surface $r = a$ can be treated likewise using the variational principle (5.37) and gives for the scattering amplitude

$$f(\theta) = -\frac{1}{ik} \sum_{l=0}^{\infty} (2l+1) \frac{j_l(ka)}{h_l(ka)} P_l(\cos \theta) \tag{5.61}$$

which can also be derived directly from formula (5.52) for $\psi(\mathbf{r})$.

By setting

$$j_l(ka) = B_l \sin \eta_l, \qquad h_l(ka) = iB_l e^{-i\eta_l} \tag{5.62}$$

we may rewrite (5.61) in the convenient form

$$f(\theta) = \frac{1}{k} \sum_{l=0}^{\infty} (2l+1) e^{i\eta_l} \sin \eta_l P_l(\cos \theta) \tag{5.63}$$

with an analogous expression replacing (5.60).

Since the function ψ is the amplitude of the wave motion we may represent the energy density of the motion by $\alpha |\psi|^2$ where α

is a constant. It then follows that the flux of energy scattered through an element of area dS perpendicular to the radial direction is given by

$$v\alpha |A|^2 |f(\theta)|^2 \, dS/r^2$$

where A is the amplitude of the incident plane wave and v is the phase velocity of the wave motion. Since the incident flux of energy through unit area perpendicular to the direction of propagation is $v\alpha |A|^2$, we see that the power scattered into the element of solid angle $d\omega = dS/r^2$ per unit incident flux is

$$|f(\theta)|^2 \, d\omega.$$

$|f(\theta)|^2$ is called the *differential cross section*. Integrating over all solid angles then yields the total scattered power per unit incident flux

$$Q = \int |f(\theta)|^2 \, d\omega \tag{5.64}$$

called the *total cross section*. Now using formula (5.63) and the orthonormality property of Legendre polynomials, we get the result

$$Q = \frac{4\pi}{k^2} \sum_{l=0}^{\infty} (2l+1) \sin^2 \eta_l \tag{5.65}$$

which may be rewritten as

$$Q = \frac{4\pi}{k^2} \sum_{l=0}^{\infty} (2l+1) \frac{j_l^2(ka)}{j_l^2(ka) + n_l^2(ka)}$$

on using (5.62) and (5.45). Since

$$j_l(\rho) \cong \frac{\rho^l}{1 \cdot 3 \cdot 5 \ldots (2l+1)}, \quad n_l(\rho) \cong -\frac{1 \cdot 1 \cdot 3 \cdot 5 \ldots (2l-1)}{\rho^{l+1}}$$

for small ρ, we see that as $ka \to 0$,

$$Q \to 4\pi a^2.$$

Examples of the above theory are provided by the scattering of sound waves and light waves by a spherical obstacle. In the case of the scattering of electromagnetic waves, the vector character of the wave motion introduces an additional complication which we have avoided here.

Another interesting application of variational principles which we shall not discuss here has been made by Levine and Schwinger[2] who examined the diffraction of scalar waves by an aperture in an infinite plane screen.

5.3 Scattering of particles in wave mechanics

We now come to our main task in this chapter which is the application of variational principles to the theory of the scattering of particles whose behaviour may be described by a wave function satisfying the Schrödinger equation of wave mechanics. Consider a beam of particles incident along the direction of the z axis of a frame of reference on a scattering centre situated at the origin. If N particles cross unit area perpendicular to the beam per second, the number of particles deflected per second through polar angles θ and ϕ into an element of solid angle $d\omega$ may be written as

$$NI(\theta, \phi)\,d\omega.$$

Since this quantity has the dimensions of inverse time, it follows that $I(\theta, \phi)$ has the dimensions of area. It is called the differential cross section. Integrating over all solid angles yields the total cross section

$$Q = \int I(\theta, \phi)\,d\omega \qquad (5.66)$$

which is just the total number of particles scattered per second from an incident beam of unit flux.

If all the particles of the incident beam have the same mass m and the same initial speed v, we can represent them by a plane wave

$$A\,e^{ikz} \qquad (5.67)$$

where the wave number k is given in terms of the de Broglie wave length λ by

$$k = mv/\hbar = 2\pi/\lambda. \qquad (5.68)$$

We have seen in section 3.12 on the Schrödinger equation that the current density $\mathbf{j}$ for a wave function ψ is given by

$$\mathbf{j} = \frac{\hbar}{2mi}(\bar{\psi}\nabla\psi - \psi\nabla\bar{\psi}), \qquad (5.69)$$

the electric charge e being omitted here. It follows that the flux of incident particles is $v|A|^2$ and so

$$A = \sqrt{\frac{N}{v}}.$$

For large values of the distance r from the origin, the scattered particles can be represented by an outgoing spherical wave

$$A\frac{e^{ikr}}{r}f(\theta, \phi) \qquad (5.70)$$

where the function $f(\theta, \phi)$ is called the scattering amplitude as in the preceding section. This produces a radial flux of magnitude

$$v|A|^2|f(\theta, \phi)|^2/r^2$$

and so

$$I(\theta, \phi) = |f(\theta, \phi)|^2. \qquad (5.71)$$

Thus we see that the formal relationships between the scattering amplitude, the differential cross section and the total cross section for the scattering of particles by a centre of force are identical to those for the scattering of waves at a surface examined in the previous section.

Combining the incident wave (5.67) with the scattered wave (5.70) yields

$$A\left\{e^{ikz} + \frac{1}{r}e^{ikr}f(\theta, \phi)\right\} \qquad (5.72)$$

for the asymptotic form of the wave function describing the scattering of the particles in the region of large r. If v particles cross unit area per second in the incident beam, the amplitude A is equal to unity. For convenience we shall suppose that this is the case in all the future discussions.

Integral equation for the scattering amplitude

In wave mechanics, the scattering of particles having energy $E = \hbar^2 k^2/2m$ by a potential $V(\mathbf{r})$ is determined by the Schrödinger equation

$$\left\{\nabla^2 + k^2 - \frac{2m}{\hbar^2} V(\mathbf{r})\right\}\psi(\mathbf{r}) = 0 \tag{5.73}$$

where for large values of r

$$\psi(\mathbf{r}) \sim e^{i\mathbf{k}_i \cdot \mathbf{r}} + \frac{1}{r} e^{ikr} f(\theta, \phi), \tag{5.74}$$

$\mathbf{k}_i$ being a vector of magnitude k pointing in the direction of the incident beam of particles.

Now a particular solution of the equation

$$(\nabla^2 + k^2)\psi(\mathbf{r}) = F(\mathbf{r}), \tag{5.75}$$

where F is a given function of $\mathbf{r}$, can be expressed in the form

$$\psi(\mathbf{r}) = \int G(\mathbf{r}, \mathbf{r}') F(\mathbf{r}') \, d\mathbf{r}' \tag{5.76}$$

where the Green's function $G(\mathbf{r}, \mathbf{r}')$ satisfies equation (5.21) and, as in section 5.2, takes the form

$$G(\mathbf{r}, \mathbf{r}') = -\frac{1}{4\pi} \frac{e^{ik|\mathbf{r}-\mathbf{r}'|}}{|\mathbf{r} - \mathbf{r}'|}$$

if $\psi(\mathbf{r})$ is to behave as an outgoing spherical wave for large values of r. Hence, using this Green's function, we may write the solution of the wave equation (5.73) having asymptotic behaviour (5.74) in the form

$$\psi(\mathbf{r}) = e^{i\mathbf{k}_i \cdot \mathbf{r}} - \frac{1}{4\pi} \int \frac{e^{ik|\mathbf{r}-\mathbf{r}'|}}{|\mathbf{r} - \mathbf{r}'|} U(\mathbf{r}')\psi(\mathbf{r}') \, d\mathbf{r}' \tag{5.77}$$

where

$$U(\mathbf{r}) = \frac{2m}{\hbar^2} V(\mathbf{r}). \tag{5.78}$$

If $\mathbf{k}_s$ is a vector of magnitude k pointing in the direction of scattering given by $\mathbf{r}$, then
$$k|\mathbf{r} - \mathbf{r}'| \sim kr - \mathbf{k}_s \cdot \mathbf{r}'$$
for large r, and so we see that
$$f(\theta, \phi) = -\frac{1}{4\pi} \int e^{-i\mathbf{k}_s \cdot \mathbf{r}'} U(\mathbf{r}')\psi(\mathbf{r}') \, d\mathbf{r}' \tag{5.79}$$
which provides an integral equation for the scattering amplitude.

Born expansion

A useful approach to the problem of evaluating the scattering amplitude is to regard the potential $V(\mathbf{r})$ as a small perturbation acting on the incident plane wave. To carry this through we replace $V(\mathbf{r})$ by $\eta V(\mathbf{r})$ as in the case of stationary state perturbation theory, and then expand the wave function and the scattering amplitude in the form
$$\psi(\mathbf{r}) = \sum_{n=0}^{\infty} \eta^n \psi_n(\mathbf{r}) \tag{5.80}$$
and
$$f(\theta, \phi) = \sum_{n=1}^{\infty} \eta^n f_{Bn}(\theta, \phi) \tag{5.81}$$
respectively. Now equating the coefficients of equal powers of η on both sides of (5.77) and (5.79) gives
$$\psi_n(\mathbf{r}) = -\frac{1}{4\pi} \int \frac{e^{ik|\mathbf{r}-\mathbf{r}'|}}{|\mathbf{r} - \mathbf{r}'|} U(\mathbf{r}')\psi_{n-1}(\mathbf{r}') \, d\mathbf{r}' \qquad (n \neq 0) \tag{5.82}$$
and
$$f_{Bn}(\theta, \phi) = -\frac{1}{4\pi} \int e^{-i\mathbf{k}_s \cdot \mathbf{r}'} U(\mathbf{r}')\psi_{n-1}(\mathbf{r}') \, d\mathbf{r}'. \tag{5.83}$$
We now revert to the original problem by allowing the parameter η to approach unity. Noting that the unperturbed wave function ψ_0 is the incident plane wave
$$\psi_0(\mathbf{r}) = e^{i\mathbf{k}_i \cdot \mathbf{r}}, \tag{5.84}$$

SCATTERING THEORY

we see that
$$f_{B1}(\theta, \phi) = -\frac{1}{4\pi} \int e^{i(\mathbf{k}_i - \mathbf{k}_s)\cdot\mathbf{r}'} U(\mathbf{r}') \, d\mathbf{r}' \qquad (5.85)$$
and
$$\psi_1(\mathbf{r}) = -\frac{1}{4\pi} \int \frac{e^{ik|\mathbf{r}-\mathbf{r}'|}}{|\mathbf{r}-\mathbf{r}'|} U(\mathbf{r}') \, e^{i\mathbf{k}_i \cdot \mathbf{r}'} \, d\mathbf{r}'$$
where f_{B1} is the *first Born approximation* to the scattering amplitude. Hence
$$f_{B2}(\theta, \phi) = \left(\frac{1}{4\pi}\right)^2 \int\int e^{-i\mathbf{k}_s \cdot \mathbf{r}_1} U(\mathbf{r}_1) \frac{e^{ik|\mathbf{r}_1-\mathbf{r}_2|}}{|\mathbf{r}_1-\mathbf{r}_2|} U(\mathbf{r}_2) \, e^{i\mathbf{k}_i \cdot \mathbf{r}_2} \, d\mathbf{r}_1 \, d\mathbf{r}_2, \qquad (5.86)$$
the second Born approximation being given by
$$f_B^{(2)} = f_{B1} + f_{B2}. \qquad (5.87)$$
In general, we see that the $(p + 1)$th Born approximation to the scattering amplitude
$$f_B^{(p+1)}(\theta, \phi) = \sum_{n=1}^{p+1} f_{Bn}(\theta, \phi) \qquad (5.88)$$
may be expressed in terms of the wave function to the pth order
$$\psi^{(p)}(\mathbf{r}) = \sum_{n=0}^{p} \psi_n(\mathbf{r}) \qquad (5.89)$$
by means of the formula
$$f_B^{(p+1)}(\theta, \phi) = -\frac{1}{4\pi} \int e^{-i\mathbf{k}_s \cdot \mathbf{r}'} U(\mathbf{r}') \psi^{(p)}(\mathbf{r}') \, d\mathbf{r}'. \qquad (5.90)$$

5.4 Variational principles for the scattering amplitude

There are two well-known forms of variational principle for the scattering amplitude. They are due to Schwinger and to Kohn and we shall examine them in turn.

Schwinger's variational principle

To establish the variational principle due to Schwinger we introduce the adjoint function $\tilde{\psi}$ satisfying the integral equation
$$\tilde{\psi}(\mathbf{r}) = e^{-i\mathbf{k}_s \cdot \mathbf{r}} - \frac{1}{4\pi} \int \frac{e^{ik|\mathbf{r}-\mathbf{r}'|}}{|\mathbf{r}-\mathbf{r}'|} U(\mathbf{r}') \tilde{\psi}(\mathbf{r}') \, d\mathbf{r}' \qquad (5.91)$$

which is the solution of the Schrödinger equation (5.73) associated with a plane wave incident along the direction of $-\mathbf{k}_s$. Multiplying both sides of (5.77) by $\tilde{\psi}(\mathbf{r})U(\mathbf{r})$ and integrating over $\mathbf{r}$ space, we obtain

$$\int \tilde{\psi}(\mathbf{r})U(\mathbf{r})\psi(\mathbf{r})\,d\mathbf{r} = \int \tilde{\psi}(\mathbf{r})U(\mathbf{r})\,e^{i\mathbf{k}_i\cdot\mathbf{r}}\,d\mathbf{r}$$
$$-\frac{1}{4\pi}\int\int \tilde{\psi}(\mathbf{r})U(\mathbf{r})\frac{e^{ik|\mathbf{r}-\mathbf{r}'|}}{|\mathbf{r}-\mathbf{r}'|}U(\mathbf{r}')\psi(\mathbf{r}')\,d\mathbf{r}\,d\mathbf{r}'$$

and so the scattering amplitude

$$f(\mathbf{k}_s, \mathbf{k}_i) = -\frac{1}{4\pi}\int e^{-i\mathbf{k}_s\cdot\mathbf{r}'}U(\mathbf{r}')\psi(\mathbf{r}')\,d\mathbf{r}' \qquad (5.92)$$

may be rewritten in the form

$$f(\mathbf{k}_s, \mathbf{k}_i) =$$
$$-\frac{\dfrac{1}{4\pi}\displaystyle\int \tilde{\psi}(\mathbf{r})U(\mathbf{r})\,e^{i\mathbf{k}_i\cdot\mathbf{r}}\,d\mathbf{r}\int e^{-i\mathbf{k}_s\cdot\mathbf{r}'}U(\mathbf{r}')\psi(\mathbf{r}')\,d\mathbf{r}'}{\displaystyle\int \tilde{\psi}(\mathbf{r})U(\mathbf{r})\psi(\mathbf{r})\,d\mathbf{r} + \frac{1}{4\pi}\int\int \tilde{\psi}(\mathbf{r})U(\mathbf{r})\frac{e^{ik|\mathbf{r}-\mathbf{r}'|}}{|\mathbf{r}-\mathbf{r}'|}U(\mathbf{r}')\psi(\mathbf{r}')\,d\mathbf{r}\,d\mathbf{r}'}$$
(5.93)

which is homogeneous in the functions ψ and $\tilde{\psi}$.

It can be readily verified that the right-hand side of (5.93) is stationary with respect to infinitesimal variations in ψ and $\tilde{\psi}$ provided they satisfy the integral equations (5.77) and (5.91) respectively. Moreover the validity of the reciprocity theorem

$$f(\mathbf{k}_s, \mathbf{k}_i) = f(-\mathbf{k}_i, -\mathbf{k}_s) \qquad (5.94)$$

is evident.

Once again our expression for the scattering amplitude has the form originated by Schwinger and its stationary property is known as the Schwinger variational principle.

Inserting the plane wave trial functions

$$\psi_T(\mathbf{r}) = e^{i\mathbf{k}_i\cdot\mathbf{r}}, \qquad \tilde{\psi}_T(\mathbf{r}) = e^{-i\mathbf{k}_s\cdot\mathbf{r}} \qquad (5.95)$$

into the right-hand side of (5.93) yields the approximate expression for the scattering amplitude

$$f(\mathbf{k}_s, \mathbf{k}_i) = \frac{\{f_{B1}(\mathbf{k}_s, \mathbf{k}_i)\}^2}{f_{B1}(\mathbf{k}_s, \mathbf{k}_i) - f_{B2}(\mathbf{k}_s, \mathbf{k}_i)}, \qquad (5.96)$$

SCATTERING THEORY

where f_{B1} is the first Born approximation to the scattering amplitude given by (5.85) and f_{B2} is the second Born approximation correction term (5.86). To the second order in the perturbation this reduces to

$$f(\mathbf{k}_s, \mathbf{k}_i) = f_{B1}(\mathbf{k}_s, \mathbf{k}_i) + f_{B2}(\mathbf{k}_s, \mathbf{k}_i)$$

which is just the second Born approximation to the scattering amplitude obtained previously.

Screened Coulomb potential. As a simple example of the application of the Schwinger variational method for the determination of the scattering amplitude, we consider the screened Coulomb potential

$$V(r) = -\frac{Be^{-\lambda r}}{r} \qquad (5.97)$$

which was proposed as a possible nucleon–nucleon interaction potential by Yukawa. We shall take the calculation to the second order in B by employing the Schwinger variational formula (5.93) together with plane wave trial functions, which necessitates the evaluation of f_{B1} and f_{B2} given by (5.85) and (5.86) respectively.

For the first Born approximation to the scattering amplitude we have

$$f_{B1} = \frac{\beta^2}{4\pi} \int e^{i\mathbf{K}\cdot\mathbf{r}} \frac{e^{-\lambda r}}{r} d\mathbf{r} \qquad (5.98)$$

where $\mathbf{K} = \mathbf{k}_i - \mathbf{k}_s$ and $\beta^2 = 2mB/\hbar^2$. Choosing the polar axis in the direction of $\mathbf{K}$ and carrying out the integrations over the polar angles ϑ, φ we get

$$f_{B1} = \frac{\beta^2}{K} \int_0^\infty \sin Kr \, e^{-\lambda r} \, dr = \frac{\beta^2}{K^2 + \lambda^2} \qquad (5.99)$$

where $K = 2k \sin \theta/2$ and θ is the scattering angle. As the range $1/\lambda$ becomes large so that $\lambda \to 0$, the screened Coulomb potential approaches the Coulombic form $-B/r$ and the differential cross section becomes

$$|f_{B1}(\theta)|^2 = \frac{1}{4}\left(\frac{B}{mv^2}\right)^2 \operatorname{cosec}^4 \theta/2 \qquad (5.100)$$

which is identical to the classical Rutherford scattering formula as well as to the result obtained with an exact wave mechanical treatment.

In order to evaluate f_{B2} we express $e^{ik|\mathbf{r}_1 - \mathbf{r}_2|}/4\pi|\mathbf{r}_1 - \mathbf{r}_2|$ in the form of the Fourier integral

$$\lim_{\epsilon \to 0} \frac{1}{(2\pi)^3} \int \frac{e^{i\mathbf{q} \cdot (\mathbf{r}_1 - \mathbf{r}_2)}}{q^2 - k^2 - i\epsilon} d\mathbf{q} \tag{5.101}$$

which enables us to separate the variables $\mathbf{r}_1$ and $\mathbf{r}_2$ and thus to write

$$\begin{aligned} f_{B2} &= \lim_{\epsilon \to 0} \frac{\beta^4}{32\pi^4} \int \frac{d\mathbf{q}}{q^2 - k^2 - i\epsilon} \int e^{i(\mathbf{q} - \mathbf{k}_s) \cdot \mathbf{r}_1} \frac{e^{-\lambda r_1}}{r_1} d\mathbf{r}_1 \\ &\quad \times \int e^{i(\mathbf{k}_i - \mathbf{q}) \cdot \mathbf{r}_2} \frac{e^{-\lambda r_2}}{r_2} d\mathbf{r}_2 \\ &= \lim_{\epsilon \to 0} \frac{\beta^4}{2\pi^2} \int \frac{d\mathbf{q}}{q^2 - k^2 - i\epsilon} \frac{1}{\{(\mathbf{q} - \mathbf{k}_s)^2 + \lambda^2\}\{(\mathbf{q} - \mathbf{k}_i)^2 + \lambda^2\}}. \end{aligned} \tag{5.102}$$

Now using the integral relation derived by Dalitz[3]

$$\lim_{\epsilon \to 0} \int \frac{d\mathbf{q}}{q^2 - p^2 - i\epsilon} \frac{1}{[(\mathbf{q} - \mathbf{P})^2 + \Lambda^2]^2} = \frac{\pi^2}{\Lambda(P^2 + \Lambda^2 - p^2 - 2p\Lambda i)} \tag{5.103}$$

and the Feynman identity

$$\frac{1}{ab} = \frac{1}{2} \int_{-1}^{1} \frac{dZ}{[\frac{1}{2}a(1 + Z) + \frac{1}{2}b(1 - Z)]^2} \tag{5.104}$$

with

$$a = (\mathbf{q} - \mathbf{k}_s)^2 + \lambda^2, \qquad b = (\mathbf{q} - \mathbf{k}_i)^2 + \lambda^2$$

and

$$\tfrac{1}{2}a(1 + Z) + \tfrac{1}{2}b(1 - Z) = (\mathbf{q} - \mathbf{P})^2 + \Lambda^2$$

we get, on performing the elementary integrations over Z,

$$\begin{aligned} f_{B2}(\theta) &= \frac{\beta^4}{2kA \sin \theta/2} \\ &\quad \times \left[\tan^{-1}\left(\frac{\lambda k \sin \theta/2}{A}\right) + \frac{i}{2} \ln\left(\frac{A + 2k^2 \sin \theta/2}{A - 2k^2 \sin \theta/2}\right) \right] \end{aligned} \tag{5.105}$$

where

$$A^2 = \lambda^4 + 4\lambda^2 k^2 + 4k^4 \sin^2 \theta/2. \tag{5.106}$$

As $\lambda \to 0$ we see that the real part of f_{B2} vanishes and so there is a zero contribution to the differential cross section for a Coulomb field from terms of order B^3. Since the differential cross section (5.100) given by the first Born approximation is identical to the exact result, all terms of order higher than the second must vanish. The term of next highest order in the differential cross section is the B^4 term which arises not only from $|f_{B2}|^2$ but also from the product of f_{B1} and the real part of f_{B3}, the third Born approximation contribution to the scattering amplitude. It has been verified by Kacser[4] that these two contributions to the term of order B^4 do indeed exactly cancel each other for the case of the Coulomb field as they should from the remark made earlier. Evidently the blind application of the second Born approximation scattering amplitude to the calculation of the differential cross section leads to false results unless the Born expansion is terminated at the appropriate order in the perturbation. The same cautionary remark applies to the use of the Schwinger variational principle. Thus although the scattering amplitude given by (5.96) used in conjunction with (5.99) and (5.105) leads to terms of all orders in B, only those up to the third order should be retained in the differential cross section $I(\theta)$. Consequently, both the second Born approximation and the Schwinger variational method employing plane wave trial functions, yield finally

$$I(\theta) = f_{B1}(\theta)[f_{B1}(\theta) + 2\mathscr{R}\{f_{B2}(\theta)\}] \qquad (5.107)$$

where the real part of f_{B2} is denoted by $\mathscr{R}\{f_{B2}\}$.

Kohn's variational principle

An alternative variational principle for the scattering amplitude due to Kohn[5] can be derived by considering the functional

$$I(-\mathbf{k}_2, \mathbf{k}_1) = \int \chi_2(\mathbf{r})\{\nabla^2 + k^2 - U(\mathbf{r})\}\chi_1(\mathbf{r}) \, d\mathbf{r} \qquad (5.108)$$

where

$$\chi_j(\mathbf{r}) \sim e^{i\mathbf{k}_j \cdot \mathbf{r}} + \frac{1}{r} e^{ikr} f_j(\mathbf{k}, \mathbf{k}_j) \qquad (j = 1, 2) \qquad (5.109)$$

for large r, $\mathbf{k}$ being a vector in the direction of $\mathbf{r}$ and $|\mathbf{k}| = |\mathbf{k}_j| = k$. We see at once that I vanishes if $\chi_1(\mathbf{r})$ is an exact solution of the Schrödinger equation (5.73).

Now let $\delta I(-\mathbf{k}_2, \mathbf{k}_1)$ be the change in $I(-\mathbf{k}_2, \mathbf{k}_1)$ arising from infinitesimal variations $\delta\chi_j$ in the χ_j such that for large r

$$\delta\chi_1 \sim \frac{1}{r} e^{ikr} \delta f_1(\mathbf{k}, \mathbf{k}_1). \tag{5.110}$$

Then to the first order of small quantities we have

$$\begin{aligned}\delta I(-\mathbf{k}_2, \mathbf{k}_1) &= \int \delta\chi_2(\mathbf{r})\{\nabla^2 + k^2 - U(\mathbf{r})\}\chi_1(\mathbf{r})\, d\mathbf{r} \\ &\quad + \int \chi_2(\mathbf{r})\{\nabla^2 + k^2 - U(\mathbf{r})\}\, \delta\chi_1(\mathbf{r})\, d\mathbf{r} \\ &= \int \delta\chi_2(\mathbf{r})\{\nabla^2 + k^2 - U(\mathbf{r})\}\chi_1(\mathbf{r})\, d\mathbf{r} \\ &\quad + \int \delta\chi_1(\mathbf{r})\{\nabla^2 + k^2 - U(\mathbf{r})\}\chi_2(\mathbf{r})\, d\mathbf{r} \\ &\quad + \oint_S \left\{\chi_2 \frac{\partial}{\partial n}\delta\chi_1 - \delta\chi_1 \frac{\partial}{\partial n}\chi_2\right\} dS \end{aligned} \tag{5.111}$$

on using Green's theorem, S being the boundary surface of the volume of integration and $\partial/\partial n$ denoting the normal derivative. Hence

$$\delta I(-\mathbf{k}_2, \mathbf{k}_1) = \oint_S \left\{\psi_2 \frac{\partial}{\partial n}\delta\psi_1 - \delta\psi_1 \frac{\partial}{\partial n}\psi_2\right\} dS \tag{5.112}$$

if we take χ_1 and χ_2 to be the exact solutions ψ_1 and ψ_2 of the Schrödinger equation having the asymptotic behaviour for large r given by

$$\psi_j(\mathbf{r}) \sim e^{i\mathbf{k}_j \cdot \mathbf{r}} + \frac{1}{r} e^{ikr} f(\mathbf{k}, \mathbf{k}_j) \qquad (j = 1, 2). \tag{5.113}$$

Choosing S to be a large spherical surface of radius r and writing

$$\delta\psi_1 \sim \frac{1}{r} e^{ikr} \delta f_1(\mathbf{k}, \mathbf{k}_1), \tag{5.114}$$

SCATTERING THEORY 237

enables us to express equation (5.112) in the form
$$\delta I(-\mathbf{k}_2, \mathbf{k}_1) = \oint_S \left\{ e^{i\mathbf{k}_2 \cdot \mathbf{r}} \frac{\partial}{\partial r} \left(\frac{e^{ikr}}{r} \right) - \frac{e^{ikr}}{r} \frac{\partial}{\partial r} (e^{i\mathbf{k}_2 \cdot \mathbf{r}}) \right\} \delta f_1(\mathbf{k}, \mathbf{k}_1) \, dS. \tag{5.115}$$

Taking the polar axis of a spherical polar system of coordinates r, ϑ, φ in the direction of $-\mathbf{k}_2$, we find that

$$e^{i\mathbf{k}_2 \cdot \mathbf{r}} \frac{\partial}{\partial r} \left(\frac{e^{ikr}}{r} \right) - \frac{e^{ikr}}{r} \frac{\partial}{\partial r} (e^{i\mathbf{k}_2 \cdot \mathbf{r}}) = \left(\frac{ik}{r} - \frac{1}{r^2} \right) e^{ikr(1-\cos\vartheta)}$$
$$+ \frac{ik \cos \vartheta}{r} e^{ikr(1-\cos\vartheta)}$$
$$\simeq \frac{ik}{r} (1 + \cos \vartheta) e^{ikr(1-\cos\vartheta)}$$

if terms of order r^{-2} are neglected, and hence

$$\delta I(-\mathbf{k}_2, \mathbf{k}_1) = ikr \int_0^{2\pi} d\varphi \int_0^{\pi} (1 + \cos \vartheta) e^{ikr(1-\cos\vartheta)}$$
$$\times \delta f_1(\mathbf{k}, \mathbf{k}_1) \sin \vartheta \, d\vartheta$$
$$= 2\pi \left\{ [(1 + \cos \vartheta) e^{ikr(1-\cos\vartheta)} \delta f_1(\mathbf{k}, \mathbf{k}_1)]_0^{\pi} \right.$$
$$\left. - \int_0^{\pi} e^{ikr(1-\cos\vartheta)} \frac{\partial}{\partial \vartheta} [(1 + \cos \vartheta) \delta f_1(\mathbf{k}, \mathbf{k}_1)] \, d\vartheta \right\}$$

on performing an integration by parts. Since the second term is of order r^{-1}, it follows that

$$\delta I(-\mathbf{k}_2, \mathbf{k}_1) = -4\pi \, \delta f_1(-\mathbf{k}_2, \mathbf{k}_1) \tag{5.116}$$

which is *Kohn's variational principle*.

Let us now insert trial functions $\psi_{jT}(\mathbf{r})$ into (5.108) where for large r

$$\psi_{1T}(\mathbf{r}) \sim e^{i\mathbf{k}_1 \cdot \mathbf{r}} + \frac{1}{r} e^{ikr} f_{1T}(\mathbf{k}, \mathbf{k}_1), \tag{5.117}$$

and denote the resulting expression for $I(-\mathbf{k}_2, \mathbf{k}_1)$ by $I_T(-\mathbf{k}_2, \mathbf{k}_1)$. Then it follows from the variational principle (5.116) that

$$f(-\mathbf{k}_2, \mathbf{k}_1) = f_{1T}(-\mathbf{k}_2, \mathbf{k}_1) + \frac{1}{4\pi} I_T(-\mathbf{k}_2, \mathbf{k}_1) \tag{5.118}$$

to the first order of small quantities.

The trial functions
$$\psi_{1T}(\mathbf{r}) = e^{i\mathbf{k}_i \cdot \mathbf{r}}, \qquad \psi_{2T}(\mathbf{r}) = e^{-i\mathbf{k}_s \cdot \mathbf{r}} \qquad (5.119)$$
yield
$$f(\mathbf{k}_s, \mathbf{k}_i) = -\frac{1}{4\pi} \int e^{-i\mathbf{k}_s \cdot \mathbf{r}} U(\mathbf{r}) e^{i\mathbf{k}_i \cdot \mathbf{r}} d\mathbf{r} \qquad (5.120)$$

which is just the first Born approximation to the scattering amplitude. The second Born approximation may be obtained by taking ψ_{1T} as in (5.119) but

$$\psi_{2T}(\mathbf{r}) = e^{-i\mathbf{k}_s \cdot \mathbf{r}} - \frac{1}{4\pi} \int \frac{e^{ik|\mathbf{r}-\mathbf{r}'|}}{|\mathbf{r}-\mathbf{r}'|} U(\mathbf{r}') e^{-i\mathbf{k}_s \cdot \mathbf{r}'} d\mathbf{r}'. \qquad (5.121)$$

More generally if we take

$$\psi_{1T}(\mathbf{r}) = \sum_{n=0}^{p} \psi_n(\mathbf{r}) = \psi^{(p)}(\mathbf{r}) \qquad (5.122)$$

and

$$\psi_{2T}(\mathbf{r}) = \sum_{n=0}^{q} \tilde{\psi}_n(\mathbf{r}) = \tilde{\psi}^{(q)}(\mathbf{r}) \qquad (5.123)$$

where $\psi^{(p)}(\mathbf{r})$ is the pth order Born approximation to the solution of the Schrödinger equation (5.73) satisfying the asymptotic condition (5.74) for large r and $\tilde{\psi}^{(q)}(\mathbf{r})$ is the adjoint function of $\psi^{(q)}(\mathbf{r})$, we see that

$$\begin{aligned} f_{1T}(\mathbf{k}_s, \mathbf{k}_i) &= -\frac{1}{4\pi} \int e^{-i\mathbf{k}_s \cdot \mathbf{r}} U(\mathbf{r}) \psi^{(p-1)}(\mathbf{r}) \, d\mathbf{r} \\ &= f_B^{(p)}(\mathbf{k}_s, \mathbf{k}_i) \end{aligned} \qquad (5.124)$$

where $f_B^{(p)}(\mathbf{k}_s, \mathbf{k}_i)$ is the pth Born approximation to the scattering amplitude and

$$\begin{aligned} I_T(\mathbf{k}_s, \mathbf{k}_i) &= \int \tilde{\psi}^{(q)}(\mathbf{r}) \{\nabla^2 + k^2 - U(\mathbf{r})\} \psi^{(p)}(\mathbf{r}) \, d\mathbf{r} \\ &= -\int \tilde{\psi}^{(q)}(\mathbf{r}) U(\mathbf{r}) \psi_p(\mathbf{r}) \, d\mathbf{r}. \end{aligned} \qquad (5.125)$$

Hence (5.118) gives
$$f(\mathbf{k}_s, \mathbf{k}_i) = f_B^{(p)}(\mathbf{k}_s, \mathbf{k}_i) - \frac{1}{4\pi} \int \tilde{\psi}^{(q)}(\mathbf{r}) U(\mathbf{r}) \psi_p(\mathbf{r}) \, d\mathbf{r} \qquad (5.126)$$

SCATTERING THEORY

and since the quantity which has been neglected

$$\int \delta\psi_2(\nabla^2 + k^2 - U)\,\delta\psi_1\,d\mathbf{r}$$

is of order $p + q + 2$, this means that expression (5.126) for the scattering amplitude is correct to order $p + q + 1$. Thus

$$f_B^{(p+q+1)}(\mathbf{k}_s, \mathbf{k}_i) = f_B^{(p)}(\mathbf{k}_s, \mathbf{k}_i) - \frac{1}{4\pi}\int \tilde{\psi}^{(q)}(\mathbf{r})U(\mathbf{r})\psi_p(\mathbf{r})\,d\mathbf{r} \quad (5.127)$$

so that if the solution of the Schrödinger equation (5.73) is known to order p, the scattering amplitude can be found to order $2p + 1$. An analogous result was derived in section 4.15 for stationary state problems.

Reformulation of Kohn's variational principle. Another stationary expression for the scattering amplitude can be obtained by introducing functions $\phi_j(\mathbf{r})$ given by

$$\psi_{jT}(\mathbf{r}) = e^{i\mathbf{k}_j \cdot \mathbf{r}} + \frac{1}{r} e^{ikr}\phi_j(\mathbf{r}) \qquad (j = 1, 2) \quad (5.128)$$

where for large r

$$\phi_j(\mathbf{r}) \sim f_{jT}(\mathbf{k}, \mathbf{k}_j) \qquad (j = 1, 2) \quad (5.129)$$

and then substituting formula (5.128) into (5.118). The resulting expression can be simplified in the following manner. Using Green's theorem we have

$$\int e^{i\mathbf{k}_2 \cdot \mathbf{r}}(\nabla^2 + k^2)\left(\frac{e^{ikr}}{r}\phi_1\right)d\mathbf{r}$$

$$= \oint_S \left\{ e^{i\mathbf{k}_2 \cdot \mathbf{r}} \frac{\partial}{\partial r}\left(\frac{e^{ikr}}{r}\right) - \frac{e^{ikr}}{r} \frac{\partial}{\partial r}(e^{i\mathbf{k}_2 \cdot \mathbf{r}}) \right\} f_{1T}(\mathbf{k}, \mathbf{k}_1)\,dS$$

where the surface integral over the large spherical surface S can be evaluated by employing the same technique used to derive (5.116) and yields $-4\pi f_{1T}(-\mathbf{k}_2, \mathbf{k}_1)$. Also, noting that

$$(\nabla^2 + k^2)\frac{e^{ikr}}{r} = -4\pi\,\delta(\mathbf{r})$$

and using Green's theorem again, we find that

$$\int \frac{e^{ikr}}{r} \phi_2 (\nabla^2 + k^2) \left(\frac{e^{ikr}}{r} \phi_1 \right) d\mathbf{r}$$
$$= -4\pi \int \frac{e^{ikr}}{r} \phi_2 \phi_1 \, \delta(\mathbf{r}) \, d\mathbf{r} - \int \left(\frac{e^{ikr}}{r} \right)^2 \nabla \phi_2 \cdot \nabla \phi_1 \, d\mathbf{r}$$
$$+ \oint_S \left(\frac{e^{ikr}}{r} \right)^2 \phi_2 \frac{\partial \phi_1}{\partial r} \, dS.$$

Since ϕ_1 is zero and $(e^{ikr}/r)\phi_2$ is finite at the origin, the first integral on the right-hand side of this equation vanishes. Further, the last integral also vanishes since $\partial \phi_1/\partial r$ tends to zero for large r, and so only the second integral is non-vanishing. Collecting all the various terms together, it follows therefore that

$$I_T(-\mathbf{k}_2, \mathbf{k}_1) + 4\pi f_{1T}(-\mathbf{k}_2, \mathbf{k}_1)$$
$$= -\int \left(e^{i\mathbf{k}_2 \cdot \mathbf{r}} + \frac{e^{ikr}}{r} \phi_2 \right) \left(e^{i\mathbf{k}_1 \cdot \mathbf{r}} + \frac{e^{ikr}}{r} \phi_1 \right) U(\mathbf{r}) \, d\mathbf{r}$$
$$- \int \left(\frac{e^{ikr}}{r} \right)^2 \nabla \phi_2 \cdot \nabla \phi_1 \, d\mathbf{r} \qquad (5.130)$$

and hence, setting $\mathbf{k}_1 = \mathbf{k}_i$ and $\mathbf{k}_2 = -\mathbf{k}_s$, we obtain as our variational approximation to the scattering amplitude

$$f(\mathbf{k}_s, \mathbf{k}_i) = -\frac{1}{4\pi} \int \left(e^{-i\mathbf{k}_s \cdot \mathbf{r}} + \frac{e^{ikr}}{r} \phi_s \right) \left(e^{i\mathbf{k}_i \cdot \mathbf{r}} + \frac{e^{ikr}}{r} \phi_i \right) U(\mathbf{r}) \, d\mathbf{r}$$
$$- \frac{1}{4\pi} \int \left(\frac{e^{ikr}}{r} \right)^2 \nabla \phi_s \cdot \nabla \phi_i \, d\mathbf{r}, \qquad (5.131)$$

where for large r

$$\phi_i(\mathbf{r}) \sim f_{iT}(\mathbf{k}, \mathbf{k}_i), \qquad \phi_s(\mathbf{r}) \sim f_{sT}(\mathbf{k}, -\mathbf{k}_s). \qquad (5.132)$$

The form of the above expression for the scattering amplitude is a generalization of that derived by Tamm for the tangent of the scattering phase shift which we will discuss in a later section (p. 248 ff.).

5.5 Scattering phase shifts

We now make the assumption that the potential $V(\mathbf{r})$ is spherically symmetrical. In this case the wave function must have cylindrical symmetry about the direction of motion of the incident particles, chosen as the polar axis of a spherical polar system of coordinates r, θ, ϕ, and so we may expand $\psi(\mathbf{r})$ in Legendre polynomials according to the formula

$$\psi(\mathbf{r}) = \sum_{l=0}^{\infty} A_l R_l(r) P_l(\cos \theta) \qquad (5.133)$$

which is commonly referred to as an *expansion in partial waves*. Substituting into the Schrödinger equation (5.73) this gives

$$\frac{1}{r^2} \frac{d}{dr}\left(r^2 \frac{dR_l}{dr}\right) + \left\{k^2 - U(r) - \frac{l(l+1)}{r^2}\right\} R_l = 0 \qquad (5.134)$$

which is an equation of the Sturm–Liouville type (3.151). On setting

$$R_l(r) = \frac{u_l(r)}{r} \qquad (5.135)$$

we obtain the radial equation

$$\frac{d^2 u_l}{dr^2} + \left\{k^2 - U(r) - \frac{l(l+1)}{r^2}\right\} u_l = 0 \qquad (5.136)$$

where $u_l(r)$ must vanish at the origin $r = 0$ since $R_l(r)$ is necessarily finite at all points.

If we now suppose that $V(r)$ falls off more rapidly than r^{-2} as $r \to \infty$, we find for large r

$$u_l(r) \sim r\{a_l j_l(kr) + b_l n_l(kr)\}, \qquad (5.137)$$

where j_l and n_l are the spherical Bessel and Neumann functions given by formulae (5.42) and (5.43) respectively and having the asymptotic forms given by (5.44). Hence we may choose a_l and b_l so that

$$u_l(r) \sim k^{-1} \sin(kr - \tfrac{1}{2}l\pi + \eta_l) \qquad (5.138)$$

as $r \to \infty$, where

$$\tan \eta_l = -b_l/a_l. \qquad (5.139)$$

η_l is known as the scattering *phase shift* for the lth partial wave. It vanishes for the case of a null potential.

We have already noted on p. 222 that a plane wave may be expanded in the form

$$e^{ikz} = \sum_{l=0}^{\infty} i^l(2l+1) j_l(kr) P_l(\cos\theta)$$

and so, putting

$$f(\theta) = \sum_{l=0}^{\infty} C_l P_l(\cos\theta), \qquad (5.140)$$

we see that for large r

$$A_l u_l(r) \sim i^l(2l+1) k^{-1} \sin(kr - \tfrac{1}{2}l\pi) + C_l e^{ikr}. \qquad (5.141)$$

Equating the coefficients of e^{ikr} and of e^{-ikr} on the two sides of (5.141) yields

$$A_l = i^l(2l+1) e^{i\eta_l} \qquad (5.142)$$

and

$$C_l = \frac{1}{2ik}(2l+1)(e^{2i\eta_l} - 1). \qquad (5.143)$$

Hence the scattering amplitude can be expressed in the form

$$f(\theta) = \frac{1}{k} \sum_{l=0}^{\infty} (2l+1) e^{i\eta_l} \sin\eta_l \, P_l(\cos\theta) \qquad (5.144)$$

and so the total cross section is given by

$$Q = \frac{2\pi}{k^2} \int_0^{\pi} \left| \sum_{l=0}^{\infty} (2l+1) e^{i\eta_l} \sin\eta_l \, P_l(\cos\theta) \right|^2 \sin\theta \, d\theta$$

where we have used (5.66) and (5.71) as well as the formula $d\omega = \sin\theta \, d\theta \, d\phi$. On performing the integration over θ and using the orthonormality property of Legendre polynomials, we find that the total cross section may be expressed in the form

$$Q = \sum_{l=0}^{\infty} Q_l, \qquad (5.145)$$

where

$$Q_l = \frac{4\pi}{k^2}(2l+1) \sin^2\eta_l \qquad (5.146)$$

is called the lth partial cross section. The analogy between the scattering of particles by a centre of force and the scattering of scalar waves by a spherical surface is now transparent since formulae (5.63) and (5.144) for the scattering amplitude and formulae (5.65) and (5.145) for the total cross section are identical in form.

It is of some interest to note that the total cross section may be rewritten as

$$Q = \frac{4\pi}{k} \mathscr{I}\{f(0)\}, \qquad (5.147)$$

known as the *optical theorem*, which expresses the total cross section in terms of the imaginary part of the forward scattering amplitude $\mathscr{I}\{f(0)\}$. Returning briefly to the example of the scattering by the screened Coulomb potential treated on pp. 233 to 235, we see that the first Born approximation gives

$$\begin{aligned} Q_{B1} &= 2\pi \int_0^\pi |f_{B1}(\theta)|^2 \sin\theta \, d\theta \\ &= \frac{2\pi\beta^4}{k^2} \int_0^{2k} \frac{KdK}{(K^2 + \lambda^2)^2} \\ &= \frac{4\pi\beta^4}{\lambda^2(4k^2 + \lambda^2)} \end{aligned} \qquad (5.148)$$

while the imaginary part of the forward scattering amplitude according to the second Born approximation is

$$\mathscr{I}\{f_B^{(2)}(0)\} = \frac{\beta^4}{\lambda^2} \frac{k}{\lambda^2 + 4k^2} \qquad (5.149)$$

so that

$$Q_{B1} = \frac{4\pi}{k} \mathscr{I}\{f_B^{(2)}(0)\}. \qquad (5.150)$$

In general, the total cross section derived from a knowledge of the nth Born approximation $f_B^{(n)}(\theta)$ to the scattering amplitude, can also be obtained by employing the optical theorem together with the $(n+1)$th Born approximation $f_B^{(n+1)}(0)$ for the forward scattering direction.

Integral equation for the phase shifts

An integral equation for the phase shifts may be readily derived by inserting into (5.79)

$$e^{-i\mathbf{k}_s \cdot \mathbf{r}'} = \sum_{l=0}^{\infty} (-i)^l (2l+1) j_l(kr') P_l(\cos \Theta')$$

where

$$\cos \Theta' = \cos \theta \cos \theta' + \sin \theta \sin \theta' \cos (\phi - \phi')$$

and

$$\psi(\mathbf{r}') = \frac{1}{r'} \sum_{l=0}^{\infty} i^l (2l+1) e^{i\eta_l} u_l(r') P_l(\cos \theta').$$

This yields

$$f(\theta) = -\sum_{l=0}^{\infty} (2l+1) e^{i\eta_l} P_l(\cos \theta) \int_0^{\infty} r' j_l(kr') u_l(r') U(r') \, dr'$$

and so, making use of (5.144), we obtain the integral equation

$$\sin \eta_l = -k \int_0^{\infty} r j_l(kr) u_l(r) U(r) \, dr \quad (5.151)$$

where

$$u_l(r) \sim k^{-1} \sin(kr - \tfrac{1}{2}l\pi + \eta_l)$$

for large r.

The first Born approximation replaces $u_l(r)$ by $r j_l(kr)$ and results in the formula

$$\sin \eta_{lB} = -k \int_0^{\infty} r^2 \{j_l(kr)\}^2 U(r) \, dr \quad (5.152)$$

for the phase shifts.

It is sometimes appropriate to choose the asymptotic behaviour of the radial function $u_l(r)$ to be given by

$$u_l(r) \sim \sin(kr - \tfrac{1}{2}l\pi) + \tan \eta_l \cos(kr - \tfrac{1}{2}l\pi) \quad (5.153)$$

for large r, in which case the integral equation for the phase shifts becomes

$$\tan \eta_l = -\int_0^{\infty} r j_l(kr) u_l(r) U(r) \, dr \quad (5.154)$$

and the first Born approximation takes the alternative form

$$\tan \eta_{lB} = -k \int_0^\infty r^2 \{j_l(kr)\}^2 U(r)\, dr. \tag{5.155}$$

5.6 Variational principles for the phase shifts

Having established the fundamental formulae of scattering theory arising from an expansion of the wave function in partial waves, we are in a position to direct our attention to the role played by variational principles in such an approach. We have seen that the basic quantities occurring in a partial wave expansion are the scattering phase shifts. As a consequence, a considerable amount of effort has been devoted to the problem of obtaining variational principles for them and several different forms of variational principle satisfied by the phase shifts have been derived by various investigators. We commence this section by considering the variational principle originated by Hulthén.[6]

Hulthén's variational principle

It is convenient to introduce the differential operator L given by

$$L \equiv \frac{d^2}{dr^2} + k^2 - U(r) - \frac{l(l+1)}{r^2} \tag{5.156}$$

since the radial equation (5.136) may then be rewritten in the succinct form

$$Lu = 0 \tag{5.157}$$

where we have dropped the subscript l to avoid cumbersome notation. Now, defining the functional

$$I[v] = \int_0^\infty vLv\, dr \tag{5.158}$$

where v is an arbitrary quadratically integrable function of r, we see that

$$I[u] = 0 \tag{5.159}$$

where u is the solution of equation (5.157) which vanishes at the origin and has the asymptotic behaviour for large r:

$$u \sim k^{-1} \sin(kr - \tfrac{1}{2}l\pi + \eta). \tag{5.160}$$

Next we consider the function $u + \delta u$ differing infinitesimally from u, vanishing at $r = 0$ and having the asymptotic form

$$u + \delta u \sim k^{-1} \sin(kr - \tfrac{1}{2}l\pi + \eta + \delta\eta) \tag{5.161}$$

as $r \to \infty$. Then δu also vanishes at the origin and has the asymptotic behaviour

$$\delta u \sim k^{-1} \cos(kr - \tfrac{1}{2}l\pi + \eta)\,\delta\eta \tag{5.162}$$

for large r.

We now let $\delta I[u]$ denote the change in $I[u]$ due to the infinitesimal variation δu in u so that

$$\delta I[u] = \int_0^\infty \delta u L u \, dr + \int_0^\infty u L \delta u \, dr + \int_0^\infty \delta u L \delta u \, dr. \tag{5.163}$$

Since

$$\int_0^\infty u \frac{d^2}{dr^2} \delta u \, dr = \int_0^\infty \delta u \frac{d^2 u}{dr^2} dr + \left[u \frac{d}{dr} \delta u - \delta u \frac{du}{dr} \right]_0^\infty \tag{5.164}$$

$$= \int_0^\infty \delta u \frac{d^2 u}{dr^2} dr - k^{-1}\,\delta\eta, \tag{5.165}$$

it follows that

$$\delta I[u] = 2 \int_0^\infty \delta u L u \, dr - k^{-1}\,\delta\eta, \tag{5.166}$$

neglecting quantities of the second order of smallness. But $Lu = 0$ and thus we arrive at the result

$$\delta I[u] = -k^{-1}\,\delta\eta \tag{5.167}$$

which is known as *Hulthén's variational principle*.

If we choose only those variations of u for which $\delta I[u] = 0$, then I remains equal to zero and $\delta\eta = 0$. This stationary property of the phase shift may be used to find an approximate value for η and an approximation to the radial function $u(r)$. We choose a trial function $u_T(r)$, which is continuous and bounded for all r, satisfying the boundary conditions

$$u_T(0) = 0 \qquad (5.168)$$

and

$$u_T(r) \sim k^{-1} \sin(kr - \tfrac{1}{2}l\pi + \eta_T) \qquad (5.169)$$

for large r, and which depends upon n parameters c_r ($r = 1, \ldots, n$) as well as upon the phase shift parameter η_T. The function $u_T(r)$ is then substituted for v in $I[v]$ and the $n + 1$ parameters are determined from the $n + 1$ equations

$$I_T = 0, \quad \frac{\partial I_T}{\partial c_r} = 0 \quad (r = 1, \ldots, n) \qquad (5.170)$$

where

$$I_T = I[u_T], \qquad (5.171)$$

the phase shift being approximated by η_T. This is known as *Hulthén's variational method*.

An alternative procedure which may be adopted depends upon the stationary property of $I + k^{-1}\eta$ as given by (5.167). In this method the parameters c_r and η_T are determined by the equations

$$\frac{\partial I_T}{\partial \eta_T} = -k^{-1}, \quad \frac{\partial I_T}{\partial c_r} = 0 \quad (r = 1, \ldots, n), \qquad (5.172)$$

the phase shift now being given by

$$\eta_V = \eta_T + kI_T \qquad (5.173)$$

since I_T does not necessarily vanish in the present case. This method of obtaining a variational approximation to the phase shift is referred to as *Kohn's method*.

So far we have chosen the asymptotic behaviour of the radial function $u(r)$ to be given by (5.160). However, it is often more convenient to let $u(r)$ have the asymptotic form for large r

$$u(r) \sim \sin(kr - \tfrac{1}{2}l\pi) + \alpha \cos(kr - \tfrac{1}{2}l\pi) \qquad (5.174)$$

where

$$\alpha = \tan \eta. \qquad (5.175)$$

If we then take the infinitesimal change δu to have the asymptotic behaviour

$$\delta u \sim \delta\alpha \cos(kr - \tfrac{1}{2}l\pi) \qquad (5.176)$$

for large r, we see from (5.163) and (5.164) that Hulthén's variational principle now assumes the form

$$\delta I[u] = -k\delta\alpha. \qquad (5.177)$$

Suppose $u_T(r)$ is a trial function satisfying the boundary conditions (5.168) at the origin and

$$u_T(r) \sim \sin(kr - \tfrac{1}{2}l\pi) + \alpha_T \cos(kr - \tfrac{1}{2}l\pi) \qquad (5.178)$$

for large r. Then Hulthén's variational method makes use of the equations (5.170) as before, while Kohn's variational method employs the equations

$$\frac{\partial I_T}{\partial \alpha_T} = -k, \quad \frac{\partial I_T}{\partial c_r} = 0 \quad (r = 1, \ldots, n) \qquad (5.179)$$

to determine the parameters α_T and c_r ($r = 1, \ldots, n$) with the phase shift η now being approximated by η_V where

$$\tan \eta_V = \alpha_V = \alpha_T + k^{-1}I_T. \qquad (5.180)$$

Although our discussion here has been entirely concerned with short range potentials, it can be generalized without undue difficulty to the very important case of the Coulomb potential.[7]

Tamm's variational method

An interesting approach to the problem of obtaining a variational approximation to the phase shift has been originated by Tamm.[8] His method is based on the introduction of a function $g(r)$ defined by the formula

$$u(r) = kr\{j_l(kr) - \alpha g(r)n_l(kr)\}. \qquad (5.181)$$

Since it is clear from the form of the radial equation (5.136) that as $r \to 0$ we have

$$u(r) \sim r^{l+1} \qquad (5.182)$$

SCATTERING THEORY 249

provided the potential is less singular than r^{-2} near the origin, and since also $n_l(r) \sim r^{-l-1}$ for small r, it follows that as $r \to 0$ we must have

$$g(r) \sim r^{2l+1}, \qquad (5.183)$$

while for large r it is evident from (5.174) and the asymptotic forms of j_l and n_l given by (5.44) that

$$g(r) \sim 1. \qquad (5.184)$$

Following Tamm's procedure we therefore take a trial function u_T having the general form

$$u_T(r) = kr\{j_l(kr) - \alpha_T g_T(r) n_l(kr)\}, \qquad (5.185)$$

where for small r

$$g_T(r) \sim r^{2l+1} \qquad (5.186)$$

while for large r

$$g_T(r) \sim 1. \qquad (5.187)$$

Now substituting $v = u_T$ into the functional $I[v]$, remembering that $rj_l(kr)$ and $rn_l(kr)$ satisfy equation (5.136) with the potential U placed equal to zero, performing an integration by parts and using the formula

$$\rho j_l(\rho) \frac{d}{d\rho} \{\rho n_l(\rho)\} - \rho n_l(\rho) \frac{d}{d\rho} \{\rho j_l(\rho)\} = 1, \qquad (5.188)$$

we find that

$$\alpha_T + k^{-1} I_T = -k\left[\int_0^\infty r^2 \{j_l(kr) - \alpha_T g_T(r) n_l(kr)\}^2 U(r)\, dr \right.$$
$$\left. + \alpha_T^2 \int_0^\infty r^2 \{n_l(kr)\}^2 \left(\frac{dg_T}{dr}\right)^2 dr \right] \qquad (5.189)$$

which is equivalent to the result originally derived by Tamm. Using the Kohn approximation to $\tan \eta$ given by (5.180) and noting that

$$\tan \eta_B = -k \int_0^\infty r^2 \{j_l(kr)\}^2 U(r)\, dr$$

where η_B denotes the Born approximation to the phase shift, we obtain

$$\tan \eta_V = \tan \eta_B + k\left[2\alpha_T \int_0^\infty r^2 j_l(kr)n_l(kr)g_T(r)U(r)\,dr \right.$$
$$\left. - \alpha_T^2 \int_0^\infty r^2\{n_l(kr)\}^2\left\{g_T^2(r)U(r) + \left(\frac{dg_T}{dr}\right)^2\right\}dr\right]. \quad (5.190)$$

The parameter α_T may be determined by employing the first of the Kohn formulae (5.179) which, on differentiating both sides of (5.189) with respect to α_T, yields

$$\alpha_T = \frac{\int_0^\infty r^2 j_l(kr)n_l(kr)g_T(r)U(r)\,dr}{\int_0^\infty r^2\{n_l(kr)\}^2\left\{g_T^2(r)U(r) + \left(\frac{dg_T}{dr}\right)^2\right\}dr} \quad (5.191)$$

enabling us to rewrite (5.190) as

$$\tan \eta_V = \tan \eta_B + \alpha_T k \int_0^\infty r^2 j_l(kr)n_l(kr)g_T(r)U(r)\,dr. \quad (5.192)$$

Variational methods and the integral equation for the phase shifts

We now concern ourselves with the question of whether the trial function $u_T(r)$ and the variational approximation to the phase shift satisfy the integral equation (5.154) established on p. 244. Returning to the preceding analysis dealing with Tamm's variational method and remembering that g_T is given in terms of u_T by the relation (5.185), we see at once that expression (5.192) for $\tan \eta_V$ can be simplified still further to the form

$$\tan \eta_V = -\int_0^\infty r j_l(kr)u_T(r)U(r)\,dr \quad (5.193)$$

which is just the integral equation (5.154) for the phase shift with the exact radial function $u(r)$ replaced by the trial function $u_T(r)$. Thus Kohn's variational method as well as Tamm's variational method automatically produce a phase shift η_V and a trial function $u_T(r)$ satisfying the integral equation (5.154).

On the other hand we can readily show that Hulthén's variational method does not have this satisfactory feature. Differentiating equation (5.189) with respect to α_T and then using (5.189) to eliminate the integrals involving g_T^2 and $(dg_T/dr)^2$, we find the relation

$$I_T + \tfrac{1}{2}\alpha_T\left(k - \frac{\partial I_T}{\partial \alpha_T}\right) = -k \int_0^\infty rj_l(kr)u_T(r)U(r)\,dr. \quad (5.194)$$

If we now employ the Kohn equations (5.179) and (5.180) we once more obtain the integral equation (5.193), whereas if we use instead the Hulthén equations $I_T = 0$ and $\tan \eta_V = \alpha_T$, we see that the integral equation (5.193) for the phase shift can only be satisfied provided the Kohn equation $\partial I_T/\partial \alpha_T = -k$ is also satisfied, a conclusion first reached by Demkov and Shepelenko.[9]

Kato's variational principle

A more general form of variational principle than the one derived on p. 246 due to Hulthén has been established by Kato.[10] He chooses the solution $u(r)$ of the radial equation (5.157) to have the asymptotic behaviour for large r given by

$$u(r) \sim \cos(kr - \tfrac{1}{2}l\pi + \theta) + \lambda \sin(kr - \tfrac{1}{2}l\pi + \theta) \quad (5.195)$$

where θ is a constant at our disposal and

$$\lambda = \cot(\eta - \theta), \quad (5.196)$$

the phase shift being denoted by η.

Let us now choose the infinitesimal change δu in u to have the asymptotic form

$$\delta u \sim \delta\lambda \sin(kr - \tfrac{1}{2}l\pi + \theta) \quad (5.197)$$

for large r. Then it follows immediately from (5.163) and (5.164) that

$$\delta I[u] = k\delta\lambda + \int_0^\infty \delta u L \delta u\,dr \quad (5.198)$$

and thus

$$F[v] = k\zeta - \int_0^\infty vLv\,dr, \quad (5.199)$$

where $v(0) = 0$ and for large r

$$v(r) \sim \cos(kr - \tfrac{1}{2}l\pi + \theta) + \zeta \sin(kr - \tfrac{1}{2}l\pi + \theta), \quad (5.200)$$

is stationary for $v = u$.

Since $Lu = 0$ we see that

$$F[u] = k\lambda \quad (5.201)$$

and so, if v differs infinitesimally from u, we have

$$k\lambda = k\zeta - \int_0^\infty vLv\,dr \quad (5.202)$$

to the first order of small quantities.

Kato's variational principle

$$\delta F[u] = 0 \quad (5.203)$$

which may be written alternatively as

$$\delta I[u] = k\delta\lambda, \quad (5.204)$$

converts to the Hulthén form (5.177) if we put $\theta = \pi/2$ since then $\lambda = -\tan\eta$.

Monotonic property of the phase shift. We now come to an interesting application of the Kato variational principle. We shall demonstrate that if η_a and η_b are phase shifts associated with the same order partial wave for the scattering of particles having wave number k by the potentials $V_a(r)$ and $V_b(r)$ respectively, then $\eta_a \leq \eta_b$ if $V_a(r) \geq V_b(r)$ for all values of r.

To begin with we shall assume that $V_a(r) - V_b(r)$ is an infinitesimal positive quantity for all r. Let us denote the solutions of the radial equations for the potentials $V_a(r)$ and $V_b(r)$ by $u_a(r)$ and $u_b(r)$ respectively. Putting $v(r) = u_b(r)$ in the Kato formula (5.202) for the scattering by the potential $V_a(r)$ yields to the first order of small quantities

$$k\cot(\eta_a - \theta) = k\cot(\eta_b - \theta) - \int_0^\infty u_b L_a u_b\,dr, \quad (5.205)$$

where L_a denotes the operator defined by (5.156) with the potential V replaced by V_a. Now

$$L_a u_b = (U_b - U_a)u_b$$

where U and V are related by formula (5.78), and so

$$k \cot(\eta_a - \theta) = k \cot(\eta_b - \theta) + \int_0^\infty (U_a - U_b)u_b^2 \, dr. \quad (5.206)$$

Since the integral on the right-hand side is positive due to the condition

$$U_a(r) - U_b(r) \geqslant 0 \quad (5.207)$$

for all r, it follows that

$$\cot(\eta_a - \theta) - \cot(\eta_b - \theta) \geqslant 0. \quad (5.208)$$

Hence

$$\eta_a \leqslant \eta_b. \quad (5.209)$$

This result must also hold when the requirement made initially concerning the infinitesimal nature of the potential difference $V_a - V_b$ is removed since we may compose any finite positive change in the potential out of infinitesimal positive variations each of which, as we have seen, must lead to an infinitesimal negative variation in the phase shift and thus collectively produce a finite negative change in the phase shift.

As an important corollary to the theorem proved above, we see at once that the phase shift η is positive if the potential $V(r) \leqslant 0$ for all r and that η is negative if $V(r) \geqslant 0$ for all r.

Hulthén's second variational method

An alternative variational method, originally established by Hulthén[7] for the zero order phase shift and generalized to higher order phase shifts by Feshbach and Rubinow,[11] which is akin to that due to Tamm, can be derived with the aid of Kato's variational principle by putting $\theta = 0$ so that

$$\lambda = \cot \eta \quad (5.210)$$

and by introducing the function $y(r)$ defined by the relation
$$u(r) = \cos(kr - \tfrac{1}{2}l\pi) + \lambda kr j_l(kr) - y(r), \quad (5.211)$$
where $u(r)$ satisfies the boundary conditions $u(0) = 0$ and
$$u(r) \sim \cos(kr - \tfrac{1}{2}l\pi) + \lambda \sin(kr - \tfrac{1}{2}l\pi) \quad (5.212)$$
for large values of r. Then we see that $y(0) = \cos(\tfrac{1}{2}l\pi)$ and $y(r)$ vanishes in the limit of large r.

The function $y(r)$ is often referred to as the *internal function* and characterizes the distortion of the radial function $u(r)$ for small values of r from its asymptotic behaviour for large r and in this respect resembles the function $g(r)$ introduced by Tamm. In fact Hulthén's internal function $y(r)$ is given by the difference between the radial function $u(r)$ and its asymptotic form for large r. On the other hand Tamm's function $g(r)$ does not have such a simple interpretation though it has the considerable advantage of producing less complicated final formulae. We note further that the spherical Neumann function $n_l(kr)$ which enters into the definition of $g(r)$ does not appear in the formula defining $y(r)$ being replaced by its asymptotic form for large r.

Substituting (5.211) into (5.199) yields
$$F[u] = -J + 2(k - N)\lambda - k\eta_B \lambda^2 \quad (5.213)$$
where
$$J = \int_0^\infty \left[k^2 y^2 - \left(\frac{dy}{dr}\right)^2 - \left\{ \frac{l(l+1)}{r^2} + U(r) \right\} \right. \\ \left. \times \{y - \cos(kr - \tfrac{1}{2}l\pi)\}^2 \right] dr, \quad (5.214)$$
$$N = k \int_0^\infty r j_l(kr) \{y - \cos(kr - \tfrac{1}{2}l\pi)\} U(r) \, dr, \quad (5.215)$$
and
$$\eta_B = -k \int_0^\infty r^2 \{j_l(kr)\}^2 U(r) \, dr \quad (5.216)$$
is the first Born approximation to the phase shift which may be obtained from formula (5.155) by assuming that η_B is small and consequently replacing $\tan \eta_B$ by η_B.

We now introduce a trial function given by
$$u_T(r) = \cos(kr - \tfrac{1}{2}l\pi) + \lambda_T k r j_l(kr) - y_T(r) \quad (5.217)$$
where $y_T(0) = \cos(\tfrac{1}{2}l\pi)$ and $y_T(r) \to 0$ as $r \to \infty$. Since $F[u]$ has been shown to be stationary on p. 252, we take
$$\frac{\partial F_T}{\partial \lambda_T} = 0, \quad (5.218)$$
where the subscript T denotes that the exact radial function u has been replaced by the trial function u_T. This yields
$$k\eta_B \lambda_T = k - N_T \quad (5.219)$$
and hence
$$F_T = -J_T + (k - N_T)\lambda_T = k\lambda_T - \Delta \quad (5.220)$$
where
$$\Delta = J_T + \lambda_T N_T. \quad (5.221)$$
Evidently if u_T is the exact solution u of the radial equation (5.157) then Δ vanishes and $\lambda_T = \lambda$.

We now suppose that the function $y_T(r)$ depends upon n arbitrary parameters c_r ($r = 1, \ldots, n$). Because $F[u]$ is stationary we take
$$\frac{\partial F_T}{\partial c_r} = 0 \quad (r = 1, \ldots, n) \quad (5.222)$$
and so
$$\frac{\partial J_T}{\partial c_r} + 2\lambda_T \frac{\partial N_T}{\partial c_r} = 0 \quad (r = 1, \ldots, n). \quad (5.223)$$
Then the $n + 1$ equations (5.219) and (5.223) determine the $n + 1$ parameters λ_T and c_r ($r = 1, \ldots, n$). Finally, using the stationary property of $F[u]$ once again, we see that the variational approximation to the phase shift η is given by the formula
$$\cot \eta_V = \lambda_T - \frac{\Delta}{k}. \quad (5.224)$$

A different way of regarding the above variational problem

follows by substituting expression (5.211) into the radial equation (5.157) which yields

$$\frac{d^2y}{dr^2} + \left\{k^2 - U(r) - \frac{l(l+1)}{r^2}\right\}y$$
$$+ \left\{U(r) + \frac{l(l+1)}{r^2}\right\}\cos(kr - \tfrac{1}{2}l\pi) + \lambda krj_l(kr)U(r) = 0. \quad (5.225)$$

Now it can be readily verified that this may be rewritten in the form of the Euler–Lagrange equation

$$\frac{\partial \mathscr{L}'}{\partial y} - \frac{d}{dr}\frac{\partial \mathscr{L}'}{\partial(\partial y/\partial r)} = 0 \qquad (5.226)$$

where

$$\int_0^\infty \mathscr{L}'\left(r, y, \frac{dy}{dr}\right) dr = J + 2\lambda N \qquad (5.227)$$

and so we see that the variational problem discussed above is equivalent to taking J to be stationary subject to the auxiliary condition that N be constant with 2λ as Lagrange multiplier.

Schwinger's variational principle

The variational principle for the phase shift due to Schwinger takes the form of a stationary expression for $k \cot \eta$. Following a similar treatment to that used in the case of the variational principle for the scattering amplitude derived by Schwinger, it may be obtained by rewriting expression (5.154) in the form of an integral equation which is homogeneous in $u(r)$. To do this we introduce the Green's function

$$G(r, r') = -kr_< r_> j_l(kr_<)n_l(kr_>), \qquad (5.228)$$

where $r_<, r_>$ are the lesser and greater of r, r', which satisfies the equation

$$\left\{\frac{\partial^2}{\partial r^2} + k^2 - \frac{l(l+1)}{r^2}\right\}G(r, r') = -\delta(r - r'). \qquad (5.229)$$

Choosing the radial function $u(r)$ to have the asymptotic behaviour for large r

$$u(r) \sim kr\{j_l(kr) - \tan \eta \, n_l(kr)\}, \qquad (5.230)$$

SCATTERING THEORY

we may write
$$u(r) = krj_l(kr) - \int_0^\infty G(r, r')U(r')u(r')\, dr' \quad (5.231)$$

and so the tangent of the phase shift η is given by the integral equation
$$\tan \eta = -\int_0^\infty r'j_l(kr')U(r')u(r')\, dr' \quad (5.232)$$

which is identical to the formula (5.154) derived previously in an alternative way. Now multiplying both sides of (5.231) by $U(r)u(r)$ and integrating over r we get

$$\int_0^\infty krj_l(kr)U(r)u(r)\, dr$$
$$= \int_0^\infty U(r)\{u(r)\}^2\, dr + \int_0^\infty U(r)u(r)\, dr \int_0^\infty G(r,r')U(r')u(r')\, dr' \quad (5.233)$$

from which it follows that we may rewrite (5.232) in the form

$$k \cot \eta =$$
$$-\frac{\int_0^\infty U(r)\{u(r)\}^2\, dr + \int_0^\infty U(r)u(r)\, dr \int_0^\infty G(r,r')U(r')u(r')\, dr'}{\left\{\int_0^\infty rj_l(kr)U(r)u(r)\, dr\right\}^2} \quad (5.234)$$

which is the expression for $k \cot \eta$ deduced by Schwinger.

For simplicity of notation we define the quantity
$$(g, \Omega f) = \int_0^\infty \bar{g}\Omega f\, dr \quad (5.235)$$

where Ω is a differential or integral operator. Then we may express $k \cot \eta$ as
$$k \cot \eta = -\frac{(u, Uu) + (u, UKu)}{(u, Uf)^2} \quad (5.236)$$

where K denotes the integral operator
$$K = \int_0^\infty dr' G(r, r')U(r') \quad (5.237)$$

and
$$f(r) = rj_l(kr). \quad (5.238)$$

Now defining the real Hermitian operator
$$\Omega = -U(1 + K) \tag{5.239}$$
and introducing the functional of a real function v given by
$$J[v] = \frac{(v, \Omega v)}{(v, Uf)^2}, \tag{5.240}$$
we see that formula (5.234) may be rewritten in the succinct form
$$k \cot \eta = J[u] \tag{5.241}$$
where u is the exact solution of the integral equation (5.231) which we may express in the symbolic form
$$u = kf - Ku,$$
or alternatively as
$$\Omega u = -kUf. \tag{5.242}$$

Stationary property of the Schwinger functional. Our next concern is to show that the Schwinger functional $J[v]$ is stationary with respect to infinitesimal variations of the function v when v is the exact radial function u. Because expression (5.240) is homogeneous in the function v, it is permissible to normalize v so that
$$(v, Uf) = 1 \tag{5.243}$$
without any loss of generality. Then we have
$$J[v] = (v, \Omega v). \tag{5.244}$$
With the same normalization the exact function u satisfies the equation
$$\Omega u = \mu Uf \tag{5.245}$$
where
$$\mu = (u, \Omega u) = k \cot \eta. \tag{5.246}$$
Now suppose that $\delta J[v]$ is the change in $J[v]$ resulting from an infinitesimal variation δv in v. Remembering that v is real and that Ω is an Hermitian operator, we find that
$$\delta J[v] = 2(\delta v, \Omega v) + (\delta v, \Omega \delta v). \tag{5.247}$$

Further, since the variation is subject to the supplementary condition (5.243) being satisfied, we have also

$$(\delta v, Uf) = 0. \tag{5.248}$$

It follows that, to the first order of small quantities,

$$\delta J[v] = 0 \tag{5.249}$$

if v is the solution u of equation (5.245). This establishes the stationary property of $J[v]$ for $v = u$.

Extremum property of the Schwinger functional. So far in this chapter we have examined only the stationary properties of certain functional expressions, that is we have been concerned just with variational principles. However, we shall now turn our attention to the problem of obtaining extremum principles rather than variational principles. In fact in this subsection we shall establish the interesting result due to Davison[12] and to Kato[13] that Schwinger's variational procedure not only provides a stationary expression for $k \cot \eta$ but that, under certain general conditions, it also yields a variational bound to $k \cot \eta$.

The proof we shall present here follows the lines of that due to Sugar and Blankenbecler[14] and is based upon the generalization to non-local potentials of the monotonicity theorem established on pp. 252 and 253. This theorem states that if two non-local potentials V_a and V_b satisfy

$$(\phi, V_a \phi) \geqslant (\phi, V_b \phi) \tag{5.250}$$

for all functions ϕ, then the corresponding phase shifts η_a and η_b satisfy

$$\eta_a \leqslant \eta_b. \tag{5.251}$$

Let us suppose that $V(r) \geqslant 0$ for all r. Then by Schwarz's inequality we know that

$$(\phi, V\phi) \geqslant (\phi, V_s \phi) \tag{5.252}$$

where

$$(\phi, V_s \phi) = \frac{(\phi, Vv)(v, V\phi)}{(v, Vv)} \tag{5.253}$$

so that V_s is the integral operator

$$V_s = \frac{V(r)v(r)\int_0^\infty dr'v(r')V(r')}{(v, Vv)}. \tag{5.254}$$

Hence we have

$$\eta \leqslant \eta_s \tag{5.255}$$

where η_s is the phase shift corresponding to the non-local potential V_s. Now the radial equation describing the scattering by V_s takes the form

$$\left\{\frac{d^2}{dr^2} + k^2 - \frac{l(l+1)}{r^2}\right\}q(r) - \frac{U(r)v(r)(v, Uq)}{(v, Uv)} = 0 \tag{5.256}$$

where $U(r) = (2m/\hbar^2)V(r)$ and $q(r)$ has the asymptotic behaviour for large r

$$q(r) \sim kr\{j_l(kr) - \tan\eta_s n_l(kr)\}. \tag{5.257}$$

Using the straightforward generalization of the integral equation (5.232) for the tangent of the phase shift appropriate to non-local potentials of the form (5.254), we obtain

$$\tan\eta_s = -\frac{(v, Uq)}{(v, Uv)} \tag{5.258}$$

where v is normalized according to equation (5.243). But the solution q of equation (5.256) may also be expressed in the integral equation form

$$q = kf - K_s q \tag{5.259}$$

where K_s is the integral operator given by

$$K_s = \frac{\int_0^\infty G(r, r')U(r')v(r')dr' \int_0^\infty dr''v(r'')U(r'')}{(v, Uv)}. \tag{5.260}$$

Therefore

$$(v, Uq) = k - \frac{(v, UKv)(v, Uq)}{(v, Uv)} \tag{5.261}$$

and so

$$k\frac{(v, Uv)}{(v, Uq)} = -(v, \Omega v) \tag{5.262}$$

where Ω is defined by (5.239). Hence we may rewrite equation (5.258) in the form

$$k \cot \eta_s = J[v] \qquad (5.263)$$

which is just the approximation to $k \cot \eta$ given by the Schwinger variational method. Consequently it follows from (5.255) that if $V(r) \geq 0$ for all r, corresponding to the case of a *repulsive potential*, the Schwinger variational method provides an *upper bound* to the exact phase shift η. Clearly we also have that if $V(r) \leq 0$ for all r, which is the condition satisfied by an *attractive potential*, the Schwinger variational phase shift η_s is a *lower bound* to η.

Generalizations of the above results to multichannel scattering have been established by Sugar and Blankenbecler.[14]

5.7 Scattering length and effective range

We now turn our attention to the scattering of low energy particles by a short range centre of force since we shall find in the succeeding section that a variational upper bound can be obtained to a quantity known as the scattering length which determines the cross section at vanishing energy.

Consider the particular case of the zero order partial wave associated with particles having zero angular momentum. In the limit $k \to 0$ corresponding to vanishing energy of incidence, the radial wave function for $l = 0$ satisfies the equation

$$\frac{d^2 u}{dr^2} = U(r) u(r) \qquad (5.264)$$

with the boundary conditions $u(0) = 0$ and

$$u(r) \sim 1 - \frac{r}{a} \qquad (5.265)$$

for large r. The quantity a is called the *scattering length* and is given by

$$\lim_{k \to 0} k \cot \eta = -\frac{1}{a}. \qquad (5.266)$$

It follows from (5.146) that the zero order partial cross section for particles of vanishing incident energy is given by

$$Q_0(k = 0) = 4\pi a^2. \qquad (5.267)$$

The resemblance to the cross section obtained on p. 226 for the scattering of waves at a spherical surface of radius a in the limit as $ka \to 0$ is worth noting.

For particles having low incident energy, $k \cot \eta$ may be conveniently expanded in the form of a power series in k^2. The leading term of this series is obviously $-a^{-1}$. The next term may be found by using Schwinger's variational principle with the zero energy function $u(r)$, the solution of equation (5.264) having the asymptotic form (5.265) for large r, as trial function.[15]

Introducing the internal function

$$y(r) = 1 - \frac{r}{a} - u(r) \qquad (5.268)$$

we see that $y(0) = 1$, $y(r) \to 0$ as $r \to \infty$, and

$$\frac{d^2y}{dr^2} = -\frac{d^2u}{dr^2} = -U(r)u(r). \qquad (5.269)$$

Then the numerator of (5.234) may be expressed in the form

$$-\int_0^\infty U(r)u(r)\left\{u(r) + \int_0^\infty G(r, r')U(r')u(r')\,dr'\right\}dr$$

$$= \int_0^\infty \frac{d^2y}{dr^2}\{u(r) + A(r)\}\,dr$$

where

$$A(r) = \int_0^\infty G(r, r')U(r')u(r')\,dr' = -\int_0^\infty G(r, r')\frac{d^2y}{dr'^2}\,dr'$$

$$= \left[y(r')\frac{\partial}{\partial r'}G(r, r') - G(r, r')\frac{d}{dr'}y(r')\right]_0^\infty$$

$$- \int_0^\infty y(r')\frac{\partial^2}{\partial r'^2}G(r, r')\,dr'$$

on integrating by parts. But
$$G(r, r') = k^{-1} \sin kr_< \cos kr_>$$
and
$$\left(\frac{\partial^2}{\partial r'^2} + k^2\right)G(r, r') = -\delta(r - r'),$$
and so
$$A(r) = -\cos kr + y(r) + k^2 \int_0^\infty y(r')G(r, r')\, dr'.$$

Hence the numerator of (5.234) may be rewritten as
$$\int_0^\infty \frac{d^2y}{dr^2}\left\{1 - \frac{r}{a} - \cos kr + k^2 \int_0^\infty y(r')G(r, r')\, dr'\right\} dr$$
$$= -\frac{1}{a} + k^2 \int_0^\infty \{2y \cos kr - y^2\}\, dr$$
$$- k^4 \int_0^\infty \int_0^\infty y(r)y(r')G(r, r')\, dr'$$
on integrating by parts again.

The denominator of (5.234) is the square of the integral
$$-\frac{1}{k}\int_0^\infty U(r)u(r) \sin kr\, dr = \frac{1}{k}\int_0^\infty \frac{d^2y}{dr^2} \sin kr\, dr$$
$$= \frac{1}{k}\left[\frac{dy}{dr} \sin kr - ky \cos kr\right]_0^\infty - k \int_0^\infty y \sin kr\, dr$$
$$= 1 - k \int_0^\infty y \sin kr\, dr.$$

Hence to order k^2 we obtain
$$k \cot \eta = -\frac{1}{a} + \frac{1}{2} r_0 k^2 \qquad (5.270)$$
where
$$r_0 = 2\int_0^\infty \left[2\left(1 - \frac{r}{a}\right)y(r) - \{y(r)\}^2\right] dr$$
$$= 2\int_0^\infty \left[\left(1 - \frac{r}{a}\right)^2 - \{u(r)\}^2\right] dr. \qquad (5.271)$$

r_0 is called the *effective range*.

It is important to notice that formula (5.270) is valid only for short range potentials such as those which fall off exponentially with increasing r. In the case of the long range potential given by

$$U(r) \sim -\frac{\beta^2}{r^4} \tag{5.272}$$

it has been shown that formula (5.270) must be replaced by

$$k \cot \eta = -\frac{1}{a} + \frac{\pi\beta^2}{3a^2}k + \frac{4\beta^2}{3a}k^2 \ln\frac{\beta k}{4} \tag{5.273}$$

where terms of order k^2 and higher have been neglected.[16]

Upper bounds to the scattering length

We now look at the problem of obtaining upper bounds to the scattering length defined in the previous section. The scattering of particles having zero energy and zero angular momentum is described by the solution of the radial equation

$$Lu = 0 \tag{5.274}$$

where

$$L = \frac{d^2}{dr^2} - U(r). \tag{5.275}$$

In the present section we choose the radial function $u(r)$ to have the asymptotic behaviour for large r

$$u(r) \sim a - r \tag{5.276}$$

where a is the scattering length, rather than the form (5.265) which we shall adopt in the succeeding section. Introducing a trial function $u_T(r)$ satisfying $u_T(0) = 0$ and

$$u_T(r) \sim a_T - r \tag{5.277}$$

for large r, and defining the function

$$w(r) = u_T(r) - u(r), \tag{5.278}$$

we obtain

$$\int_0^\infty u_T L u_T \, dr = \left[u\frac{dw}{dr} - w\frac{du}{dr}\right]_0^\infty + \int_0^\infty wLw \, dr$$

SCATTERING THEORY

which yields the result
$$a = a_T - \int_0^\infty u_T L u_T \, dr + \int_0^\infty w L w \, dr. \tag{5.279}$$

If the potential $V(r)$ has no bound states then
$$\int \chi H \chi \, d\mathbf{r} \geq 0, \tag{5.280}$$
where
$$H = -\frac{\hbar^2}{2m} \nabla^2 + V(r) \tag{5.281}$$

is the Hamiltonian operator for the particle, provided the function $\chi(r)$ satisfies the condition
$$\int \chi^2 \, d\mathbf{r} < \infty. \tag{5.282}$$

Hence, setting
$$\chi(r) = r^{-1} w(r), \tag{5.283}$$
we see that
$$\int_0^\infty w L w \, dr \leq 0 \tag{5.284}$$

provided w is quadratically integrable, that is
$$\int_0^\infty w^2 \, dr < \infty. \tag{5.285}$$

However, it can be verified at once from (5.278) that $w \sim a_T - a$ for large r and is therefore not quadratically integrable. On the other hand $w(r, \lambda) = w(r) e^{-\lambda r}$ is quadratically integrable for $\lambda > 0$ and so
$$K(\lambda) = \int_0^\infty w(r, \lambda) L w(r, \lambda) \, dr \leq 0. \tag{5.286}$$
Now
$$K(\lambda) - K(0) = \int_0^\infty w(r)(e^{-2\lambda r} - 1) L w(r) \, dr$$
$$- 2\lambda \int_0^\infty w(r) \frac{dw(r)}{dr} e^{-2\lambda r} \, dr + \lambda^2 \int_0^\infty \{w(r)\}^2 e^{-2\lambda r} \, dr \tag{5.287}$$

and since $|w(r)|$ is bounded from above for all r and $dw(r)/dr$ vanishes in the limit of large r, we see that all three terms on the right-hand side of (5.287) tend to zero as $\lambda \to 0$. Hence $K(\lambda)$ is continuous at $\lambda = 0$ and so it follows from (5.286) that $w(r)$ satisfies the inequality (5.284). If we had chosen $u(r)$ to have the alternative asymptotic behaviour given by (5.265), we would have had $w \sim \{(1/a) - (1/a_T)\}r$ which is unbounded and then $K(\lambda)$ would not have been continuous at $\lambda = 0$, preventing further progress along the present lines.

Now using (5.284) together with (5.279) we obtain the result first established by Spruch and Rosenberg[17]:

$$a \leqslant a_T - \int_0^\infty u_T L u_T \, dr \qquad (5.288)$$

which provides an upper bound to the scattering length.

For the zero energy case, Hulthén's variational principle (5.177) takes the form

$$\delta I[u] = \delta a \qquad (5.289)$$

where $u(r)$ has the asymptotic behaviour (5.276) for large r and

$$I[v] = \int_0^\infty v L v \, dr.$$

If a_T and the parameters c_r $(r = 1, \ldots, n)$ upon which the trial function u_T depends are determined by employing the Hulthén equations (5.170):

$$I_T = 0, \qquad \frac{\partial I_T}{\partial c_r} = 0 \qquad (r = 1, \ldots, n)$$

the right-hand side of (5.288) becomes a_T and does not provide the least upper bound to a for the given form of trial function. The least upper bound to the scattering length for the given trial function can be found, however, by using the Kohn equations

$$\frac{\partial I_T}{\partial a_T} = 1, \qquad \frac{\partial I_T}{\partial c_r} = 0 \qquad (r = 1, \ldots, n). \qquad (5.290)$$

It follows that for zero energy particles, Kohn's method is superior to that due to Hulthén.

A bound to the scattering length can also be obtained by employing Tamm's variational method or Hulthén's second variational method which is equivalent to it in the present case. To do this we define the function $g_T(r)$ given by

$$u_T(r) = a_T g_T(r) - r \qquad (5.291)$$

where $g_T(r) \sim r$ for small r and $g_T(r) \sim 1$ for large r. It then follows from the analysis given on pp. 248 to 250 that

$$\int_0^\infty u_T L u_T \, dr = a_T - a_B + 2Ba_T - Ca_T^2 \qquad (5.292)$$

where

$$B = \int_0^\infty r g_T(r) U(r) \, dr, \qquad (5.293)$$

$$C = \int_0^\infty \left[\left(\frac{dg_T}{dr} \right)^2 + \{g_T(r)\}^2 U(r) \right] dr, \qquad (5.294)$$

and

$$a_B = \int_0^\infty r^2 U(r) \, dr \qquad (5.295)$$

is the scattering length given by the Born approximation.

Hence by the inequality (5.288) we see that

$$a \leqslant a_B - 2Ba_T + Ca_T^2. \qquad (5.296)$$

Using the first of the Kohn equations (5.290) yields B/C as the optimum value of a_T and therefore

$$a \leqslant a_B - \frac{B^2}{C}. \qquad (5.297)$$

Since the right-hand side of (5.296) is bounded from below, C must be positive and so $a_B - B^2/C$ provides a more accurate value for the scattering length than the Born approximation.

Now suppose that there exists a single bound state for the potential $V(r)$ with eigenenergy E_1 and normalized eigenfunction $\psi_1(\mathbf{r})$ satisfying the Schrödinger equation

$$H\psi_1 = E_1 \psi_1 \qquad (5.298)$$

where H is the Hamiltonian operator (5.281). Since

$$\chi = \phi - \psi_1 \int \bar{\psi}_1 \phi \, d\mathbf{r}, \tag{5.299}$$

where ϕ is an arbitrary normalized function, is orthogonal to ψ_1 we have from (4.138) that

$$\int \bar{\chi} H \chi \, d\mathbf{r} \geqslant 0 \tag{5.300}$$

provided

$$\int |\chi|^2 \, d\mathbf{r} < \infty, \tag{5.301}$$

and so

$$\int \bar{\phi} H \phi \, d\mathbf{r} \geqslant E_1 \left| \int \bar{\phi} \psi_1 \, d\mathbf{r} \right|^2. \tag{5.302}$$

For the zero angular momentum case we may write

$$\phi(\mathbf{r}) = \frac{1}{\sqrt{(4\pi)}} \frac{w(r)}{r} \tag{5.303}$$

and

$$\psi_1(\mathbf{r}) = \frac{1}{\sqrt{(4\pi)}} \frac{u_1(r)}{r} \tag{5.304}$$

where $w(r)$ and $u_1(r)$ are normalized radial functions. Then we have

$$\int_0^\infty wLw \, dr \leqslant \epsilon_1 \left(\int_0^\infty w u_1 \, dr \right)^2 \tag{5.305}$$

with $\epsilon_1 = -(2m/\hbar^2)E_1$. Although the difference function given by (5.278) is not quadratically integrable we may, following an analogous argument to that which established the validity of (5.284), replace the function $w(r)$ entering into the inequality (5.305) by the difference function (5.278). Then we see that

$$\int_0^\infty w u_1 \, dr = \int_0^\infty u_T u_1 \, dr \tag{5.306}$$

since u and u_1 are orthogonal, and so it follows from equation (5.279) that

$$a \leqslant a_T - \int_0^\infty u_T L u_T \, dr + \epsilon_1 \left\{ \int_0^\infty u_T u_1 \, dr \right\}^2. \tag{5.307}$$

As it stands this inequality is of reduced practical use because the exact eigenfunction ψ_1 and eigenenergy E_1 which determine u_1 and ϵ_1 are unknown in general. However, we will now show that we may replace the right-hand side of (5.307) by an upper bound to the scattering length which depends only upon approximations to u_1 and ϵ_1 determined by employing the variational method for bound states.

To carry this out we apply the result (4.155) with

$$\chi_1(\mathbf{r}) = \frac{1}{\sqrt{(4\pi)}} \frac{u_{1T}(r)}{r} \tag{5.308}$$

and

$$\chi_2(\mathbf{r}) = \frac{1}{\sqrt{(4\pi)}} \frac{w(r)}{r} \tag{5.309}$$

where $u_{1T}(r)$ is a variational approximation to $u_1(r)$. Then we get

$$\int_0^\infty wLw\, dr \leqslant \frac{1}{\epsilon_{1T}} \left\{ \int_0^\infty u_{1T} Lw\, dr \right\}^2 \tag{5.310}$$

with

$$\epsilon_{1T} = \int_0^\infty u_{1T} L u_{1T}\, dr. \tag{5.311}$$

Since $Lw = Lu_T$ it follows that

$$a \leqslant a_T - \int_0^\infty u_T L u_T\, dr + \frac{1}{\epsilon_{1T}} \left\{ \int_0^\infty u_{1T} L u_T\, dr \right\}^2 \tag{5.312}$$

which is the desired inequality for the case in which the potential $V(r)$ is sufficiently strong to have a single bound state.[18]

The variational approximation to the scattering length given by the right-hand side of (5.312) can also be obtained directly by employing the trial function

$$u'_T(r) = u_T(r) + d_1 u_{1T}(r)$$

where d_1 is an adjustable parameter. We have

$$a_V = a_T - \int_0^\infty u'_T L u'_T\, dr$$

$$= a_T - \int_0^\infty u_T L u_T\, dr - 2d_1 \int_0^\infty u_{1T} L u_T\, dr - d_1^2 \epsilon_{1T}$$

and thus, on selecting the optimum value of d_1 given by

$$d_1 = -\frac{1}{\epsilon_{1T}} \int_0^\infty u_{1T} L u_T \, dr$$

which comes from putting $\partial I[u_T']/\partial d_1 = 0$, we obtain the required formula.

This result can be generalized to cases in which the potential possesses N bound states. Thus the trial function

$$u_T'(r) = u_T(r) + \sum_{i=1}^N d_i u_{iT}(r),$$

where $u_{1T}, u_{2T}, \ldots, u_{NT}$ form a set of linearly independent functions such that the matrix with elements $\int_0^\infty u_{iT} L u_{jT} \, dr$ has positive eigenvalues only, leads to a variational expression for the scattering length which is an absolute upper bound to the true scattering length provided the linear parameters d_i are taken to have their optimum values. If the functions are chosen to satisfy the conditions

$$\int_0^\infty u_{iT} u_{jT} \, dr = \delta_{ij}, \qquad \int_0^\infty u_{iT} L u_{jT} \, dr = \epsilon_{iT} \delta_{ij} \qquad (\epsilon_{iT} > 0),$$

the optimum values of the parameters d_i are given by

$$d_i = -\frac{1}{\epsilon_{iT}} \int_0^\infty u_{iT} L u_T \, dr$$

and the variational approximation to the scattering length becomes

$$a_V = a_T - \int_0^\infty u_T L u_T \, dr + \sum_{i=1}^N \frac{1}{\epsilon_{iT}} \left\{ \int_0^\infty u_{iT} L u_T \, dr \right\}^2.$$

Lower bound to the reciprocal of the scattering length

In this subsection we shall show that when the potential $V(r)$ satisfies certain conditions, a lower bound to the reciprocal of the scattering length can be obtained by taking the boundary con-

SCATTERING THEORY

ditions satisfied by the radial function $u(r)$ to be $u(r) \sim r$ for small r and

$$u(r) \sim 1 - \frac{r}{a}$$

for large r, as in section 5.7.

Choosing a variation δu satisfying the boundary conditions $\delta u = 0$ at the origin and

$$\delta u = -r\delta\left(\frac{1}{a}\right)$$

for large r, we find that the change $\delta I[u]$ in $I[u]$ is given by

$$\delta I[u] = -\delta\left(\frac{1}{a}\right) + \int_0^\infty \delta u L \delta u \, dr. \quad (5.313)$$

Now taking a trial function $u_T(r)$ having the behaviour $u_T(r) \sim r$ for small r and having the asymptotic form

$$u_T(r) \sim 1 - \frac{r}{a_T} \quad (5.314)$$

for large r, it immediately follows from (5.313) that

$$\frac{1}{a} = \frac{1}{a_V} - \int_0^\infty \delta u L \delta u \, dr \quad (5.315)$$

where the variational method approximation a_V to the scattering length is given by

$$\frac{1}{a_V} = \frac{1}{a_T} + I[u_T]. \quad (5.316)$$

Expressing the trial function in terms of the function $g_T(r)$ defined according to the formula

$$u_T(r) = g_T(r) - \frac{r}{a_T} \quad (5.317)$$

where $g_T(r) \sim r$ for small r and $g_T(r) \sim 1$ for large r, we find that

$$I[u_T] = \frac{1}{a_T} - \int_0^\infty \left\{g_T(r) - \frac{r}{a_T}\right\}^2 U(r) \, dr - \int_0^\infty \left(\frac{dg_T}{dr}\right)^2 dr. \quad (5.318)$$

Next we determine the parameter a_T by using the Kohn variational equation

$$\frac{\partial I[u_T]}{\partial (1/a_T)} = -1 \qquad (5.319)$$

which yields

$$\frac{1}{a_T} = \frac{1}{a_B}\left[\int_0^\infty rg_T(r)U(r)\,dr + 1\right] \qquad (5.320)$$

where a_B is the Born approximation to the scattering length. Then we may rewrite the variational approximation to the scattering length given by (5.316) in the form

$$\frac{1}{a_V} = \frac{1}{a_B}(1+B)^2 - C \qquad (5.321)$$

where B and C are given by (5.293) and (5.294) respectively.

In order to prove that $1/a_V$ provides a lower bound to the reciprocal of the scattering length when certain general conditions are satisfied by the potential $V(r)$, we now consider the second order quantity $\int_0^\infty \delta u L \delta u\, dr$. Noting that

$$\delta u = g_T(r) - g(r) - \left(\frac{1}{a_T} - \frac{1}{a}\right)r \qquad (5.322)$$

where $g(r)$ is defined in terms of the exact radial function $u(r)$ by the relation

$$u(r) = g(r) - \frac{r}{a} \qquad (5.323)$$

and satisfies $g(r) \sim r$ for small r and $g(r) \sim 1$ for large r, it can be readily verified that

$$\int_0^\infty \delta u L \delta u\, dr = -\int_0^\infty \left[\left\{\frac{d}{dr}(g_T - g)\right\}^2 + \left\{g_T(r) - g(r) - \left(\frac{1}{a_T} - \frac{1}{a}\right)r\right\}^2 U(r)\right] dr. \qquad (5.324)$$

It follows at once that

$$\int_0^\infty \delta u L \delta u\, dr \leqslant 0 \qquad (5.325)$$

SCATTERING THEORY

if $U(r) \geqslant 0$ for all r and so we see from equation (5.315) that for a repulsive potential

$$\frac{1}{a} \geqslant \frac{1}{a_V}. \tag{5.326}$$

Let us now suppose that the potential is attractive so that $U(r) \leqslant 0$ for all r, or more generally that $a_B < 0$. Using the result that a_T is given by (5.320) in Kohn's variational method and that the true scattering length a is given by

$$\frac{1}{a} = \frac{1}{a_B}\left[\int_0^\infty rg(r)U(r)\,dr + 1\right] \tag{5.327}$$

we obtain

$$\int_0^\infty \delta u L \delta u\,dr = -\int_0^\infty \left[\left\{\frac{d}{dr}(g_T - g)\right\}^2 + \{g_T(r) - g(r)\}^2 U(r)\right]dr$$
$$+ \frac{1}{a_B}\left[\int_0^\infty r\{g_T(r) - g(r)\}U(r)\,dr\right]^2. \tag{5.328}$$

We now assume that the potential $V(r)$ does not possess more than a single bound state. Then setting $w(r) = g_T(r) - g(r)$ and $u_{1T}(r) = r\,e^{-\lambda r}$ in (5.310) and allowing $\lambda \to 0$, we obtain

$$-\int_0^\infty \left[\left\{\frac{d}{dr}(g_T - g)\right\}^2 + \{g_T(r) - g(r)\}^2 U(r)\right]dr$$
$$+ \frac{1}{a_B}\left[\int_0^\infty r\{g_T(r) - g(r)\}U(r)\,dr\right]^2 \leqslant 0 \tag{5.329}$$

by virtue of the fact that $g_T(r) - g(r)$ vanishes at the origin and tends to zero sufficiently rapidly as $r \to \infty$. Consequently the inequality (5.325) is again satisfied and so $1/a_V$ provides a lower bound to the reciprocal of the scattering length.

As long as we exclude from consideration those cases for which $a > 0$ but $a_V < 0$, it also follows that a_V provides an upper bound to a.

Finally we note that a minimum principle for $k \cot \eta$ at non-zero energies has been established by Hahn, O'Malley and Spruch.[19]

5.8 Elastic scattering of electrons by hydrogen atoms

The most extensive application of variational methods in scattering theory has been to elastic collisions between electrons and hydrogen atoms which provides the simplest example of an atomic collision process.

The wave function $\Psi(\mathbf{r}_1, \mathbf{r}_2)$ of the system of two electrons with position vectors $\mathbf{r}_1$ and $\mathbf{r}_2$ relative to the proton satisfies the Schrödinger equation

$$\left[-\frac{\hbar^2}{2m}(\nabla_1^2 + \nabla_2^2) - \frac{e^2}{r_1} - \frac{e^2}{r_2} + \frac{e^2}{r_{12}} - E\right]\Psi(\mathbf{r}_1, \mathbf{r}_2) = 0 \quad (5.330)$$

where E is the total energy of the system and r_{12} is the distance between the electrons.

This equation may be solved by expanding Ψ in terms of the orthogonal and normalized set of hydrogen atom wave functions $\psi_n(\mathbf{r})$ satisfying the equation

$$\left\{-\frac{\hbar^2}{2m}\nabla^2 - \frac{e^2}{r} - E_n\right\}\psi_n = 0 \quad (5.331)$$

where E_n is the eigenenergy of the nth state of the atom.

Because the two electrons are indistinguishable, the total wave function of the system must be antisymmetric with respect to interchange of both the space and spin coordinates of the electrons as required by the Pauli principle. Thus the space functions $\Psi^+(\mathbf{r}_1, \mathbf{r}_2)$ and $\Psi^-(\mathbf{r}_1, \mathbf{r}_2)$ associated with singlet and triplet spin functions for the two electrons respectively may be expressed in the symmetrized forms

$$\Psi^\pm(\mathbf{r}_1, \mathbf{r}_2) = \frac{1}{\sqrt{2}}\{\Psi(\mathbf{r}_1, \mathbf{r}_2) \pm \Psi(\mathbf{r}_2, \mathbf{r}_1)\} \quad (5.332)$$

with

$$\Psi(\mathbf{r}_1, \mathbf{r}_2) = \mathop{\mathsf{S}}_{m} F_m^\pm(\mathbf{r}_1)\psi_m(\mathbf{r}_2), \quad (5.333)$$

where the summation denoted by the symbol S is over all states of the hydrogen atom including the continuum states.

SCATTERING THEORY

Substituting into (5.330), multiplying across by $\bar{\psi}_n(\mathbf{r}_2)$ and integrating with respect to $\mathbf{r}_2$, we obtain the Hartree–Fock equations for continuum states

$$[\nabla_1^2 + k_n^2]F_n^\pm(\mathbf{r}_1)$$
$$= \frac{2m}{\hbar^2} \sum_m \left\{ V_{nm}(\mathbf{r}_1)F_m^\pm(\mathbf{r}_1) \mp \int K_{nm}(\mathbf{r}_1, \mathbf{r}_2)F_m^\pm(\mathbf{r}_2)\, d\mathbf{r}_2 \right\} \quad (5.334)$$

where the matrix element V_{nm} and the kernel K_{nm} are given by

$$V_{nm}(\mathbf{r}_1) = \int \bar{\psi}_n(\mathbf{r}_2)\left(\frac{e^2}{r_{12}} - \frac{e^2}{r_1}\right)\psi_m(\mathbf{r}_2)\, d\mathbf{r}_2 \quad (5.335)$$

and

$$K_{nm}(\mathbf{r}_1, \mathbf{r}_2) = \bar{\psi}_n(\mathbf{r}_2)\psi_m(\mathbf{r}_1)\left\{E - E_n - E_m - \frac{e^2}{r_{12}}\right\}, \quad (5.336)$$

and the wave number k_n is given by

$$k_n^2 = \frac{2m}{\hbar^2}(E - E_n). \quad (5.337)$$

If the hydrogen atom is initially in its ground state which we shall denote by $n = 0$, we have for large r

$$F_0^\pm(\mathbf{r}) \sim e^{i\mathbf{k}_0 \cdot \mathbf{r}} + \frac{1}{r} e^{ik_0 r} f_0^\pm(\theta, \phi) \quad (5.338)$$

and

$$F_n^\pm(\mathbf{r}) \sim \frac{1}{r} e^{ik_n r} f_n^\pm(\theta, \phi) \; (n \neq 0). \quad (5.339)$$

Let us now consider the elastic scattering of electrons by hydrogen atoms in the ground $1s$ state. If we neglect the effect of all states other than the $n = 0$ state, the infinite set of coupled integro-differential equations (5.334) reduce to the pair of equations corresponding to positive and negative symmetry

$$\left[\nabla_1^2 + k_0^2 - \frac{2m}{\hbar^2} V_{00}(r_1)\right]F_0^\pm(\mathbf{r}_1) \pm \frac{2m}{\hbar^2} \int K_{00}(\mathbf{r}_1, \mathbf{r}_2)F_0^\pm(\mathbf{r}_2)\, d\mathbf{r}_2 = 0 \quad (5.340)$$

where $F_0^\pm(\mathbf{r})$ have the asymptotic behaviour (5.338) for large r. Assuming that the single antisymmetric spin function

$$(1/\sqrt{2})\{\sigma_1(s_1)\sigma_2(s_2) - \sigma_1(s_2)\sigma_2(s_1)\}$$

associated with the singlet state of the system and the three symmetric spin functions

$$\sigma_1(s_1)\sigma_1(s_2), \quad \sigma_2(s_1)\sigma_2(s_2),$$
$$(1/\sqrt{2})\{\sigma_1(s_1)\sigma_2(s_2) + \sigma_1(s_2)\sigma_2(s_1)\}$$

associated with the triplet state of the system have equal weight, it follows that the differential cross section is given by

$$I(\theta, \phi) = \tfrac{1}{4}\{|f_0^+(\theta, \phi)|^2 + 3|f_0^-(\theta, \phi)|^2\}. \quad (5.341)$$

Expanding $F_0^\pm(\mathbf{r})$ in the form

$$F_0^\pm(\mathbf{r}) = \frac{1}{r} \sum_{l=0}^{\infty} i^l (2l+1) e^{i\eta_l^\pm} u_l^\pm(r) P_l(\cos\theta) \quad (5.342)$$

we obtain

$$\left\{\frac{d^2}{dr^2} + k_0^2 - \frac{2m}{\hbar^2} V_{00}(r) - \frac{l(l+1)}{r^2}\right\} u_l^\pm(r)$$
$$\pm \frac{2m}{\hbar^2} \int_0^\infty \kappa_l(r, r') u_l^\pm(r') \, dr' = 0 \quad (5.343)$$

where

$$\kappa_l(r, r') = \frac{4\pi}{2l+1} rr' \psi_0(r)\psi_0(r')\{(E - 2E_0)\delta_{l0} - e^2\gamma_l(r, r')\} \quad (5.344)$$

and $\gamma_l(r, r')$ is given by (4.180), the asymptotic behaviour of $u_l^\pm(r)$ for large r being

$$u_l^\pm(r) \sim \frac{1}{k_0} \sin(k_0 r - \tfrac{1}{2}l\pi + \eta_l^\pm). \quad (5.345)$$

The total elastic cross section Q is given in terms of the phase shifts $\eta_l^\pm$ by the formula

$$Q = \sum_{l=0}^{\infty} Q_l \quad (5.346)$$

where

$$Q_l = \frac{\pi}{k_0^2} (2l+1)\{\sin^2 \eta_l^+ + 3\sin^2 \eta_l^-\}. \quad (5.347)$$

The zero order partial wave $l = 0$ has been studied in considerable detail by a number of investigators. We first consider the static field approximation in which the kernel $\kappa_0(r, r')$ is put equal to zero and which is equivalent to using the unsymmetrized form (5.333) for the space function describing the two electrons. The scattering is then entirely due to the static field potential $V_{00}(r)$ of the hydrogen atom. A suitable trial function has the form

$$u_T(r) = \sin k_0 r + (\alpha_T + c_1 e^{-r/a_0})(1 - e^{-r/a_0}) \cos k_0 r \quad (5.348)$$

where α_T and c_1 are arbitrary parameters. It can be readily verified that this trial function satisfies the boundary conditions $u_T(0) = 0$ and for large r

$$u_T(r) \sim \sin k_0 r + \alpha_T \cos k_0 r \quad (5.349)$$

which is identical to (5.178) for $l = 0$.

A comparison between the values of the phase shift η_0 obtained by the numerical integration of the differential equation

$$\left\{\frac{d^2}{dr^2} + k_0^2 - \frac{2m}{\hbar^2} V_{00}(r)\right\} u_0(r) = 0 \quad (5.350)$$

and by using the Hulthén variational method equations (5.179) together with the trial function (5.348) is given in Table 5.1. The agreement is very satisfactory. We may also use Tamm's

Table 5.1. Elastic scattering of electrons by hydrogen atoms in the ground state

Wave number k_0 (in a_0^{-1})	Static field approximation		Exchange approximation			
	η_0		η_0^+		η_0^-	
	V	N	V	N	V	N
0	$(-9\cdot44)$*	$(-9\cdot44)$*	$(9\cdot03)$*	$(8\cdot06)$*	$(2\cdot35)$*	$(2\cdot35)$*
0·2	0·972	0·973	1·819	1·870	2·679	2·679
0·5	1·044	1·045	1·074	1·031	2·070	2·070
1·0	0·904	0·906	0·645	0·543	1·390	1·391

* Scattering length in a_0.
N = numerical integration, V = variational method (Hulthén).[20]

variational method, described on pp. 248 to 250, in conjunction with the nonlinear trial function

$$g_T(r) = 1 - e^{-\beta r/a_0} \tag{5.351}$$

where β is an adjustable parameter. For $k_0 = 1$ it is found that the optimum value of β is 2·5 for which we obtain 0·9055 for η_0, which is in very good agreement with the value 0·9057 obtained by numerical integration, despite the simple form of trial function employed.

We now turn our attention to the case in which the integral term in equation (5.343) is retained. This is called the exchange approximation since the symmetrized form (5.332) allows for the possibility of the incident electron exchanging roles with the atomic electron. Equation (5.343) has been numerically integrated for $l = 0$ and the two cases of different symmetry by several investigators. The values of the zero order phase shift $\eta_0^\pm$ calculated in this way are displayed in Table 5.1 where they are compared with the values of $\eta_0^\pm$ obtained by employing a generalization of Hulthén's variational method in which the operator L given by expression (5.156) is replaced by the integro-differential operator

$$\frac{d^2}{dr^2} + k_0^2 - \frac{2m}{\hbar^2} V_{00}(r) \pm \frac{2m}{\hbar^2} \int_0^\infty dr' \kappa_0(r, r')$$

and again using the trial function $u_T(r)$ given by (5.348). Although excellent results are obtained with the variational method for the triplet spin function case associated with an antisymmetric space function, the values for the singlet spin function case corresponding to a symmetric space function are considerably less satisfactory.

We have already noted on p. 176 in connection with two electron atomic systems, that in the zero angular momentum case the wave function describing the system may be expressed entirely in terms of the coordinates r_1, r_2, r_{12} and does not depend upon any of the angular coordinates. This means that we may write the $l = 0$ space wave function in the form $\Psi(r_1, r_2, r_{12})$ where for large r_1

$$\Psi(r_1, r_2, r_{12}) \sim \frac{1}{r_1}(\sin k_0 r_1 + \alpha \cos k_0 r_1)\psi_0(r_2). \tag{5.352}$$

SCATTERING THEORY

A variational treatment which involves wave functions having the above form clearly necessitates a generalization of the variational principle due to Hulthén derived on p. 246. To this end we consider the functional

$$I[X] = \int X(L - \epsilon)X \, d\tau \qquad (5.353)$$

where

$$L = -\frac{2m}{\hbar^2} H = \nabla_1^2 + \nabla_2^2 + \frac{2}{a_0}\left(\frac{1}{r_1} + \frac{1}{r_2} - \frac{1}{r_{12}}\right) \qquad (5.354)$$

and

$$\epsilon = -\frac{2m}{\hbar^2} E, \qquad (5.355)$$

H being the total Hamiltonian and E being the total energy of the system of incident electron and target hydrogen atom. Then it can be shown without difficulty that

$$\delta I[\Psi] = -4\pi k_0 \, \delta\alpha \qquad (5.356)$$

which is the required generalization of Hulthén's variational principle.

The static field approximation corresponds to the choice of trial function having the separable form

$$\Psi_T(r_1, r_2, r_{12}) = \frac{1}{r_1} u_T(r_1)\psi_0(r_2) \qquad (5.357)$$

which neglects the dependence upon the r_{12} coordinate. Allowance for the correlation between the two electrons may be made by employing a function having the form

$$\Psi_T(r_1, r_2, r_{12}) = \frac{1}{r_1} u_T(r_1, r_{12})\psi_0(r_2) \qquad (5.358)$$

which depends explicitly on the interelectron distance r_{12}. A suitable trial function is obtained by taking

$$u_T(r_1, r_{12}) = \sin k_0 r_1 + \{\alpha_T + (c_1 + c_2 r_{12}) e^{-r_1/a_0}\}$$
$$\times (1 - e^{-r_1/a_0}) \cos k_0 r_1 \qquad (5.359)$$

which reduces to the static field approximation trial function (5.348) if c_2 is chosen to vanish.

As can be seen from figure 2, where the zero order phase shifts calculated by Massey and Moiseiwitsch[20] using the variational method together with different forms of trial function are illustrated, the effect of the introduction of the term depending upon r_{12} is to make the phase shift tend to π radians as k_0 becomes small. This behaviour of the correlation approximation phase shift contrasts with that found by using the static field approximation for which η_0 tends to zero as k_0 becomes small but is similar to that obtained with the exchange approximation which corresponds to taking the symmetrized forms of trial function

$$\Psi_T^\pm(r_1, r_2, r_{12}) = \frac{1}{\sqrt{2}} \left\{ \frac{1}{r_1} u_T(r_1)\psi_0(r_2) \pm \frac{1}{r_2} u_T(r_2)\psi_0(r_1) \right\}, \quad (5.360)$$

the outside factor $1/\sqrt{2}$ ensuring that the variational principle (5.356) holds without any modification. Since the integer multiple of π to which the phase shift tends in the limit of vanishing energy of incidence is, according to Levinson's theorem, equal to the number of bound states in the field of the target atom of an electron having the prescribed angular momentum, it follows that the static field of the hydrogen atom without any allowance for correlation or exchange is unable to produce a bound state of the negative ion H^-.

Both exchange and correlation may be taken into account by employing functions having the symmetrized forms

$$\Psi_T^\pm(r_1, r_2, r_{12}) = \frac{1}{\sqrt{2}} \{\Psi_T(r_1, r_2, r_{12}) \pm \Psi_T(r_2, r_1, r_{12})\}. \quad (5.361)$$

If we choose $\Psi_T(r_1, r_2, r_{12})$ to be given by (5.358) together with (5.359) it is found that the effect of introducing the dependence upon the interelectron distance is negligible for the triplet spin function case but produces a small increase in the phase shift η_0^+ for the singlet spin function case as can be seen by referring to figure 2.

Much attention has been paid to the special case of zero energy incident particles since it is more readily dealt with than a non-zero energy case, and because the scattering lengths resulting from the application of Hulthén's and Kohn's variational methods given

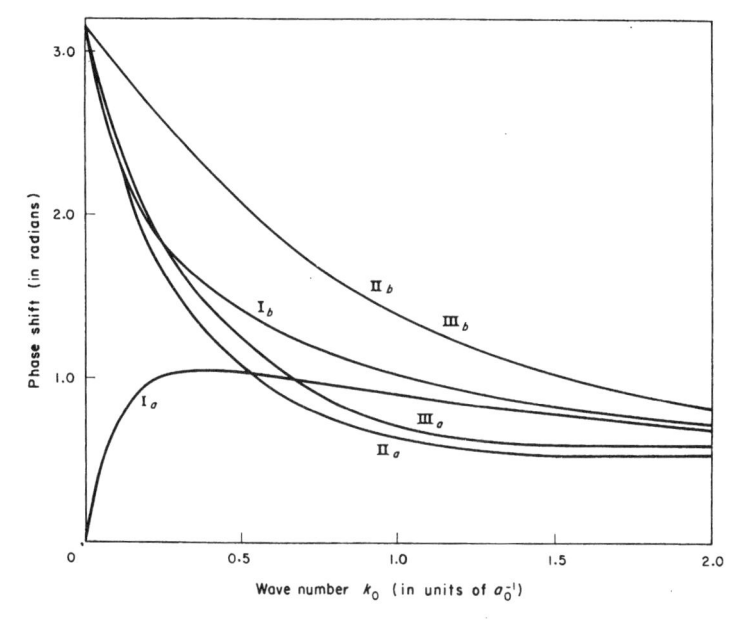

Fig. 2. Zero order phase shifts for the elastic scattering of electrons by hydrogen atoms calculated by Massey and Moiseiwitsch[20] using Hulthén's variational method together with the trial functions (5.348) and (5.359). I_a, static field approximation for η_0; I_b, correlation approximation for η_0; II_a, exchange approximation for η_0^+; II_b, exchange approximation for η_0^-; III_a, exchange–correlation approximation for η_0^+; III_b, exchange–correlation approximation for η_0^-.

on p. 266 provide upper bounds to the true scattering lengths.* Values of the scattering length a for the static field and exchange approximations calculated by numerical integration and by employing variational methods are given in Table 5.1. More involved calculations have been performed by Rosenberg, Spruch

* The proof of this for potential scattering given on pp. 264 to 266 is readily generalizable to scattering by a compound system such as a hydrogen atom or a deuteron.[21]

and O'Malley[22] using the symmetrized forms (5.361) together with

$$\Psi_T(r_1, r_2, r_{12}) = \frac{1}{r_1} \Big[\{a_T(1 - e^{-br_1/a_0}) - r_1\} e^{-r_2/a_0}$$
$$+ c_1 r_1 e^{-(d_1 r_1 + f_1 r_2)/a_0} + c_2 r_1 e^{-(d_2 r_1 + f_2 r_2 + g r_{12})/a_0}\Big].$$
(5.362)

They obtained the values $a_+ = 6.23 a_0$ and $a_- = 1.93 a_0$ for the singlet and triplet scattering lengths both of which are upper bounds to the actual values of the scattering lengths. Since the total elastic cross section for incident particles of zero energy is given by

$$Q(k_0 = 0) = \pi(a_+^2 + 3a_-^2), \qquad (5.363)$$

it follows that $50\pi a_0^2$ must be an upper bound to the zero energy limit of the elastic cross section. In fact the very elaborate and accurate variational calculations of Schwartz[23] have resulted in the values $a_+ = 5.965 \pm 0.003 a_0$ and $a_- = 1.7686 \pm 0.0002 a_0$ yielding $Q(k_0 = 0) = 44.97 \pi a_0^2$.

Elastic scattering of positrons by hydrogen atoms

Because it is not necessary to satisfy the Pauli exclusion principle in the case of the scattering of positrons by hydrogen atoms, the space function describing the system need not be symmetrized. On the other hand it is necessary to make allowance for virtual positronium formation, that is for the possibility of virtually forming a system consisting of an electron and a positron bound together, in the trial function which is employed with the variational method.

Considering the special case of the scattering of positrons with vanishing energy by atomic hydrogen, it has been found by Massey and Moussa[24] that a trial function having the simple form

$$\Psi_T(r_1, r_2, r_{12}) = \frac{1}{r_1} [(a_T + c_1 e^{-r_1/a_0})(1 - e^{-r_1/a_0}) - r_1] e^{-r_2/a_0}$$
(5.364)

gives rise to a scattering length having the value $0 \cdot 582 a_0$ while the inclusion of the effect of correlation by the use of the trial function

$$\Psi_T(r_1, r_2, r_{12})$$
$$= \frac{1}{r_1} [\{a_T + (c_1 + c_2 r_{12}) e^{-r_1/a_0}\}(1 - e^{-r_1/a_0}) - r_1] e^{-r_2/a_0}$$
(5.365)

yields $0 \cdot 512 a_0$ for the scattering length. One might be led to suppose from these results that the effect of the introduction of a dependence upon the electron–positron separation is small and that the value $0 \cdot 512 a_0$ for the scattering length is close to the true value. However, this conclusion is incorrect as has been demonstrated by Spruch and Rosenberg.[25] Employing the trial function

$$\Psi_T(r_1, r_2, r_{12}) = \frac{1}{r_1} [\{a_T(1 - e^{-r_1/a_0}) - r_1\} e^{-r_2/a_0}$$
$$+ c_1 r_1 e^{-(dr_1 + f_1 r_2)/a_0} + c_2 r_1 e^{-(f_2 r_2 + g r_{12})/a_0}]$$
(5.366)

which makes allowance for virtual positronium formation through the term involving the factor $e^{-g r_{12}/a_0}$, they obtained the value $-1 \cdot 356 a_0$ for the scattering length which is larger in absolute magnitude and opposite in sign to that found using the trial function (5.365). Thus despite the fact that the force acting on the positron due to the static potential field of the hydrogen atom is repulsive, we see that the over-all force acting on the positron is attractive since it produces a negative scattering length. Remembering that the value $-1 \cdot 356 a_0$ for the scattering length is an upper bound to the true scattering length and noting that it is negative in sign, we see that it must provide a lower bound $7 \cdot 36 \pi a_0^2$ to the zero energy limit of the elastic cross section.

Elaborate variational calculations by Schwartz[23] have yielded the value $-2 \cdot 10 a_0$ for the scattering length giving $17 \cdot 6 \pi a_0^2$ for $Q (k_0 = 0)$.

5.9 Elastic scattering of neutrons by deuterons

The analogous problem to the scattering of electrons by hydrogen atoms in the theory of collisions between light nuclei is the scattering of neutrons by deuterons. Distinguishing between the three nucleons involved in the collision by the numerals 1, 2, 3 we may write the Schrödinger equation in the centre of mass system of coordinates as

$$[T + V(12) + V(23) + V(31) - E]\Psi(123) = 0 \quad (5.367)$$

where T is the kinetic energy operator, $V(ij)$ is the potential of interaction between the nucleons i and j, E is the total energy and $\Psi(123)$ is the total wave function of the system. Let $\mathbf{r}_1$, $\mathbf{r}_2$, $\mathbf{r}_3$ denote the position vectors of the three nucleons relative to some fixed origin, particles 1 and 2 being respectively the proton and neutron forming the target deuteron and particle 3 being the incident neutron. Now introducing the vectors

$$\mathbf{R} = \mathbf{r}_2 - \mathbf{r}_1, \quad \mathbf{r} = \mathbf{r}_3 - \tfrac{1}{2}(\mathbf{r}_1 + \mathbf{r}_2) \quad (5.368)$$

where $\mathbf{R}$ is the position vector of the bound neutron relative to the proton and $\mathbf{r}$ is the position vector of the incident neutron relative to the centre of mass of the deuteron, we may write the kinetic energy operator in the form

$$T = -\frac{\hbar^2}{2M}(2\nabla_{\mathbf{R}}^2 + \tfrac{3}{2}\nabla_{\mathbf{r}}^2), \quad (5.369)$$

M being the mass of a neutron or proton. The first term of T corresponds to the kinetic energy of the internal motion of the deuteron while the second term corresponds to the kinetic energy of relative motion of the incident neutron and the deuteron.

Unlike the electrostatic potential between two charged particles, the precise form of the nucleon–nucleon interaction potential is not known. We shall assume that the nuclear interaction between a neutron and a proton and between a pair of neutrons are the same and that they may be represented by the potential corresponding to the so-called ordinary force given by

$$V(ij) = -A\{\tfrac{1}{2}(1 + x) + \tfrac{1}{2}(1 - x)B_{ij}\} e^{-(r_{ij}/\lambda)^2} \quad (5.370)$$

where r_{ij} is the distance between particles i and j, B_{ij} is the Bartlett operator interchanging the spin coordinates of particles i and j, and A, x and λ are constants. The ratio x of the ^1S to ^3S depths of the potential interaction between a neutron and a proton is chosen to be equal to 0·652. Further we take $A = 86·4$ Mev and the range $\lambda = 1·332 \times 10^{-13}$ cm.

Since the spin of the deuteron is 1 and the spin of the incident neutron is $\frac{1}{2}$, the total spin of the whole system may be either $\frac{1}{2}$, corresponding to a doublet state, or $\frac{3}{2}$, corresponding to a quartet state of the system. For convenience we shall confine our discussion to the latter case. Remembering that the total wave function of the system must be antisymmetric with respect to the interchange of the space and spin coordinates of the two neutrons in order to satisfy the Pauli principle, we write the wave function describing quartet scattering in the form

$$\Psi_q(123) = \frac{1}{\sqrt{2}} \{\phi(12, 3) - \phi(13, 2)\}\sigma_q(123) \qquad (5.371)$$

where $\phi(ij, k)$ is a function of the space coordinates of the particles which is symmetrical with respect to the interchange of the coordinates of particles i and j, and $\sigma_q(123)$ is one of the four quartet spin functions all of which are symmetrical with respect to the interchange of the spin coordinates of any pair of particles. Neglecting the polarization of the deuteron by the incident neutron we may put

$$\phi(12, 3) = \psi_D(R)F_q(\mathbf{r}) \qquad (5.372)$$

where $\psi_D(R)$ is the unperturbed deuteron ground state wave function and $F_q(\mathbf{r})$ is the function describing the motion of the incident neutron relative to the centre of mass of the deuteron. For the elastic scattering of neutrons by deuterons with zero energy of relative motion, we take the asymptotic behaviour of $F_q(\mathbf{r})$ for large r to be

$$F_q(\mathbf{r}) \sim \frac{1}{r}(a_q - r) \qquad (5.373)$$

where a_q is the quartet scattering length.

An upper bound to the scattering length may be determined by the application of the Kohn variational method. Thus defining the operator

$$L = -\frac{4M}{3\hbar^2}[T + V(12) + V(23) + V(31)] \quad (5.374)$$

and putting $\epsilon_D = -(4M/3\hbar^2)E_D$ where $-E_D$ ($= 2\cdot2$ Mev) is the binding energy of the deuteron, it can be shown without undue difficulty that

$$a_q \leq a_{qT} - \frac{1}{4\pi}\int \Psi_{qT}(L - \epsilon_D)\Psi_{qT}\, d\tau \quad (5.375)$$

since there does not exist a quartet bound state of a neutron in the field of a deuteron, that is a quartet bound state of the triton. The asymptotic form of the trial function Ψ_{qT} is determined by the behaviour of F_{qT} for large r:

$$F_{qT}(\mathbf{r}) \sim \frac{1}{r}(a_{qT} - r). \quad (5.376)$$

Using a deuteron wave function having the form

$$\psi_D(R) = \sum_{i=1}^{3} c_i\, e^{-\mu_i R^2} \quad (5.377)$$

together with the trial function

$$F_{qT}(\mathbf{r}) = \frac{1}{r}\left[1 - \frac{r}{a_{qT}} - \exp\{-br^2 - d(s^2 - R^2)\}\right] \quad (5.378)$$

where $\mathbf{s} = \mathbf{R} - \mathbf{r}$ and b, d are variable parameters, Sartori and Rubinow obtained $6\cdot9 \times 10^{-13}$ cm as a variational approximation to the scattering length.[26] Although their trial function (5.378) does not possess the asymptotic behaviour given by (5.376) it can be easily renormalized to do so by taking the trial function $F'_{qT}(r) = a_{qT}F_{qT}(r)$ in which case the inequality (5.375) becomes

$$a_q \leq a_{qT} - \frac{a_{qT}^2}{4\pi}\int \Psi_{qT}(L - \epsilon_D)\Psi_{qT}\, d\tau \quad (5.379)$$

and yields the improved value $5\cdot8 \times 10^{-13}$ cm as an upper bound to a_q.[21] This may be compared with the value $5\cdot26 \times 10^{-13}$ cm

SCATTERING THEORY

determined by numerical integration of the appropriate integro-differential equation and the value $6 \cdot 2 \pm 0 \cdot 2 \times 10^{-13}$ cm determined from the experimental data on low energy neutron–deuteron scattering. Our variational upper bound need not, of course, be an upper bound to the experimentally determined scattering length since the potential (5.370) is only an approximation to the actual potential.

5.10 Inelastic scattering of electrons by hydrogen atoms

In order to deal with inelastic scattering of electrons by hydrogen atoms, we require a generalization of the variational principle derived on p. 279. To end this we introduce the functional

$$I(\mathbf{k}_f, \mathbf{k}_i) = \int \Psi_f (L - \epsilon) \Psi_i \, d\tau \qquad (5.380)$$

where the operator L is given by (5.354) and ϵ by (5.355), and the wave functions Ψ_i and Ψ_f correspond to electrons incident with wave vectors $\mathbf{k}_i$ and $-\mathbf{k}_f$ upon atoms in different states i and f respectively. We now expand in the forms

$$\Psi_i(\mathbf{r}_1, \mathbf{r}_2) = \frac{1}{\sqrt{2}} \left[\underset{m}{S} F_m^i(\mathbf{r}_1)\psi_m(\mathbf{r}_2) \pm \underset{m}{S} F_m^i(\mathbf{r}_2)\psi_m(\mathbf{r}_1) \right] \quad (5.381)$$

and

$$\Psi_f(\mathbf{r}_1, \mathbf{r}_2) = \frac{1}{\sqrt{2}} \left[\underset{n}{S} F_n^f(\mathbf{r}_1)\bar{\psi}_n(\mathbf{r}_2) \pm \underset{n}{S} F_n^f(\mathbf{r}_2)\bar{\psi}_n(\mathbf{r}_1) \right], \quad (5.382)$$

where the functions F_m^i and F_n^f have the asymptotic behaviour in the limit of large r

$$F_m^i(\mathbf{r}) \sim e^{i\mathbf{k}_i \cdot \mathbf{r}} + \frac{e^{ik_i r}}{r} f_i(\mathbf{k}, \mathbf{k}_i) \qquad (m = i)$$

$$\sim \frac{e^{ik_m r}}{r} f_{im}(\mathbf{k}, \mathbf{k}_i) \qquad (m \neq i), \quad (5.383)$$

$\mathbf{k}$ being a vector in the direction of $\mathbf{r}$ such that $|\mathbf{k}| = k_m$; and

$$F_n^f(\mathbf{r}) \sim e^{-i\mathbf{k}_f \cdot \mathbf{r}} + \frac{e^{ik_f r}}{r} f_f(\mathbf{k}, -\mathbf{k}_f) \qquad (n = f)$$

$$\sim \frac{e^{ik_n r}}{r} f_{fn}(\mathbf{k}, -\mathbf{k}_f) \qquad (n \neq f), \quad (5.384)$$

k being a vector in the direction of **r** such that $|\mathbf{k}| = k_n$. Thus $f_{im}(\mathbf{k}, \mathbf{k}_i)$ is the scattering amplitude for the excitation of the state m from the state i by electrons incident in the direction of $\mathbf{k}_i$ and scattered in the direction of **k**.

Substituting (5.381) and (5.382) into (5.380) we find that I may be expressed as

$$I(\mathbf{k}_f, \mathbf{k}_i) = \sum_n \int F_n^f(\mathbf{r}_1)\Bigg[(\nabla_1^2 + k_n^2)F_n^i(\mathbf{r}_1) \\ - \frac{2m}{\hbar^2}\sum_m \left\{V_{nm}(\mathbf{r}_1)F_m^i(\mathbf{r}_1) \mp \int K_{nm}(\mathbf{r}_1, \mathbf{r}_2)F_m^i(\mathbf{r}_2)\, d\mathbf{r}_2\right\}\Bigg] d\mathbf{r}_1 \tag{5.385}$$

where V_{nm} and K_{nm} are given by (5.335) and (5.336) respectively.

Now let $\delta I(\mathbf{k}_f, \mathbf{k}_i)$ be the change in $I(\mathbf{k}_f, \mathbf{k}_i)$ due to infinitesimal variations δF_n^f and δF_m^i in the functions F_n^f and F_m^i respectively. Then we get

$$\delta I(\mathbf{k}_f, \mathbf{k}_i) = \sum_n \int \delta F_n^f(\mathbf{r}_1)\Bigg[(\nabla_1^2 + k_n^2)F_n^i(\mathbf{r}_1) \\ - \frac{2m}{\hbar^2}\sum_m \left\{V_{nm}(\mathbf{r}_1)F_m^i(\mathbf{r}_1) \mp \int K_{nm}(\mathbf{r}_1, \mathbf{r}_2)F_m^i(\mathbf{r}_2)\, d\mathbf{r}_2\right\}\Bigg] d\mathbf{r}_1 \\ + \sum_n \int \delta F_n^i(\mathbf{r}_1)\Bigg[(\nabla_1^2 + k_n^2)F_n^f(\mathbf{r}_1) \\ - \frac{2m}{\hbar^2}\sum_m \left\{V_{mn}(\mathbf{r}_1)F_m^f(\mathbf{r}_1) \mp \int K_{mn}(\mathbf{r}_2, \mathbf{r}_1)F_m^f(\mathbf{r}_2)\, d\mathbf{r}_2\right\}\Bigg] d\mathbf{r}_1 \\ + \sum_n \oint_S \{F_n^f(\mathbf{r})\nabla\, \delta F_n^i(\mathbf{r}) - \delta F_n^i(\mathbf{r})\nabla F_n^f(\mathbf{r})\}\cdot d\mathbf{S}. \tag{5.386}$$

Taking S to be a large spherical surface of radius r, the surface integral becomes

$$\oint_S \left\{e^{-i\mathbf{k}_f \cdot \mathbf{r}}\frac{\partial}{\partial r}\left(\frac{e^{ik_f r}}{r}\right) - \frac{e^{ik_f r}}{r}\frac{\partial}{\partial r}(e^{-i\mathbf{k}_f \cdot \mathbf{r}})\right\}\delta f_{if}(\mathbf{k}, \mathbf{k}_i)\, dS$$

which can be shown to be equal to $-4\pi\, \delta f_{if}(\mathbf{k}_f, \mathbf{k}_i)$ by following an analogous procedure to that employed on p. 236 ff. Hence, if the F_n^i and F_n^f satisfy the Hartree–Fock equations

$$(\nabla_1^2 + k_n^2)F_n(\mathbf{r}_1) - \frac{2m}{\hbar^2} \sum_m \left\{ V_{nm}(\mathbf{r}_1)F_m(\mathbf{r}_1) \right.$$
$$\left. \mp \int K_{nm}(\mathbf{r}_1, \mathbf{r}_2)F_m(\mathbf{r}_2)\, d\mathbf{r}_2 \right\} = 0 \quad (5.387)$$

and their conjugate equations respectively, we have the variational principle

$$\delta I(\mathbf{k}_f, \mathbf{k}_i) = -4\pi\, \delta f_{if}(\mathbf{k}_f, \mathbf{k}_i). \quad (5.388)$$

Conversely if $I(\mathbf{k}_f, \mathbf{k}_i) + 4\pi f_{if}(\mathbf{k}_f, \mathbf{k}_i)$ is stationary for arbitrary small variations of F_m^i and F_n^f, the Hartree–Fock equations must hold.

5.11 Virial theorem

We now proceed to establish the virial theorem for the scattering of particles by a potential $V(\mathbf{r})$ employing the approach introduced by Demkov.[27] Putting $U(\mathbf{r}) = (2m/\hbar^2)V(\mathbf{r})$, the relevant Schrödinger equation takes the form

$$\{\nabla^2 + k^2 - U(\mathbf{r})\}\psi(\mathbf{r}) = 0, \quad (5.389)$$

the asymptotic behaviour of the solution $\psi_j(\mathbf{r}, k)$ of this equation associated with particles incident in the direction of the vector $\mathbf{k}_j$ being given by

$$\psi_j(\mathbf{r}, k) \sim e^{i\mathbf{k}_j \cdot \mathbf{r}} + \frac{e^{ikr}}{kr} f(\mathbf{k}, \mathbf{k}_j) \quad (5.390)$$

where $\mathbf{k}$ is a vector in the direction of $\mathbf{r}$ such that $|\mathbf{k}| = |\mathbf{k}_j| = k$.

The method we shall employ to derive the virial theorem is based upon the Kohn variational principle for the scattering amplitude which we write in the form

$$\delta I_k(\psi_2, \psi_1) = -\frac{4\pi}{k}\, \delta f_1(-\mathbf{k}_2, \mathbf{k}_1) \quad (5.391)$$

where

$$I_k(\chi_2, \chi_1) = \int \chi_2(\mathbf{r}, k)\{\nabla^2 + k^2 - U(\mathbf{r})\}\chi_1(\mathbf{r}, k)\, d\mathbf{r}, \quad (5.392)$$

the functions $\chi_j(\mathbf{r}, \mathbf{k})$ having the asymptotic behaviour for large r

$$\chi_j(\mathbf{r}, k) \sim e^{i\mathbf{k}_j \cdot \mathbf{r}} + \frac{e^{ikr}}{kr} f_j(\mathbf{k}, \mathbf{k}_j). \tag{5.393}$$

We begin by introducing a scaling parameter μ and turn to the consideration of the scattering of particles with wave number μk. The exact solutions for such particles are denoted by $\psi_j(\mathbf{r}, \mu k)$. However, as our varied functions in the Kohn variational method we take

$$\chi_j(\mathbf{r}, \mu k) = \psi_j(\mu \mathbf{r}, k) \tag{5.394}$$

which are just the exact solutions for particles with wave number k scaled by means of the parameter μ. Substituting into (5.392) we obtain

$$I_{\mu k}(\chi_2, \chi_1) = \int \psi_2(\mu \mathbf{r}, k)\{\nabla^2 + (\mu k)^2 - U(\mathbf{r})\}\psi_1(\mu \mathbf{r}, k)\, d\mathbf{r} \tag{5.395}$$

and so, putting $\mu \mathbf{r} = \boldsymbol{\rho}$, we get

$$I_{\mu k}(\chi_2, \chi_1) = \int \psi_2(\boldsymbol{\rho}, k)\left\{\frac{1}{\mu}(\nabla_\rho^2 + k^2) - \frac{1}{\mu^3} U(\boldsymbol{\rho}/\mu)\right\}\psi_1(\boldsymbol{\rho}, k)\, d\boldsymbol{\rho}.$$

Hence, using equation (5.389) we find

$$I_{\mu k}(\chi_2, \chi_1) = \int \psi_2(\boldsymbol{\rho}, k)\left\{\frac{1}{\mu} U(\boldsymbol{\rho}) - \frac{1}{\mu^3} U(\boldsymbol{\rho}/\mu)\right\}\psi_1(\boldsymbol{\rho}, k)\, d\boldsymbol{\rho}.$$

Taking $\mu = 1 + \epsilon$ where ϵ is a small quantity we see that

$$\frac{1}{\mu} U(\boldsymbol{\rho}) - \frac{1}{\mu^3} U(\boldsymbol{\rho}/\mu)$$
$$\cong (1 - \epsilon)U(\boldsymbol{\rho}) - (1 - 3\epsilon)\{U(\boldsymbol{\rho}) - \epsilon \boldsymbol{\rho} \cdot \nabla_\rho U(\boldsymbol{\rho})\}$$
$$\cong \epsilon\{2U(\boldsymbol{\rho}) + \boldsymbol{\rho} \cdot \nabla_\rho U(\boldsymbol{\rho})\}$$

and thus, to the first order in ϵ, we have

$$I_{k+\epsilon k}(\chi_2, \chi_1) = \epsilon \int \psi_2(\mathbf{r}, k)\{2U(\mathbf{r}) + \mathbf{r} \cdot \nabla U(\mathbf{r})\}\psi_1(\mathbf{r}, k)\, d\mathbf{r} \tag{5.396}$$

where we have replaced $\boldsymbol{\rho}$ by the vector $\mathbf{r}$. Since

$$\delta f_1(-\mathbf{k}_2, \mathbf{k}_1) = -\epsilon k \frac{\partial}{\partial k} f(-\mathbf{k}_2, \mathbf{k}_1)$$

SCATTERING THEORY

it follows from the Kohn variational principle (5.391) that

$$\int \psi_2(\mathbf{r}, k)\{2U(\mathbf{r}) + \mathbf{r}\cdot\nabla U(\mathbf{r})\}\psi_1(\mathbf{r}, k)\, d\mathbf{r} = 4\pi \frac{\partial}{\partial k} f(-\mathbf{k}_2, \mathbf{k}_1) \tag{5.397}$$

which is the virial theorem for scattering.

A similar result may be obtained for the individual terms of the partial wave expansion if the potential is spherically symmetrical. In this case we consider the radial equation

$$\left\{\frac{d^2}{dr^2} + k^2 - U(r)\right\}u(r, k) = 0 \tag{5.398}$$

whose solution has the asymptotic behaviour for large r given by

$$u(r, k) \sim \sin(kr + \eta). \tag{5.399}$$

For convenience we have absorbed the centrifugal term $l(l + 1)/r^2$ into $U(r)$. We now introduce the functional

$$I_k[v] = \int_0^\infty v(r, k)\left\{\frac{d^2}{dr^2} + k^2 - U(r)\right\}v(r, k)\, dr \tag{5.400}$$

where for large r the function $v(r, k)$ has the asymptotic behaviour

$$v(r, k) \sim \sin(kr + \theta), \tag{5.401}$$

and employ the Hulthén variational principle in the form

$$\delta I_k[u] = -k\delta\eta. \tag{5.402}$$

We again take a scaling parameter μ and consider particles with wave numbers μk. Choosing as our varied function

$$v(r, \mu k) = u(\mu r, k) \tag{5.403}$$

we obtain

$$I_{\mu k}[v] = \int_0^\infty u(\mu r, k)\left\{\frac{d^2}{dr^2} + (\mu k)^2 - U(r)\right\}u(\mu r, k)\, dr \tag{5.404}$$

which, or putting $\rho = \mu r$, becomes

$$I_{\mu k}[v] = \int_0^\infty u(\rho, k)\left\{\mu\left(\frac{d^2}{d\rho^2} + k^2\right) - \frac{1}{\mu}U(\rho/\mu)\right\}u(\rho, k)\, d\rho$$

$$= \int_0^\infty u(\rho, k)\left\{\mu U(\rho) - \frac{1}{\mu}U(\rho/\mu)\right\}u(\rho, k)\, d\rho.$$

Taking $\mu = 1 + \epsilon$ we see that to the first order in ϵ

$$\mu U(\rho) - \frac{1}{\mu} U(\rho/\mu) = \epsilon\left\{2U(\rho) + \rho \frac{d}{d\rho} U(\rho)\right\}$$

and so to the same order

$$I_{\mu k}[v] = \epsilon \int_0^\infty u(r, k)\left\{2U(r) + r\frac{d}{dr} U(r)\right\}u(r, k)\, dr \quad (5.405)$$

where we have substituted r for ρ.

Since the change in η due to the use of the varied function $v(r, \mu k) = u(\mu r, k)$ instead of the exact function $u(r, \mu k)$ is given by

$$\delta\eta = -\epsilon k \frac{\partial \eta}{\partial k},$$

it follows from (5.402) that

$$\int_0^\infty u(r, k)\left\{2U(r) + r\frac{d}{dr} U(r)\right\}u(r, k)\, dr = k^2 \frac{\partial \eta}{\partial k} \quad (5.406)$$

which is the one-dimensional form of the virial theorem for scattering. It is of interest to note that the virial theorem provides a formula for the collision lifetime which is defined as $T = 2\hbar \partial \eta/\partial E$, where $E = \hbar^2 k^2/2m$ is the total energy of the incident particle. In fact we find that

$$T = \frac{2m}{\hbar k^3} \int_0^\infty u(r, k)\left\{2U(r) + r\frac{d}{dr} U(r)\right\}u(r, k)\, dr.$$

5.12 Time-dependent scattering theory

Thus far in this chapter we have examined the theory of scattering from an entirely time-independent point of view. However, it is also of considerable importance to investigate the evolution with time of a scattering process and we shall now turn our attention to this aspect of collision theory. We consider a pair of interacting quantum mechanical systems described by a total Hamiltonian operator H which can be expressed as a sum of an unperturbed Hamiltonian operator H_0 and an interaction potential V which vanishes in the limit when the interacting systems are at an

SCATTERING THEORY

infinite separation. The time-dependent Schrödinger equation which determines the scattering is then

$$i\hbar \frac{\partial}{\partial t} \Psi(t) = (H_0 + V)\Psi(t) \qquad (5.407)$$

where we have denoted the state vector of the entire system by $\Psi(t)$ rather than by the notation $|t\rangle$ used in section 2.7 on quantum mechanics, which is less convenient for our present purpose. The time dependence arising from the unperturbed Hamiltonian H_0 can now be removed by introducing the unitary transformation to the so-called *interaction representation*

$$\Phi(t) = e^{iH_0 t/\hbar} \Psi(t), \qquad (5.408)$$

since we may rewrite (5.407) in the alternative form

$$i\hbar \frac{\partial}{\partial t} \Phi(t) = V(t)\Phi(t) \qquad (5.409)$$

where

$$V(t) = e^{iH_0 t/\hbar} V e^{-iH_0 t/\hbar}. \qquad (5.410)$$

We now express the state vector $\Phi(t)$ representing the dynamical system at time t in terms of the initial state vector $\Phi(-\infty)$ corresponding to $t = -\infty$ by the formula

$$\Phi(t) = U_+(t)\Phi(-\infty) \qquad (5.411)$$

where $U_+(t)$ is a unitary operator which satisfies the terminal condition

$$U_+(-\infty) = I. \qquad (5.412)$$

By substituting (5.411) into (5.409) we see at once that the operator $U_+(t)$ is a solution of the equation

$$i\hbar \frac{\partial}{\partial t} U_+(t) = V(t)U_+(t). \qquad (5.413)$$

Remembering that $U_+(t)$ obeys the terminal condition (5.412), it follows that this operator may be expressed in the integral equation form

$$U_+(t) = I - \frac{i}{\hbar} \int_{-\infty}^{t} V(t')U_+(t')\, dt' \qquad (5.414)$$

which can be rewritten as
$$U_+(t) = I - \frac{i}{\hbar} \int_{-\infty}^{\infty} \eta(t - t')V(t')U_+(t')\,dt' \tag{5.415}$$
by introducing the step function
$$\eta(x) = \begin{cases} 0 & (x < 0) \\ 1 & (x > 0). \end{cases} \tag{5.416}$$

The state vector $\Phi(\infty)$ representing the total system in its final state corresponding to $t = \infty$ may be written
$$\Phi(\infty) = S\Phi(-\infty) \tag{5.417}$$
where
$$S = U_+(\infty) \tag{5.418}$$
is known as the *collision operator*. Using (5.414) we may express S in the integral equation form
$$S = I - \frac{i}{\hbar} \int_{-\infty}^{\infty} V(t)U_+(t)\,dt. \tag{5.419}$$

An alternative development can be obtained by writing the state vector $\Phi(t)$ in terms of the final state vector $\Phi(\infty)$ by the formula
$$\Phi(t) = U_-(t)\Phi(\infty) \tag{5.420}$$
where $U_-(t)$ is a unitary operator which obeys the terminal condition
$$U_-(\infty) = I \tag{5.421}$$
and which can be readily shown to be a solution of the differential equation
$$i\hbar \frac{\partial}{\partial t} U_-(t) = V(t)U_-(t) \tag{5.422}$$
as well as the integral equation
$$U_-(t) = I + \frac{i}{\hbar} \int_t^{\infty} V(t')U_-(t')\,dt'$$
$$= I + \frac{i}{\hbar} \int_{-\infty}^{\infty} \eta(t' - t)V(t')U_-(t')\,dt'. \tag{5.423}$$

SCATTERING THEORY

Making use of (5.420) and (5.417), and then comparing the resulting formula with (5.411) we obtain

$$U_+(t) = U_-(t)S \qquad (5.424)$$

which yields, on allowing $t \to -\infty$, the following integral equation:

$$S^{-1} = U_-(-\infty) = I + \frac{i}{\hbar}\int_{-\infty}^{\infty} V(t)U_-(t)\,dt. \qquad (5.425)$$

We are now in a position to establish the variational principles for the collision operator first obtained by Lippmann and Schwinger.[28] To this end we consider the functional of the operators $U_+(t)$ and $U_-^*(t)$ given by

$$S[U_-^*, U_+] = U_+(\infty) - \int_{-\infty}^{\infty} U_-^*(t)\left\{\frac{\partial}{\partial t} + \frac{i}{\hbar}V(t)\right\}U_+(t)\,dt \qquad (5.426)$$

where $U_+(t)$ and $U_-^*(t)$ are chosen to obey the terminal conditions specified by (5.412) and (5.421) respectively. Then if $\delta U_+(t)$ and $\delta U_-^*(t)$ are small independent variations of these operators satisfying

$$\delta U_+(-\infty) = 0, \qquad \delta U_-^*(\infty) = 0 \qquad (5.427)$$

the ensuing variation in $S[U_-^*, U_+]$ is given by

$$\delta S = -\int_{-\infty}^{\infty} \delta U_-^*(t)\left\{\frac{\partial}{\partial t} + \frac{i}{\hbar}V(t)\right\}U_+(t)\,dt$$
$$+ \int_{-\infty}^{\infty} \left[\left\{\frac{\partial}{\partial t} + \frac{i}{\hbar}V(t)\right\}U_-(t)\right]^* \delta U_+(t)\,dt \qquad (5.428)$$

and thus we see that $S[U_-^*, U_+]$ is stationary if $U_+(t)$ and $U_-(t)$ are solutions of the differential equations (5.413) and (5.422) respectively, and furthermore that the stationary value of $S[U_-^*, U_+]$ is just the collision operator S. The first Born approximation to the collision operator

$$S = I - \frac{i}{\hbar}\int_{-\infty}^{\infty} V(t)\,dt \qquad (5.429)$$

follows by substituting the approximations $U_-(t) = U_+(t) = I$ in (5.426).

An alternative variational principle for the collision operator can be established by considering the functional

$$S[U_-^*, U_+] = I - \frac{i}{\hbar} \int_{-\infty}^{\infty} \{U_-^*(t)V(t) + V(t)U_+(t)\} \, dt$$

$$+ \frac{i}{\hbar} \int_{-\infty}^{\infty} U_-^*(t)V(t)U_+(t) \, dt$$

$$+ \left(\frac{i}{\hbar}\right)^2 \int_{-\infty}^{\infty} \int_{-\infty}^{\infty} U_-^*(t)V(t)\eta(t-t')$$
$$\times V(t')U_+(t') \, dt \, dt'. \quad (5.430)$$

Since the variation in this functional due to small independent variations $\delta U_+(t)$ and $\delta U_-^*(t)$ is given by

$$\delta S = \frac{i}{\hbar} \int_{-\infty}^{\infty} \delta U_-^*(t)V(t)\left[U_+(t) - I\right.$$
$$\left. + \frac{i}{\hbar} \int_{-\infty}^{\infty} \eta(t-t')V(t')U_+(t') \, dt'\right] dt$$

$$+ \frac{i}{\hbar} \int_{-\infty}^{\infty} \left[U_-(t) - I - \frac{i}{\hbar} \int_{-\infty}^{\infty} \eta(t'-t)V(t')U_-(t') \, dt'\right]^*$$
$$\times V(t) \, \delta U_+(t) \, dt, \quad (5.431)$$

it follows that $S[U_-^*, U_+]$ is stationary if $U_+(t)$ and $U_-(t)$ satisfy the integral equations (5.415) and (5.423) respectively. In addition it can be readily verified that the stationary value of $S[U_-^*, U_+]$ is equal to the collision operator S. The second Born approximation to the collision operator

$$S = I - \frac{i}{\hbar} \int_{-\infty}^{\infty} V(t) \, dt + \left(\frac{i}{\hbar}\right)^2 \int_{-\infty}^{\infty} \int_{-\infty}^{\infty} V(t)\eta(t-t')V(t') \, dt \, dt'$$
$$(5.432)$$

can be obtained by setting $U_-(t) = U_+(t) = I$ in (5.430). Since the variational principle based on the functional (5.426) only gave the first Born approximation with the same substitution, this demonstrates the superiority of the variational formula (5.430) over (5.426).

SCATTERING THEORY

A simplification of some of the formulae derived in this section can be achieved by introducing another operator T defined by

$$T = S - I, \qquad (5.433)$$

for then we may express equations (5.419) and (5.425) in the alternative forms

$$T = -\frac{i}{\hbar} \int_{-\infty}^{\infty} V(t) U_+(t) \, dt \qquad (5.434)$$

and

$$T^* = \frac{i}{\hbar} \int_{-\infty}^{\infty} V(t) U_-(t) \, dt \qquad (5.435)$$

since $S^* = S^{-1}$.

Let us now suppose that the initial state of the whole system is represented by the eigenvector ψ_i of the unperturbed Hamiltonian H_0 associated with eigenenergy E_i. Then we see that the probability of finding the total system in a state represented by the eigenvector ψ_f associated with eigenenergy E_f is

$$P_{fi} = |(\psi_f, S\psi_i)|^2, \qquad (5.436)$$

where we have used the notation $(\psi_f, S\psi_i)$ to denote the scalar product of $S\psi_i$ and the complex conjugate of ψ_f rather than the notation used in section 2.7. P_{fi} is called the *transition probability*. It may be rewritten as

$$P_{fi} = |T_{fi}|^2, \qquad (5.437)$$

where

$$T_{fi} = (\psi_f, T\psi_i), \qquad (5.438)$$

provided the initial and final states are different so that ψ_i and ψ_f are orthogonal. Using formulae (5.434) and (5.410) we may put T_{fi} into the form

$$T_{fi} = -\frac{i}{\hbar} \int_{-\infty}^{\infty} (\psi_f, \, e^{iH_0 t/\hbar} V \, e^{-iH_0 t/\hbar} U_+(t) \psi_i) \, dt$$

which we may express equivalently as

$$T_{fi} = -\frac{i}{\hbar} (\psi_f, V\Psi_i^+(E_f)) \qquad (5.439)$$

where

$$\Psi_i^+(E) = \int_{-\infty}^{\infty} dt \, e^{i(E-H_0)t/\hbar} U_+(t) \psi_i. \qquad (5.440)$$

Now replacing $U_+(t)$ in the above formula by its integral equation form (5.415) we are led to the result

$$\Psi_i^+(E) = \int_{-\infty}^{\infty} dt\, e^{i(E-E_i)t/\hbar}\psi_i - \frac{i}{\hbar}\int_0^{\infty} d\tau\, e^{i(E-H_0)\tau/\hbar} V\Psi_i^+(E) \quad (5.441)$$

where we have set $\tau = |t - t'|$. For the purpose of evaluating the integrals in this equation for $\Psi_i^+(E)$, we introduce integrating factors and employ the formula for the Dirac δ function given by

$$\lim_{\epsilon \to 0} \int_{-\infty}^{\infty} dt\, e^{i(E-E_i)t/\hbar} e^{-\epsilon|t|/\hbar} = 2\pi\hbar\, \delta(E - E_i) \quad (5.442)$$

and the integral

$$-\frac{i}{\hbar}\int_0^{\infty} d\tau\, e^{i(E-H_0)\tau/\hbar} e^{-\epsilon\tau/\hbar} = \frac{1}{E - H_0 + i\epsilon}. \quad (5.443)$$

Then we find that equation (5.441) may be rewritten in the form

$$\Psi_i^+(E) = 2\pi\hbar\, \delta(E - E_i)\psi_i + \lim_{\epsilon \to 0} \frac{1}{E - H_0 + i\epsilon}\, V\Psi_i^+(E). \quad (5.444)$$

In a similar fashion, starting from the formula (5.435), it can be readily verified that

$$T_{if} = -\frac{i}{\hbar}(\Psi_i^-(E_f), V\psi_f) \quad (5.445)$$

where

$$\Psi_i^-(E) = \int_{-\infty}^{\infty} dt\, e^{i(E-H_0)t/\hbar} U_-(t)\psi_i \quad (5.446)$$

satisfies the equation

$$\Psi_i^-(E) = 2\pi\hbar\, \delta(E - E_i)\psi_i + \lim_{\epsilon \to 0} \frac{1}{E - H_0 - i\epsilon}\, V\Psi_i^-(E). \quad (5.447)$$

If we now set

$$\Psi_i^\pm(E) = 2\pi\hbar\, \delta(E - E_i)\Psi_i^\pm, \quad (5.448)$$

we arrive at the *Lippmann–Schwinger equation*

$$\Psi_i^\pm = \psi_i + G_\pm(E_i) V\Psi_i^\pm \quad (5.449)$$

where

$$G_{\pm}(E_i) = \lim_{\epsilon \to 0} \frac{1}{E_i - H_0 \pm i\epsilon} \qquad (5.450)$$

are Green's functions for the homogeneous equation $(H_0 - E_i)\psi_i = 0$, with the $+$ and $-$ signs indicating outgoing and incoming scattered waves. Thus we see that the formula for $\Psi_i^{\pm}$ given by the Lippmann–Schwinger equation is just a formal way of expressing the solution of the time-independent Schrödinger equation for the whole system:

$$(H_0 - E_i)\Psi_i = -V\Psi_i. \qquad (5.451)$$

In addition we take note of the identity

$$\lim_{\epsilon \to 0} \frac{1}{\omega \pm i\epsilon} = \lim_{\epsilon \to 0} \left\{ \frac{\omega}{\omega^2 + \epsilon^2} \mp \frac{i\epsilon}{\omega^2 + \epsilon^2} \right\} = P \frac{1}{\omega} \mp i\pi\, \delta(\omega),$$

where P indicates that the principal value is to be chosen when an integration past the ω^{-1} singularity is performed, since it reveals that the Green's functions given by (5.450) can be written alternatively as

$$G_{\pm}(E_i) = P \frac{1}{E_i - H_0} \mp i\pi\, \delta(E_i - H_0). \qquad (5.452)$$

Returning now to formulae (5.439) and (5.445) for the matrix elements T_{fi} and T_{if}, and using (5.448), we see that we can write

$$T_{fi} = -2\pi i\, \delta(E_f - E_i) \mathcal{T}_{fi} \qquad (5.453)$$

where

$$\mathcal{T}_{fi} = (\psi_f, V\Psi_i^+) = (\Psi_f^-, V\psi_i) \qquad (5.454)$$

from which it follows that the transition probability becomes

$$P_{fi} = 4\pi^2 \{\delta(E_f - E_i)\}^2 |\mathcal{T}_{fi}|^2$$

or

$$P_{fi} = \frac{2\pi}{\hbar} \delta(E_f - E_i) |\mathcal{T}_{fi}|^2 \lim_{\epsilon \to 0} \int_{-\infty}^{\infty} e^{i(E_f - E_i)t/\hbar}\, e^{-\epsilon|t|/\hbar}\, dt$$

on replacing one of the δ functions by its equivalent integral expression (5.442). Employing the characteristic property of the remaining δ function we now find that

$$P_{fi} = \frac{2\pi}{\hbar} \delta(E_f - E_i)|\mathcal{T}_{fi}|^2 \int_{-\infty}^{\infty} dt \qquad (5.455)$$

which is infinite because it corresponds to the transition probability for an infinite time interval. However, we can immediately deduce from this expression for P_{fi} that the *transition rate* from the state i to the state f must be given by

$$w_{fi} = \frac{2\pi}{\hbar} \delta(E_f - E_i)|\mathcal{T}_{fi}|^2. \qquad (5.456)$$

A variational principle for the transition matrix element $\mathcal{T}_{fi}$ can be readily obtained. Thus consider the functional

$$\mathcal{T}[\Psi_f^-, \Psi_i^+] = (\Psi_f^-, V\psi_i) + (\psi_f, V\Psi_i^+) - (\Psi_f^-, V\Psi_i^+) \\ + (\Psi_f^-, VG_+(E)V\Psi_i^+) \qquad (5.457)$$

where $E = E_i = E_f$. First we note that the change in $\mathcal{T}[\Psi_f^-, \Psi_i^+]$ resulting from small independent variations $\delta\Psi_i^+$ and $\delta\Psi_f^-$ is given by

$$\delta\mathcal{T} = (\delta\Psi_f^-, V\{\psi_i - \Psi_i^+ + G_+(E)V\Psi_i^+\}) \\ + (\{\psi_f - \Psi_f^- + G_-(E)V\Psi_f^-\}, V\,\delta\Psi_i^+) \qquad (5.458)$$

from which it is plain that $\mathcal{T}[\Psi_f^-, \Psi_i^+]$ is stationary when Ψ_i^+ and Ψ_f^- are solutions of the Lippmann–Schwinger equation. Also it can be easily seen that the stationary value of $\mathcal{T}[\Psi_f^-, \Psi_i^+]$ is the transition matrix element $\mathcal{T}_{fi}$. Consequently the stationary property of the functional (5.457) provides a variational principle for $\mathcal{T}_{fi}$.

In conclusion, we can readily derive an expression for the transition matrix element $\mathcal{T}_{fi}$ which is homogeneous in Ψ_i^+ and Ψ_f^-. Thus the Lippmann–Schwinger equation for Ψ_i^+ gives

$$(\Psi_f^-, V\psi_i) = (\Psi_f^-, V\Psi_i^+) - (\Psi_f^-, VG_+(E)V\Psi_i^+)$$

which enables us to write

$$\mathcal{T}_{fi} = \frac{(\Psi_f^-, V\psi_i)(\psi_f, V\Psi_i^+)}{(\Psi_f^-, V\Psi_i^+) - (\Psi_f^-, VG_+(E)V\Psi_i^+)}. \quad (5.459)$$

This expression for the transition matrix element has the customary form originated by Schwinger and is stationary with respect to infinitesimal variations of Ψ_i^+ and Ψ_f^-.

References

1. Watson, G. N., *Theory of Bessel Functions*, Macmillan, New York, 1944, p. 128.
2. Levine, H. and Schwinger, J., *Phys. Rev.*, **74**, 958 (1948); **75**, 1423 (1949).
3. Dalitz, R. H., *Proc. Roy. Soc.*, **A206**, 509 (1951).
4. Kacser, C., *Nuovo Cimento*, **13**, 303 (1959).
5. Kohn, W., *Phys. Rev.*, **74**, 1763 (1948).
6. Hulthén, L., *Kgl. Fysiogr. Sällsk. Lund. Förh.*, **14**, No. 21 (1944).
7. Hulthén, L., *Arkiv. Mat. Astron. Fysik*, **35A**, No. 25 (1948).
8. Tamm, I. E., *J. Exp. Theor. Phys. U.S.S.R.*, **18**, 337 (1948).
9. Demkov, Yu. N. and Shepelenko, F. P., *Soviet Phys. J.E.T.P.* **6** (33), 1144 (1958).
10. Kato, T., *Phys. Rev.*, **80**, 475 (1950).
11. Feshbach, H. and Rubinow, S. I., *Phys. Rev.*, **88**, 484 (1952).
12. Davison, B., *Phys. Rev.*, **71**, 694 (1947).
13. Kato, T., *Prog. Theor. Phys. (Kyoto)*, **6**, 295 (1951).
14. Sugar, R. and Blankenbecler, R., *Phys. Rev.*, **136**, B472 (1964).
15. Blatt, J. M. and Jackson, J. D., *Phys. Rev.*, **76**, 18 (1949).
16. O'Malley, T. F., Spruch, L. and Rosenberg, L., *J. Math. Phys.*, **2**, 491 (1961).
17. Spruch, L. and Rosenberg, L., *Phys. Rev.*, **116**, 1034 (1959).
18. Rosenberg, L., Spruch, L. and O'Malley, T. F., *Phys. Rev.*, **118**, 184 (1960).
19. Hahn, Y., O'Malley, T. F. and Spruch, L., *Phys. Rev.*, **128**, 932 (1962); **130**, 381 (1963).
20. Massey, H. S. W. and Moiseiwitsch, B. L., *Proc. Roy. Soc.*, **A205**, 483 (1951).
21. Spruch, L. and Rosenberg, L., *Phys. Rev.*, **117**, 1095 (1960).
22. Rosenberg, L., Spruch, L. and O'Malley, T. F., *Phys. Rev.*, **119**, 164 (1960).

23. Schwartz, C., *Phys. Rev.*, **124,** 1468 (1961).
24. Massey, H. S. W. and Moussa, A. H. A., *Proc. Phys. Soc.*, **71,** 38 (1958).
25. Spruch, L. and Rosenberg, L., *Phys. Rev.*, **117,** 143 (1960).
26. Sartori, L. and Rubinow, S. I., *Phys. Rev.*, **112,** 214 (1958).
27. Demkov, Yu. N., *Dokl. Akad. Nauk S.S.S.R.*, **89,** 249 (1953).
28. Lippmann, B. A. and Schwinger, J., *Phys. Rev.*, **79,** 469 (1950).

Bibliography

Additional material as well as amplification and alternative treatments of some of the topics discussed in the present volume may be found in the following books:

Courant, R. and Hilbert, D., *Methods of Mathematical Physics*, Vol. I, Interscience, New York, 1953.
Demkov, Yu. N. (trans. by Kemmer, N.), *Variational Principles in the Theory of Collisions*, Pergamon, Oxford, 1963.
Dirac, P. A. M., *The Principles of Quantum Mechanics*, 3rd ed., Oxford University Press, London, 1947.
Goldstein, H., *Classical Mechanics*, Addison-Wesley, Cambridge, Mass., 1950.
Gould, S. H., *Variational Methods for Eigenvalue Problems*, University of Toronto Press, 1957.
Lanczos, C., *The Variational Principles of Mechanics*, 2nd ed., University of Toronto Press, 1963.
Mikhlin, S. G. (trans. by Boddington, T.), *Variational Methods in Mathematical Physics*, Pergamon, Oxford, 1964.
Morse, P. M. and Feshbach, H., *Methods of Theoretical Physics*, Vols. I and II, McGraw-Hill, New York, 1953.
Schiff, L. I., *Quantum Mechanics*, 2nd ed., McGraw-Hill, New York, 1955.
Spruch, L., "Minimum principles in scattering theory", *Lectures in Theoretical Physics*, Vol. IV (Ed. by Brittin, W. E., Downs, B. W. and Downs, J.), Interscience, New York, 1962.
Temple, G. and Bickley, W. G., *Rayleigh's Principle*, Dover, London, 1933.
Wentzel, G., *Quantum Theory of Fields*, Interscience, New York, 1949.
Whittaker, E. T., *Analytical Dynamics*, 4th ed., Cambridge University Press, 1937.
Yourgrau, W. and Mandelstam, S., *Variational Principles in Dynamics and Quantum Theory*, 2nd ed., Pitman, London, 1960.

Index

Action, 23, 55
Adjoint operator, 197
Adjoint solution, 215, 219
Angular frequency, 136
   stationary property of, 140–142
Associated Legendre functions, 104
Atomic eigenenergies, 170–183
Auxiliary conditions, 15
   (*see also* Subsidiary conditions)

Be^{2+}, 172–181
Beach, J. Y., 207
Bessel functions, 221
Blankenbecler, R., 259, 261
Born approximation, first and second, 231, 233–235, 238–239, 243, 295–296
Born expansion, 230–231, 238–239

Canonical momenta, 24
Charge and current densities, 11, 105
   for complex field, 115
   for Dirac equation, 128
   for scalar meson field, 121–122
   for Schrödinger equation, 118–119
Christoffel symbols, 47–48
Collision operator, 294
Commuting operators, 67–68
Completeness, 68, 152
Complex conjugate vector, 64
Composition law of transformation functions, 68
Conservation of energy, 4–5, 12–13
Conservative force, 5
Constraints, holonomic, 6
   invariable, 12–13
   non-holonomic, 16
Contact transformations, 33–35
Continuity equation, for charged scalar meson field, 122
   for complex field, 116
   for electromagnetic field, 107
   for fluid motion, 85
   for Schrödinger equation, 119
Convergence in the mean, 152
Coolidge, A. S., 194
Coordinate representation, 71
Correlation approximation, 279–281, 283
Coulomb integral, 192
Coulomb's law, 87
Cross section, 226, 227–228

D'Alembertian operator, 90
D'Alembert's principle, 6–8
Dalgarno, A., 210
Dalitz, R. H., 234
Davison, B., 259
De Broglie wave length, 57
Degeneracy, 67, 138
Degrees of freedom, 6
Delves, L. M., 207
Demkov, Yu. N., 251, 289
Diamagnetic susceptibility, 210–211
Differential cross section, 226, 227–228, 276
Diffusion equation, 112–114
Dirac δ function, 67, 217
Dirac equation, 61–63, 126–129
Dissociation energy of hydrogen molecule, 193, 195
Dynamical principle of Schwinger, 75, 132

Effective range, 263
Eigenenergies, of atoms, 170–183
   of quantum mechanical system, 163–170
Eigenfunctions, 144
Eigenvalue equation, 66
Eigenvalues, 66, 136–138, 144
Eigenvector, 66

INDEX 305

Eigenvector—*cont.*
 simultaneous, 68
Elastic cross section, 276
Elastic scattering, of electrons by hydrogen atoms, 274–282
 of neutrons by deuterons, 283–287
 of positrons by hydrogen atoms, 282–283
Electromagnetic field, 105–111
 motion of a charged particle in, 11–12
Electron affinity, 174
 of hydrogen, 174–175
Electrostatics, 87–89
Elliptic coordinates, confocal, 189
Energy, 2–5
 conservation of, 12–13
 in relativistic mechanics, 29, 31–32
Energy density, 94, 95
Equilibrium, 7–8, 134
Equivalence principle, 44
Euclidean space, 45
Euler, 1
Euler equation of fluid flow, 96
Euler–Lagrange equations, 14–16
 for complex field, 115
 in field theory, 77–81
 for geodesic line in Riemannian space, 45–48
 for light ray *in vacuo*, 49–50
 with subsidiary conditions, 100–101
Exchange approximation, 277–278, 280, 281
Exchange-correlation approximation, 280–281
Exchange integral, 192
Excited states, of helium, 181–183
 upper bounds to eigenenergies of, 165–168
Expectation value, 69
 variational principle for, 207–211

Fermat's principle of least time, 1, 53–55
Feshbach, H., 253
Feynman, R. P., 76
Feynman identity, 234

Field components, 77
 complex, 115–117
Field equations, 78
 quantum, 131–133
Fluid flow, 84–86, 96–97
Force, 1
 applied, 7
 conservative, 5
 of constraint, 7
 external, 3
 generalized, 9
 of inertia, 8
Four-tensor expression of Maxwell's equations, 107–108
Frequency, 55
Functional derivative, 81, 83

Gauge transformation, 106
Generalized coordinates, 56
Generalized forces, 9
Generalized potentials, 10
 for charged particle in electromagnetic field, 12
Geodesic line, 42
 in Riemannian space, 45–48
Geometrical mechanics, 41–43
Geometrical optics, 51–56
Gravitational field, 43–45
 motion of particle in, 48–49
Green's function, in formal scattering theory, 299
 one-dimensional, 214, 256
 three-dimensional, 218, 229
Green's theorem, 78, 80
Ground state, 163

H^-, 172–181
Hahn, Y., 273
Hamiltonian density, 82
 for complex field, 117
 for diffusion equation, 114
 for Dirac equation, 128–129
 for electromagnetic field, 110–111
 for Klein–Gordon equation, 120
 for scalar meson field, 121
 for scalar meson in electromagnetic field, 123

for Schrödinger equation, 118
for sound waves, 99–100
for vector meson field, 125–126
for vibrating membrane, 95
for vibrating string, 94
Hamiltonian function, 25
for electromagnetic field, 110–111
in field theory, 81–83
in special relativity, 32, 60
for system of particles, 59
Hamiltonian operator, 59, 71
for hydrogen molecular ion, 188
for hydrogen molecule, 191
for hydrogenic atom, 171
for two electron systems, 172
Hamilton–Jacobi equation, 35–36
for particle in gravitational field, 49
for relativistic particle, 60–62
Hamilton's characteristic function, 37–40, 52, 55, 57, 60
Hamilton's equations, 24–25
in field theory, 81–83
variational principle for, 25–26
Hamilton's integral, 36–37
in quantum mechanics, 75
Hamilton's principal function, 36, 37, 40, 57
for a particle in gravitational field, 49
Hamilton's principle, 17–20
in field theory, 80
in quantum mechanics, 73–76
in special relativity, 30
Hartree approximation, 186
Hartree–Fock approximation, 175, 183
Hartree–Fock equations, for bound states, 183–187
for continuum states, 275, 289
He, 172–181
excited states of, 181–183
Heat conduction, 112–113
Heitler–London method, 190–193
Helmholtz equation, 101–103, 130, 156–158
Hermitian adjoint operator, 65

Hermitian operators, 65
complete set of commuting, 68
Holonomic constraints, 6
Hulthén's second variational method, 253–256, 267
Hulthén's variational method, 247–248, 266
Hulthén's variational principle, 245–248, 266
generalization of, 279
Huygen's principle, 51–52
Hydrogen molecular ion, 187–190
Hydrogen molecule, 190–195
Hydrogenic atom, 171–172
Hylleraas coordinates, 176, 178, 278
Hylleraas–Undheim method, 166–168

Incompressible fluid, 84–86
Inelastic scattering of electrons by hydrogen atoms, 287–289
Inertial frame of reference, 2, 27–28
Integral equation, for phase shifts, 244–245, 250–251
for scattering amplitude, 218–219, 229–230
Interaction representation, 293
Internal function, 254
Ionization energy, 174
for excited states of He, 183
for He, Li^+, Be^{2+}, 174–175, 181
Irrotational motion of incompressible fluid, 84–86

James, H. M., 194

Kacser, C., 235
Kato, T., 259
Kato's variational principle, 251–253
Kinetic energy, 3, 41
for irrotational fluid, 86, 99
in relativistic mechanics, 29
for vibrating membrane, 95
for vibrating string, 92–93
Klein–Gordon equation, 60–61, 119–126, 130
Kohn's variational method, for phase shifts, 247–248

Kohn's variational method—*cont.*
  for scattering length, 266
Kohn's variational principle for scattering amplitude, 235–240
Kronecker delta, 47

Lagrange undetermined multipliers, method of, 15–16, 100–101
Lagrange's equations, 8–10
  in field theory, 81
  in quantum field theory, 133
  for quantum mechanical system, 76
Lagrange's theorem, 142
Lagrangian density, 77
  for complex field, 115
  for diffusion equation, 113
  for Dirac equation, 126–127
  for electromagnetic field, 108–109
  for Helmholtz equation, 102–103
  for irrotational motion of incompressible fluid, 86
  for Klein–Gordon equation, 120
  for Laplace's equation, 84
  modified, 101, 103, 105
  for Poisson's equation, 86, 89, 90
  for scalar meson field, 120
  for scalar meson in electromagnetic field, 122
  for scalar wave equation, 91
  for Schrödinger equation, 117
  for sound waves, 99
  for Sturm–Liouville equation, 105
  for vector meson field, 124
  for vibrating membrane, 95
  for vibrating string, 94
Lagrangian function, for conservative system, 10
  for charged particle in electromagnetic field, 12
  in field theory, 80
  for geodesic line in Riemannian space, 46
  for light ray *in vacuo*, 50
  for particle in gravitational field, 48
  for quantum mechanical system, 75
  for relativistic particle, 30–31
  for small oscillations of dynamical system, 135, 140
Laplace's equation, 83–86, 130
Lcao approximation, 187
Legendre polynomials, 221
Levine, H., 227
Li^+, 172–181
Linear equations, 129–131
Linear operator, 65
Lippmann, B. A., 295
Lippman–Schwinger equation, 298
Lorentz force, 12
Lorentz transformation, 27–28
Lower bounds, to ground state eigenenergy, 168–170
  to phase shifts, 259–261
  to reciprocal of scattering length, 270–273

Mass, inertial, 1–2, 43
  gravitational, 43
  in relativistic mechanics, 28–29
Massey, H. S. W., 281, 282
Maupertuis, 1
Maxwell's equations, 11, 105
Mean square integral, 151–152
Membrane, vibrating, 94–95
Meson, scalar, 119–123
  vector, 123–126
Metric tensor, 41, 45
Minimum principle for $k \cot \eta$, 273
Moiseiwitsch, B. L., 281
Molecular energy curves, 187–195
Momentum, 1
  canonical, 24
  for particle in gravitational field, 49
  in relativistic mechanics, 28, 31
Momentum density, 82
Moussa, A. H. A., 282
Mutual potential energy, 88-90

Neutron–deuteron scattering, 283–287
Neutron diffusion, 112
Newtonian gravitation, 89–90

Newton's law of gravitation, 43, 89
Newton's laws of motion, 1–2
Noether's theorem, 116
Non-Euclidean space, 45
Normal coordinates, 138–140
Normal mode, 140

Observables, 65
O'Malley, T. F., 273, 282
Optical theorem, 243
Orthogonality and orthonormality, 67
  and the Sturm–Liouville eigenvalue problem, 148–149
Overlap integral, 192

Partial cross section, 242, 276
Partial waves expansion, 241–243, 276
Pauli exclusion principle, 175, 184, 274
Pauling, L., 207
Pekeris, C. L., 179, 183
Perimetric coordinates, 179
Perturbation theory, 198–202
Phase, 56
Phase shifts, 240–243
  integral equation for, 244–245, 257
  monotonic property of, 252–253, 259
Phase velocity, 56, 93, 95
Planck's constant, 57–58
Plane wave, 217, 227
  expansion in Legendre polynomials, 222
Poisson's equation, 86–90, 130
Polarizability, 202–205
Potential, 4–5
  generalized, 10
  scalar and vector, 11, 106–107
Potential barrier, scattering by, 213–216
Potential energy, for irrotational fluid, 98–99
  for vibrating membrane, 95
  for vibrating string, 92–93
Principle of energy, 2–5
Principle of equivalence, 43–44

Principle of least action, 22–24
  Jacobi's form of, 24, 33, 42–43, 54
  in relativistic mechanics, 33
Principle of superposition of states, 64
Principle of virtual work, 7
Probability amplitude, 69
Probability density, 118
Proper values, *see* Eigenvalues

Quantum field equations, 131–133
Quantum mechanics, 64–76

Rayleigh's principle, 142–144
  for vibrating string, 144–147
Reciprocity theorem, 219, 232
Relativistic mechanics, 28–33
Relativity, special theory of, 26–28
Representative, 68
Rest energy, 29
Rest mass, 29
Riemannian geometry, 41
Ritz variational method, 153–156, 166
Rosenberg, L., 266, 281, 283
Rubinow, S. I., 253, 286
Rutherford scattering formula, 234

Sartori, L., 286
Scalar product, 64
Scattering, by aperture in infinite plane screen, 227
  by Coulomb potential, 233–234
  of electrons by hydrogen atoms, 274–282, 287–289
  of neutrons by deuterons, 283–287
  by one-dimensional potential barrier, 213–216
  of particles in wave mechanics, 227–231
  of positrons by hydrogen atoms, 282–283
  by screened Coulomb potential, 233–235, 243
  at a spherical surface, 220–227
  at a surface, 216–220
Scattering amplitude, 217, 228–240, 242–243

Scattering amplitude—*cont.*
 integral equation for, 218, 220, 229–230
Scattering length, 261–273
Schrödinger equation, 57–60, 69–71, 117–119, 161–163
Schwartz, C., 282, 283
Schwinger, J., 227, 295
Schwinger dynamical principle, 75, 132
Schwinger functional, extremum property of, 259–261
 stationary property of, 258–259
Schwinger variational principle, for the phase shifts, 256–261, 262–263
 for the scattering amplitude, 219–220, 231–235
 for the transition matrix, 301
 for the transmission amplitude, 216
Secular equation, 136
Self-adjoint operator, 65
Shepelenko, F. P., 251
Small oscillations of a dynamical system, 134–140
Sound waves, 96–100
Special theory of relativity, 26–28
Spherical Bessel, Neumann and Hankel functions, 221, 241
Spherical wave, 217, 228
 expansion in Legendre polynomials, 224–225, 242
Spin functions, 186, 275–276
Spruch, L., 266, 273, 281, 283
Stable equilibrium, 134–135
State vector, 64
Static field approximation, 277–279, 281
Stevenson's formula, 170
Stewart, A. L., 210
String, vibrating, 91–94
Sturm–Liouville eigenvalue problem, 147–152
Sturm–Liouville equation, 103–105
Subsidiary conditions, 100–101
 (*see also* Auxiliary conditions)
Sugar, R., 259, 261

Tamm's variational method, 248–250, 267, 277–278
Temple's formula, 168–169, 179–180
Time-dependent scattering theory, 292–301
Total cross section, 226, 227, 242
Transformation function, 68
Transition probability, 297
Transition rate, 300
Transmission amplitude, 214–216
Trial function, 165, 216, 232, 237–238, 247
Two electron systems, 172–183

Unitary operator, 70
 in time dependent scattering theory, 293
Unitary transformations, 72–73
Upper bounds, to eigenenergies, 164–165
 to eigenenergies of $1s\sigma$ and $2p\sigma$ states of H_2^+, 190
 to eigenenergies of 2^1S, 2^3S, 2^1P, 2^3P states of He, 182
 to the ground state eigenenergies of H^-, He, Li^+, Be^{2+}, 174–176, 180
 to interaction energy between two hydrogen atoms, 207
 to phase shifts, 259–261
 to scattering length, 264–270

Van der Waals force, 205–207
Variational methods and integral equation for the phase shifts, 250–251
Variational principles, for arbitrary operator, 207–211
 for collision operator, 295–296
 for eigenvalue problems, 156–161
 for general eigenvalue equation, 197–198
 for Hamilton's equations, 25–26
 for Hartree–Fock equations, 185, 287–289
 for Helmholtz equation, 156–161
 for phase shifts, 245–261

for scattering amplitude, 219–220, 231–240
for Schrödinger equation, 161–164
for transition matrix, 300–301
for transmission amplitude, 216
Vector meson field, 123–126
Velocity potential, 86
Vibrating membrane, 94–95, 159–161
Vibrating string, 91–94
Virial theorem, 20–22
    for atomic systems, 177–178
    for molecular systems, 195–196
    for scattering problems, 289–292
Virtual displacement, 6
Virtual work, 7
Vorticity, 85

Wang, S. C., 193
Wave equation, 55, 90–100, 130
Wave function, 55, 58, 71
Wave length, 56
    de Broglie, 57, 227
Wave motion, 55–56, 156–161
    on a membrane, 94–95
    sound, 96–100
    on a string, 91–94
Wave number, 213
Weinstein's formula, 169–170

A CATALOG OF SELECTED
DOVER BOOKS
IN SCIENCE AND MATHEMATICS

Astronomy

BURNHAM'S CELESTIAL HANDBOOK, Robert Burnham, Jr. Thorough guide to the stars beyond our solar system. Exhaustive treatment. Alphabetical by constellation: Andromeda to Cetus in Vol. 1; Chamaeleon to Orion in Vol. 2; and Pavo to Vulpecula in Vol. 3. Hundreds of illustrations. Index in Vol. 3. 2,000pp. 6⅛ x 9¼.
Vol. I: 23567-X
Vol. II: 23568-8
Vol. III: 23673-0

EXPLORING THE MOON THROUGH BINOCULARS AND SMALL TELESCOPES, Ernest H. Cherrington, Jr. Informative, profusely illustrated guide to locating and identifying craters, rills, seas, mountains, other lunar features. Newly revised and updated with special section of new photos. Over 100 photos and diagrams. 240pp. 8¼ x 11. 24491-1

THE EXTRATERRESTRIAL LIFE DEBATE, 1750–1900, Michael J. Crowe. First detailed, scholarly study in English of the many ideas that developed from 1750 to 1900 regarding the existence of intelligent extraterrestrial life. Examines ideas of Kant, Herschel, Voltaire, Percival Lowell, many other scientists and thinkers. 16 illustrations. 704pp. 5⅜ x 8½. 40675-X

THEORIES OF THE WORLD FROM ANTIQUITY TO THE COPERNICAN REVOLUTION, Michael J. Crowe. Newly revised edition of an accessible, enlightening book recreates the change from an earth-centered to a sun-centered conception of the solar system. 242pp. 5⅜ x 8½. 41444-2

A HISTORY OF ASTRONOMY, A. Pannekoek. Well-balanced, carefully reasoned study covers such topics as Ptolemaic theory, work of Copernicus, Kepler, Newton, Eddington's work on stars, much more. Illustrated. References. 521pp. 5⅜ x 8½.
65994-1

A COMPLETE MANUAL OF AMATEUR ASTRONOMY: Tools and Techniques for Astronomical Observations, P. Clay Sherrod with Thomas L. Koed. Concise, highly readable book discusses: selecting, setting up and maintaining a telescope; amateur studies of the sun; lunar topography and occultations; observations of Mars, Jupiter, Saturn, the minor planets and the stars; an introduction to photoelectric photometry; more. 1981 ed. 124 figures. 26 halftones. 37 tables. 335pp. 6½ x 9¼.
42820-6

AMATEUR ASTRONOMER'S HANDBOOK, J. B. Sidgwick. Timeless, comprehensive coverage of telescopes, mirrors, lenses, mountings, telescope drives, micrometers, spectroscopes, more. 189 illustrations. 576pp. 5⅝ x 8¼. (Available in U.S. only.)
24034-7

STARS AND RELATIVITY, Ya. B. Zel'dovich and I. D. Novikov. Vol. 1 of *Relativistic Astrophysics* by famed Russian scientists. General relativity, properties of matter under astrophysical conditions, stars, and stellar systems. Deep physical insights, clear presentation. 1971 edition. References. 544pp. 5⅝ x 8¼. 69424-0

CATALOG OF DOVER BOOKS

Chemistry

THE SCEPTICAL CHYMIST: The Classic 1661 Text, Robert Boyle. Boyle defines the term "element," asserting that all natural phenomena can be explained by the motion and organization of primary particles. 1911 ed. viii+232pp. 5⅜ x 8½. 42825-7

RADIOACTIVE SUBSTANCES, Marie Curie. Here is the celebrated scientist's doctoral thesis, the prelude to her receipt of the 1903 Nobel Prize. Curie discusses establishing atomic character of radioactivity found in compounds of uranium and thorium; extraction from pitchblende of polonium and radium; isolation of pure radium chloride; determination of atomic weight of radium; plus electric, photographic, luminous, heat, color effects of radioactivity. ii+94pp. 5⅜ x 8½. 42550-9

CHEMICAL MAGIC, Leonard A. Ford. Second Edition, Revised by E. Winston Grundmeier. Over 100 unusual stunts demonstrating cold fire, dust explosions, much more. Text explains scientific principles and stresses safety precautions. 128pp. 5⅜ x 8½. 67628-5

THE DEVELOPMENT OF MODERN CHEMISTRY, Aaron J. Ihde. Authoritative history of chemistry from ancient Greek theory to 20th-century innovation. Covers major chemists and their discoveries. 209 illustrations. 14 tables. Bibliographies. Indices. Appendices. 851pp. 5⅜ x 8½. 64235-6

CATALYSIS IN CHEMISTRY AND ENZYMOLOGY, William P. Jencks. Exceptionally clear coverage of mechanisms for catalysis, forces in aqueous solution, carbonyl- and acyl-group reactions, practical kinetics, more. 864pp. 5⅜ x 8½. 65460-5

ELEMENTS OF CHEMISTRY, Antoine Lavoisier. Monumental classic by founder of modern chemistry in remarkable reprint of rare 1790 Kerr translation. A must for every student of chemistry or the history of science. 539pp. 5⅜ x 8½. 64624-6

THE HISTORICAL BACKGROUND OF CHEMISTRY, Henry M. Leicester. Evolution of ideas, not individual biography. Concentrates on formulation of a coherent set of chemical laws. 260pp. 5⅜ x 8½. 61053-5

A SHORT HISTORY OF CHEMISTRY, J. R. Partington. Classic exposition explores origins of chemistry, alchemy, early medical chemistry, nature of atmosphere, theory of valency, laws and structure of atomic theory, much more. 428pp. 5⅜ x 8½. (Available in U.S. only.) 65977-1

GENERAL CHEMISTRY, Linus Pauling. Revised 3rd edition of classic first-year text by Nobel laureate. Atomic and molecular structure, quantum mechanics, statistical mechanics, thermodynamics correlated with descriptive chemistry. Problems. 992pp. 5⅜ x 8½. 65622-5

FROM ALCHEMY TO CHEMISTRY, John Read. Broad, humanistic treatment focuses on great figures of chemistry and ideas that revolutionized the science. 50 illustrations. 240pp. 5⅜ x 8½. 28690-8

CATALOG OF DOVER BOOKS

Engineering

DE RE METALLICA, Georgius Agricola. The famous Hoover translation of greatest treatise on technological chemistry, engineering, geology, mining of early modern times (1556). All 289 original woodcuts. 638pp. 6¾ x 11. 60006-8

FUNDAMENTALS OF ASTRODYNAMICS, Roger Bate et al. Modern approach developed by U.S. Air Force Academy. Designed as a first course. Problems, exercises. Numerous illustrations. 455pp. 5⅜ x 8½. 60061-0

DYNAMICS OF FLUIDS IN POROUS MEDIA, Jacob Bear. For advanced students of ground water hydrology, soil mechanics and physics, drainage and irrigation engineering, and more. 335 illustrations. Exercises, with answers. 784pp. 6⅛ x 9¼. 65675-6

THEORY OF VISCOELASTICITY (Second Edition), Richard M. Christensen. Complete, consistent description of the linear theory of the viscoelastic behavior of materials. Problem-solving techniques discussed. 1982 edition. 29 figures. xiv+364pp. 6⅛ x 9¼. 42880-X

MECHANICS, J. P. Den Hartog. A classic introductory text or refresher. Hundreds of applications and design problems illuminate fundamentals of trusses, loaded beams and cables, etc. 334 answered problems. 462pp. 5⅜ x 8½. 60754-2

MECHANICAL VIBRATIONS, J. P. Den Hartog. Classic textbook offers lucid explanations and illustrative models, applying theories of vibrations to a variety of practical industrial engineering problems. Numerous figures. 233 problems, solutions. Appendix. Index. Preface. 436pp. 5⅜ x 8½. 64785-4

STRENGTH OF MATERIALS, J. P. Den Hartog. Full, clear treatment of basic material (tension, torsion, bending, etc.) plus advanced material on engineering methods, applications. 350 answered problems. 323pp. 5⅜ x 8½. 60755-0

A HISTORY OF MECHANICS, René Dugas. Monumental study of mechanical principles from antiquity to quantum mechanics. Contributions of ancient Greeks, Galileo, Leonardo, Kepler, Lagrange, many others. 671pp. 5⅜ x 8½. 65632-2

STABILITY THEORY AND ITS APPLICATIONS TO STRUCTURAL MECHANICS, Clive L. Dym. Self-contained text focuses on Koiter postbuckling analyses, with mathematical notions of stability of motion. Basing minimum energy principles for static stability upon dynamic concepts of stability of motion, it develops asymptotic buckling and postbuckling analyses from potential energy considerations, with applications to columns, plates, and arches. 1974 ed. 208pp. 5⅜ x 8½.
42541-X

METAL FATIGUE, N. E. Frost, K. J. Marsh, and L. P. Pook. Definitive, clearly written, and well-illustrated volume addresses all aspects of the subject, from the historical development of understanding metal fatigue to vital concepts of the cyclic stress that causes a crack to grow. Includes 7 appendixes. 544pp. 5⅜ x 8½. 40927-9

CATALOG OF DOVER BOOKS

ROCKETS, Robert Goddard. Two of the most significant publications in the history of rocketry and jet propulsion: "A Method of Reaching Extreme Altitudes" (1919) and "Liquid Propellant Rocket Development" (1936). 128pp. 5⅜ x 8½. 42537-1

STATISTICAL MECHANICS: Principles and Applications, Terrell L. Hill. Standard text covers fundamentals of statistical mechanics, applications to fluctuation theory, imperfect gases, distribution functions, more. 448pp. 5⅜ x 8½. 65390-0

ENGINEERING AND TECHNOLOGY 1650–1750: Illustrations and Texts from Original Sources, Martin Jensen. Highly readable text with more than 200 contemporary drawings and detailed engravings of engineering projects dealing with surveying, leveling, materials, hand tools, lifting equipment, transport and erection, piling, bailing, water supply, hydraulic engineering, and more. Among the specific projects outlined–transporting a 50-ton stone to the Louvre, erecting an obelisk, building timber locks, and dredging canals. 207pp. 8⅜ x 11¼. 42232-1

THE VARIATIONAL PRINCIPLES OF MECHANICS, Cornelius Lanczos. Graduate level coverage of calculus of variations, equations of motion, relativistic mechanics, more. First inexpensive paperbound edition of classic treatise. Index. Bibliography. 418pp. 5⅜ x 8½. 65067-7

PROTECTION OF ELECTRONIC CIRCUITS FROM OVERVOLTAGES, Ronald B. Standler. Five-part treatment presents practical rules and strategies for circuits designed to protect electronic systems from damage by transient overvoltages. 1989 ed. xxiv+434pp. 6⅛ x 9¼. 42552-5

ROTARY WING AERODYNAMICS, W. Z. Stepniewski. Clear, concise text covers aerodynamic phenomena of the rotor and offers guidelines for helicopter performance evaluation. Originally prepared for NASA. 537 figures. 640pp. 6⅛ x 9¼. 64647-5

INTRODUCTION TO SPACE DYNAMICS, William Tyrrell Thomson. Comprehensive, classic introduction to space-flight engineering for advanced undergraduate and graduate students. Includes vector algebra, kinematics, transformation of coordinates. Bibliography. Index. 352pp. 5⅜ x 8½. 65113-4

HISTORY OF STRENGTH OF MATERIALS, Stephen P. Timoshenko. Excellent historical survey of the strength of materials with many references to the theories of elasticity and structure. 245 figures. 452pp. 5⅜ x 8½. 61187-6

ANALYTICAL FRACTURE MECHANICS, David J. Unger. Self-contained text supplements standard fracture mechanics texts by focusing on analytical methods for determining crack-tip stress and strain fields. 336pp. 6⅛ x 9¼. 41737-9

STATISTICAL MECHANICS OF ELASTICITY, J. H. Weiner. Advanced, self-contained treatment illustrates general principles and elastic behavior of solids. Part 1, based on classical mechanics, studies thermoelastic behavior of crystalline and polymeric solids. Part 2, based on quantum mechanics, focuses on interatomic force laws, behavior of solids, and thermally activated processes. For students of physics and chemistry and for polymer physicists. 1983 ed. 96 figures. 496pp. 5⅜ x 8½. 42260-7

Mathematics

FUNCTIONAL ANALYSIS (Second Corrected Edition), George Bachman and Lawrence Narici. Excellent treatment of subject geared toward students with background in linear algebra, advanced calculus, physics, and engineering. Text covers introduction to inner-product spaces, normed, metric spaces, and topological spaces; complete orthonormal sets, the Hahn-Banach Theorem and its consequences, and many other related subjects. 1966 ed. 544pp. 6⅛ x 9¼. 40251-7

ASYMPTOTIC EXPANSIONS OF INTEGRALS, Norman Bleistein & Richard A. Handelsman. Best introduction to important field with applications in a variety of scientific disciplines. New preface. Problems. Diagrams. Tables. Bibliography. Index. 448pp. 5⅜ x 8½. 65082-0

VECTOR AND TENSOR ANALYSIS WITH APPLICATIONS, A. I. Borisenko and I. E. Tarapov. Concise introduction. Worked-out problems, solutions, exercises. 257pp. 5⅝ x 8¼. 63833-2

THE ABSOLUTE DIFFERENTIAL CALCULUS (CALCULUS OF TENSORS), Tullio Levi-Civita. Great 20th-century mathematician's classic work on material necessary for mathematical grasp of theory of relativity. 452pp. 5⅝ x 8¼. 63401-9

AN INTRODUCTION TO ORDINARY DIFFERENTIAL EQUATIONS, Earl A. Coddington. A thorough and systematic first course in elementary differential equations for undergraduates in mathematics and science, with many exercises and problems (with answers). Index. 304pp. 5⅜ x 8½. 65942-9

FOURIER SERIES AND ORTHOGONAL FUNCTIONS, Harry F. Davis. An incisive text combining theory and practical example to introduce Fourier series, orthogonal functions and applications of the Fourier method to boundary-value problems. 570 exercises. Answers and notes. 416pp. 5⅜ x 8½. 65973-9

COMPUTABILITY AND UNSOLVABILITY, Martin Davis. Classic graduate-level introduction to theory of computability, usually referred to as theory of recurrent functions. New preface and appendix. 288pp. 5⅜ x 8½. 61471-9

ASYMPTOTIC METHODS IN ANALYSIS, N. G. de Bruijn. An inexpensive, comprehensive guide to asymptotic methods–the pioneering work that teaches by explaining worked examples in detail. Index. 224pp. 5⅜ x 8½ 64221-6

APPLIED COMPLEX VARIABLES, John W. Dettman. Step-by-step coverage of fundamentals of analytic function theory–plus lucid exposition of five important applications: Potential Theory; Ordinary Differential Equations; Fourier Transforms; Laplace Transforms; Asymptotic Expansions. 66 figures. Exercises at chapter ends. 512pp. 5⅜ x 8½. 64670-X

INTRODUCTION TO LINEAR ALGEBRA AND DIFFERENTIAL EQUATIONS, John W. Dettman. Excellent text covers complex numbers, determinants, orthonormal bases, Laplace transforms, much more. Exercises with solutions. Undergraduate level. 416pp. 5⅜ x 8½. 65191-6

CATALOG OF DOVER BOOKS

CALCULUS OF VARIATIONS WITH APPLICATIONS, George M. Ewing. Applications-oriented introduction to variational theory develops insight and promotes understanding of specialized books, research papers. Suitable for advanced undergraduate/graduate students as primary, supplementary text. 352pp. 5⅜ x 8½.
64856-7

COMPLEX VARIABLES, Francis J. Flanigan. Unusual approach, delaying complex algebra till harmonic functions have been analyzed from real variable viewpoint. Includes problems with answers. 364pp. 5⅜ x 8½.
61388-7

AN INTRODUCTION TO THE CALCULUS OF VARIATIONS, Charles Fox. Graduate-level text covers variations of an integral, isoperimetrical problems, least action, special relativity, approximations, more. References. 279pp. 5⅜ x 8½.
65499-0

COUNTEREXAMPLES IN ANALYSIS, Bernard R. Gelbaum and John M. H. Olmsted. These counterexamples deal mostly with the part of analysis known as "real variables." The first half covers the real number system, and the second half encompasses higher dimensions. 1962 edition. xxiv+198pp. 5⅜ x 8½.
42875-3

CATASTROPHE THEORY FOR SCIENTISTS AND ENGINEERS, Robert Gilmore. Advanced-level treatment describes mathematics of theory grounded in the work of Poincaré, R. Thom, other mathematicians. Also important applications to problems in mathematics, physics, chemistry, and engineering. 1981 edition. References. 28 tables. 397 black-and-white illustrations. xvii+666pp. 6⅛ x 9¼.
67539-4

INTRODUCTION TO DIFFERENCE EQUATIONS, Samuel Goldberg. Exceptionally clear exposition of important discipline with applications to sociology, psychology, economics. Many illustrative examples; over 250 problems. 260pp. 5⅜ x 8½.
65084-7

NUMERICAL METHODS FOR SCIENTISTS AND ENGINEERS, Richard Hamming. Classic text stresses frequency approach in coverage of algorithms, polynomial approximation, Fourier approximation, exponential approximation, other topics. Revised and enlarged 2nd edition. 721pp. 5⅜ x 8½.
65241-6

INTRODUCTION TO NUMERICAL ANALYSIS (2nd Edition), F. B. Hildebrand. Classic, fundamental treatment covers computation, approximation, interpolation, numerical differentiation and integration, other topics. 150 new problems. 669pp. 5⅜ x 8½.
65363-3

THREE PEARLS OF NUMBER THEORY, A. Y. Khinchin. Three compelling puzzles require proof of a basic law governing the world of numbers. Challenges concern van der Waerden's theorem, the Landau-Schnirelmann hypothesis and Mann's theorem, and a solution to Waring's problem. Solutions included. 64pp. 5⅜ x 8½.
40026-3

THE PHILOSOPHY OF MATHEMATICS: An Introductory Essay, Stephan Körner. Surveys the views of Plato, Aristotle, Leibniz & Kant concerning propositions and theories of applied and pure mathematics. Introduction. Two appendices. Index. 198pp. 5⅜ x 8½.
25048-2

CATALOG OF DOVER BOOKS

INTRODUCTORY REAL ANALYSIS, A.N. Kolmogorov, S. V. Fomin. Translated by Richard A. Silverman. Self-contained, evenly paced introduction to real and functional analysis. Some 350 problems. 403pp. 5⅜ x 8½. 61226-0

APPLIED ANALYSIS, Cornelius Lanczos. Classic work on analysis and design of finite processes for approximating solution of analytical problems. Algebraic equations, matrices, harmonic analysis, quadrature methods, more. 559pp. 5⅜ x 8½. 65656-X

AN INTRODUCTION TO ALGEBRAIC STRUCTURES, Joseph Landin. Superb self-contained text covers "abstract algebra": sets and numbers, theory of groups, theory of rings, much more. Numerous well-chosen examples, exercises. 247pp. 5⅜ x 8½.
65940-2

QUALITATIVE THEORY OF DIFFERENTIAL EQUATIONS, V. V. Nemytskii and V.V. Stepanov. Classic graduate-level text by two prominent Soviet mathematicians covers classical differential equations as well as topological dynamics and ergodic theory. Bibliographies. 523pp. 5⅜ x 8½. 65954-2

THEORY OF MATRICES, Sam Perlis. Outstanding text covering rank, nonsingularity and inverses in connection with the development of canonical matrices under the relation of equivalence, and without the intervention of determinants. Includes exercises. 237pp. 5⅜ x 8½. 66810-X

INTRODUCTION TO ANALYSIS, Maxwell Rosenlicht. Unusually clear, accessible coverage of set theory, real number system, metric spaces, continuous functions, Riemann integration, multiple integrals, more. Wide range of problems. Undergraduate level. Bibliography. 254pp. 5⅜ x 8½. 65038-3

MODERN NONLINEAR EQUATIONS, Thomas L. Saaty. Emphasizes practical solution of problems; covers seven types of equations. ". . . a welcome contribution to the existing literature. . . . "–*Math Reviews*. 490pp. 5⅜ x 8½. 64232-1

MATRICES AND LINEAR ALGEBRA, Hans Schneider and George Phillip Barker. Basic textbook covers theory of matrices and its applications to systems of linear equations and related topics such as determinants, eigenvalues, and differential equations. Numerous exercises. 432pp. 5⅜ x 8½. 66014-1

MATHEMATICS APPLIED TO CONTINUUM MECHANICS, Lee A. Segel. Analyzes models of fluid flow and solid deformation. For upper-level math, science, and engineering students. 608pp. 5⅜ x 8½. 65369-2

ELEMENTS OF REAL ANALYSIS, David A. Sprecher. Classic text covers fundamental concepts, real number system, point sets, functions of a real variable, Fourier series, much more. Over 500 exercises. 352pp. 5⅜ x 8½. 65385-4

SET THEORY AND LOGIC, Robert R. Stoll. Lucid introduction to unified theory of mathematical concepts. Set theory and logic seen as tools for conceptual understanding of real number system. 496pp. 5⅜ x 8¼. 63829-4

CATALOG OF DOVER BOOKS

TENSOR CALCULUS, J.L. Synge and A. Schild. Widely used introductory text covers spaces and tensors, basic operations in Riemannian space, non-Riemannian spaces, etc. 324pp. 5⅜ x 8¼. 63612-7

ORDINARY DIFFERENTIAL EQUATIONS, Morris Tenenbaum and Harry Pollard. Exhaustive survey of ordinary differential equations for undergraduates in mathematics, engineering, science. Thorough analysis of theorems. Diagrams. Bibliography. Index. 818pp. 5⅜ x 8½. 64940-7

INTEGRAL EQUATIONS, F. G. Tricomi. Authoritative, well-written treatment of extremely useful mathematical tool with wide applications. Volterra Equations, Fredholm Equations, much more. Advanced undergraduate to graduate level. Exercises. Bibliography. 238pp. 5⅜ x 8½. 64828-1

FOURIER SERIES, Georgi P. Tolstov. Translated by Richard A. Silverman. A valuable addition to the literature on the subject, moving clearly from subject to subject and theorem to theorem. 107 problems, answers. 336pp. 5⅜ x 8½. 63317-9

INTRODUCTION TO MATHEMATICAL THINKING, Friedrich Waismann. Examinations of arithmetic, geometry, and theory of integers; rational and natural numbers; complete induction; limit and point of accumulation; remarkable curves; complex and hypercomplex numbers, more. 1959 ed. 27 figures. xii+260pp. 5⅜ x 8½. 42804-4

POPULAR LECTURES ON MATHEMATICAL LOGIC, Hao Wang. Noted logician's lucid treatment of historical developments, set theory, model theory, recursion theory and constructivism, proof theory, more. 3 appendixes. Bibliography. 1981 ed. ix+283pp. 5⅜ x 8½. 67632-3

CALCULUS OF VARIATIONS, Robert Weinstock. Basic introduction covering isoperimetric problems, theory of elasticity, quantum mechanics, electrostatics, etc. Exercises throughout. 326pp. 5⅜ x 8½. 63069-2

THE CONTINUUM: A Critical Examination of the Foundation of Analysis, Hermann Weyl. Classic of 20th-century foundational research deals with the conceptual problem posed by the continuum. 156pp. 5⅜ x 8½. 67982-9

CHALLENGING MATHEMATICAL PROBLEMS WITH ELEMENTARY SOLUTIONS, A. M. Yaglom and I. M. Yaglom. Over 170 challenging problems on probability theory, combinatorial analysis, points and lines, topology, convex polygons, many other topics. Solutions. Total of 445pp. 5⅜ x 8½. Two-vol. set.
Vol. I: 65536-9 Vol. II: 65537-7

INTRODUCTION TO PARTIAL DIFFERENTIAL EQUATIONS WITH APPLICATIONS, E. C. Zachmanoglou and Dale W. Thoe. Essentials of partial differential equations applied to common problems in engineering and the physical sciences. Problems and answers. 416pp. 5⅜ x 8½. 65251-3

THE THEORY OF GROUPS, Hans J. Zassenhaus. Well-written graduate-level text acquaints reader with group-theoretic methods and demonstrates their usefulness in mathematics. Axioms, the calculus of complexes, homomorphic mapping, p-group theory, more. 276pp. 5⅜ x 8½. 40922-8

Math–Decision Theory, Statistics, Probability

ELEMENTARY DECISION THEORY, Herman Chernoff and Lincoln E. Moses. Clear introduction to statistics and statistical theory covers data processing, probability and random variables, testing hypotheses, much more. Exercises. 364pp. 5⅜ x 8½. 65218-1

STATISTICS MANUAL, Edwin L. Crow et al. Comprehensive, practical collection of classical and modern methods prepared by U.S. Naval Ordnance Test Station. Stress on use. Basics of statistics assumed. 288pp. 5⅜ x 8½. 60599-X

SOME THEORY OF SAMPLING, William Edwards Deming. Analysis of the problems, theory, and design of sampling techniques for social scientists, industrial managers, and others who find statistics important at work. 61 tables. 90 figures. xvii +602pp. 5⅜ x 8½. 64684-X

LINEAR PROGRAMMING AND ECONOMIC ANALYSIS, Robert Dorfman, Paul A. Samuelson and Robert M. Solow. First comprehensive treatment of linear programming in standard economic analysis. Game theory, modern welfare economics, Leontief input-output, more. 525pp. 5⅜ x 8½. 65491-5

PROBABILITY: An Introduction, Samuel Goldberg. Excellent basic text covers set theory, probability theory for finite sample spaces, binomial theorem, much more. 360 problems. Bibliographies. 322pp. 5⅜ x 8½. 65252-1

GAMES AND DECISIONS: Introduction and Critical Survey, R. Duncan Luce and Howard Raiffa. Superb nontechnical introduction to game theory, primarily applied to social sciences. Utility theory, zero-sum games, n-person games, decision-making, much more. Bibliography. 509pp. 5⅜ x 8½. 65943-7

INTRODUCTION TO THE THEORY OF GAMES, J. C. C. McKinsey. This comprehensive overview of the mathematical theory of games illustrates applications to situations involving conflicts of interest, including economic, social, political, and military contexts. Appropriate for advanced undergraduate and graduate courses; advanced calculus a prerequisite. 1952 ed. x+372pp. 5⅜ x 8½. 42811-7

FIFTY CHALLENGING PROBLEMS IN PROBABILITY WITH SOLUTIONS, Frederick Mosteller. Remarkable puzzlers, graded in difficulty, illustrate elementary and advanced aspects of probability. Detailed solutions. 88pp. 5⅜ x 8½. 65355-2

PROBABILITY THEORY: A Concise Course, Y. A. Rozanov. Highly readable, self-contained introduction covers combination of events, dependent events, Bernoulli trials, etc. 148pp. 5⅜ x 8¼. 63544-9

STATISTICAL METHOD FROM THE VIEWPOINT OF QUALITY CONTROL, Walter A. Shewhart. Important text explains regulation of variables, uses of statistical control to achieve quality control in industry, agriculture, other areas. 192pp. 5⅜ x 8½. 65232-7

Math–Geometry and Topology

ELEMENTARY CONCEPTS OF TOPOLOGY, Paul Alexandroff. Elegant, intuitive approach to topology from set-theoretic topology to Betti groups; how concepts of topology are useful in math and physics. 25 figures. 57pp. 5⅜ x 8½. 60747-X

COMBINATORIAL TOPOLOGY, P. S. Alexandrov. Clearly written, well-organized, three-part text begins by dealing with certain classic problems without using the formal techniques of homology theory and advances to the central concept, the Betti groups. Numerous detailed examples. 654pp. 5⅜ x 8½. 40179-0

EXPERIMENTS IN TOPOLOGY, Stephen Barr. Classic, lively explanation of one of the byways of mathematics. Klein bottles, Moebius strips, projective planes, map coloring, problem of the Koenigsberg bridges, much more, described with clarity and wit. 43 figures. 210pp. 5⅜ x 8½. 25933-1

CONFORMAL MAPPING ON RIEMANN SURFACES, Harvey Cohn. Lucid, insightful book presents ideal coverage of subject. 334 exercises make book perfect for self-study. 55 figures. 352pp. 5⅝ x 8¼. 64025-6

THE GEOMETRY OF RENÉ DESCARTES, René Descartes. The great work founded analytical geometry. Original French text, Descartes's own diagrams, together with definitive Smith-Latham translation. 244pp. 5⅜ x 8½. 60068-8

PRACTICAL CONIC SECTIONS: The Geometric Properties of Ellipses, Parabolas and Hyperbolas, J. W. Downs. This text shows how to create ellipses, parabolas, and hyperbolas. It also presents historical background on their ancient origins and describes the reflective properties and roles of curves in design applications. 1993 ed. 98 figures. xii+100pp. 6½ x 9¼. 42876-1

THE THIRTEEN BOOKS OF EUCLID'S ELEMENTS, translated with introduction and commentary by Thomas L. Heath. Definitive edition. Textual and linguistic notes, mathematical analysis. 2,500 years of critical commentary. Unabridged. 1,414pp. 5⅜ x 8½. Three-vol. set. Vol. I: 60088-2 Vol. II: 60089-0 Vol. III: 60090-4

GEOMETRY OF COMPLEX NUMBERS, Hans Schwerdtfeger. Illuminating, widely praised book on analytic geometry of circles, the Moebius transformation, and two-dimensional non-Euclidean geometries. 200pp. 5⅝ x 8¼. 63830-8

DIFFERENTIAL GEOMETRY, Heinrich W. Guggenheimer. Local differential geometry as an application of advanced calculus and linear algebra. Curvature, transformation groups, surfaces, more. Exercises. 62 figures. 378pp. 5⅜ x 8½. 63433-7

CURVATURE AND HOMOLOGY: Enlarged Edition, Samuel I. Goldberg. Revised edition examines topology of differentiable manifolds; curvature, homology of Riemannian manifolds; compact Lie groups; complex manifolds; curvature, homology of Kaehler manifolds. New Preface. Four new appendixes. 416pp. 5⅜ x 8½. 40207-X

CATALOG OF DOVER BOOKS

History of Math

THE WORKS OF ARCHIMEDES, Archimedes (T. L. Heath, ed.). Topics include the famous problems of the ratio of the areas of a cylinder and an inscribed sphere; the measurement of a circle; the properties of conoids, spheroids, and spirals; and the quadrature of the parabola. Informative introduction. clxxxvi+326pp; supplement, 52pp. 5⅜ x 8½. 42084-1

A SHORT ACCOUNT OF THE HISTORY OF MATHEMATICS, W. W. Rouse Ball. One of clearest, most authoritative surveys from the Egyptians and Phoenicians through 19th-century figures such as Grassman, Galois, Riemann. Fourth edition. 522pp. 5⅜ x 8½. 20630-0

THE HISTORY OF THE CALCULUS AND ITS CONCEPTUAL DEVELOPMENT, Carl B. Boyer. Origins in antiquity, medieval contributions, work of Newton, Leibniz, rigorous formulation. Treatment is verbal. 346pp. 5⅜ x 8½. 60509-4

THE HISTORICAL ROOTS OF ELEMENTARY MATHEMATICS, Lucas N. H. Bunt, Phillip S. Jones, and Jack D. Bedient. Fundamental underpinnings of modern arithmetic, algebra, geometry, and number systems derived from ancient civilizations. 320pp. 5⅜ x 8½. 25563-8

A HISTORY OF MATHEMATICAL NOTATIONS, Florian Cajori. This classic study notes the first appearance of a mathematical symbol and its origin, the competition it encountered, its spread among writers in different countries, its rise to popularity, its eventual decline or ultimate survival. Original 1929 two-volume edition presented here in one volume. xxviii+820pp. 5⅜ x 8½. 67766-4

GAMES, GODS & GAMBLING: A History of Probability and Statistical Ideas, F. N. David. Episodes from the lives of Galileo, Fermat, Pascal, and others illustrate this fascinating account of the roots of mathematics. Features thought-provoking references to classics, archaeology, biography, poetry. 1962 edition. 304pp. 5⅜ x 8½. (Available in U.S. only.) 40023-9

OF MEN AND NUMBERS: The Story of the Great Mathematicians, Jane Muir. Fascinating accounts of the lives and accomplishments of history's greatest mathematical minds–Pythagoras, Descartes, Euler, Pascal, Cantor, many more. Anecdotal, illuminating. 30 diagrams. Bibliography. 256pp. 5⅜ x 8½. 28973-7

HISTORY OF MATHEMATICS, David E. Smith. Nontechnical survey from ancient Greece and Orient to late 19th century; evolution of arithmetic, geometry, trigonometry, calculating devices, algebra, the calculus. 362 illustrations. 1,355pp. 5⅜ x 8½. Two-vol. set. Vol. I: 20429-4 Vol. II: 20430-8

A CONCISE HISTORY OF MATHEMATICS, Dirk J. Struik. The best brief history of mathematics. Stresses origins and covers every major figure from ancient Near East to 19th century. 41 illustrations. 195pp. 5⅜ x 8½. 60255-9

CATALOG OF DOVER BOOKS

Physics

OPTICAL RESONANCE AND TWO-LEVEL ATOMS, L. Allen and J. H. Eberly. Clear, comprehensive introduction to basic principles behind all quantum optical resonance phenomena. 53 illustrations. Preface. Index. 256pp. 5⅜ x 8½. 65533-4

QUANTUM THEORY, David Bohm. This advanced undergraduate-level text presents the quantum theory in terms of qualitative and imaginative concepts, followed by specific applications worked out in mathematical detail. Preface. Index. 655pp. 5⅜ x 8½. 65969-0

ATOMIC PHYSICS: 8th edition, Max Born. Nobel laureate's lucid treatment of kinetic theory of gases, elementary particles, nuclear atom, wave-corpuscles, atomic structure and spectral lines, much more. Over 40 appendices, bibliography. 495pp. 5⅜ x 8½. 65984-4

A SOPHISTICATE'S PRIMER OF RELATIVITY, P. W. Bridgman. Geared toward readers already acquainted with special relativity, this book transcends the view of theory as a working tool to answer natural questions: What is a frame of reference? What is a "law of nature"? What is the role of the "observer"? Extensive treatment, written in terms accessible to those without a scientific background. 1983 ed. xlviii+172pp. 5⅜ x 8½. 42549-5

AN INTRODUCTION TO HAMILTONIAN OPTICS, H. A. Buchdahl. Detailed account of the Hamiltonian treatment of aberration theory in geometrical optics. Many classes of optical systems defined in terms of the symmetries they possess. Problems with detailed solutions. 1970 edition. xv+360pp. 5⅜ x 8½. 67597-1

PRIMER OF QUANTUM MECHANICS, Marvin Chester. Introductory text examines the classical quantum bead on a track: its state and representations; operator eigenvalues; harmonic oscillator and bound bead in a symmetric force field; and bead in a spherical shell. Other topics include spin, matrices, and the structure of quantum mechanics; the simplest atom; indistinguishable particles; and stationary-state perturbation theory. 1992 ed. xiv+314pp. 6⅛ x 9¼. 42878-8

LECTURES ON QUANTUM MECHANICS, Paul A. M. Dirac. Four concise, brilliant lectures on mathematical methods in quantum mechanics from Nobel Prize–winning quantum pioneer build on idea of visualizing quantum theory through the use of classical mechanics. 96pp. 5⅜ x 8½. 41713-1

THIRTY YEARS THAT SHOOK PHYSICS: The Story of Quantum Theory, George Gamow. Lucid, accessible introduction to influential theory of energy and matter. Careful explanations of Dirac's anti-particles, Bohr's model of the atom, much more. 12 plates. Numerous drawings. 240pp. 5⅜ x 8½. 24895-X

ELECTRONIC STRUCTURE AND THE PROPERTIES OF SOLIDS: The Physics of the Chemical Bond, Walter A. Harrison. Innovative text offers basic understanding of the electronic structure of covalent and ionic solids, simple metals, transition metals and their compounds. Problems. 1980 edition. 582pp. 6⅛ x 9¼. 66021-4

CATALOG OF DOVER BOOKS

HYDRODYNAMIC AND HYDROMAGNETIC STABILITY, S. Chandrasekhar. Lucid examination of the Rayleigh-Benard problem; clear coverage of the theory of instabilities causing convection. 704pp. 5⅜ x 8¼. 64071-X

INVESTIGATIONS ON THE THEORY OF THE BROWNIAN MOVEMENT, Albert Einstein. Five papers (1905–8) investigating dynamics of Brownian motion and evolving elementary theory. Notes by R. Fürth. 122pp. 5⅜ x 8½. 60304-0

THE PHYSICS OF WAVES, William C. Elmore and Mark A. Heald. Unique overview of classical wave theory. Acoustics, optics, electromagnetic radiation, more. Ideal as classroom text or for self-study. Problems. 477pp. 5⅜ x 8½. 64926-1

PHYSICAL PRINCIPLES OF THE QUANTUM THEORY, Werner Heisenberg. Nobel Laureate discusses quantum theory, uncertainty, wave mechanics, work of Dirac, Schroedinger, Compton, Wilson, Einstein, etc. 184pp. 5⅜ x 8½. 60113-7

ATOMIC SPECTRA AND ATOMIC STRUCTURE, Gerhard Herzberg. One of best introductions; especially for specialist in other fields. Treatment is physical rather than mathematical. 80 illustrations. 257pp. 5⅜ x 8½. 60115-3

AN INTRODUCTION TO STATISTICAL THERMODYNAMICS, Terrell L. Hill. Excellent basic text offers wide-ranging coverage of quantum statistical mechanics, systems of interacting molecules, quantum statistics, more. 523pp. 5⅜ x 8½. 65242-4

THEORETICAL PHYSICS, Georg Joos, with Ira M. Freeman. Classic overview covers essential math, mechanics, electromagnetic theory, thermodynamics, quantum mechanics, nuclear physics, other topics. xxiii+885pp. 5⅜ x 8½. 65227-0

PROBLEMS AND SOLUTIONS IN QUANTUM CHEMISTRY AND PHYSICS, Charles S. Johnson, Jr. and Lee G. Pedersen. Unusually varied problems, detailed solutions in coverage of quantum mechanics, wave mechanics, angular momentum, molecular spectroscopy, more. 280 problems, 139 supplementary exercises. 430pp. 6½ x 9¼. 65236-X

THEORETICAL SOLID STATE PHYSICS, Vol. I: Perfect Lattices in Equilibrium; Vol. II: Non-Equilibrium and Disorder, William Jones and Norman H. March. Monumental reference work covers fundamental theory of equilibrium properties of perfect crystalline solids, non-equilibrium properties, defects and disordered systems. Total of 1,301pp. 5⅜ x 8½. Vol. I: 65015-4 Vol. II: 65016-2

WHAT IS RELATIVITY? L. D. Landau and G. B. Rumer. Written by a Nobel Prize physicist and his distinguished colleague, this compelling book explains the special theory of relativity to readers with no scientific background, using such familiar objects as trains, rulers, and clocks. 1960 ed. vi+72pp. 23 b/w illustrations. 5⅜ x 8½.
42806-0 $6.95

A TREATISE ON ELECTRICITY AND MAGNETISM, James Clerk Maxwell. Important foundation work of modern physics. Brings to final form Maxwell's theory of electromagnetism and rigorously derives his general equations of field theory. 1,084pp. 5⅜ x 8½. Two-vol. set. Vol. I: 60636-8 Vol. II: 60637-6

CATALOG OF DOVER BOOKS

QUANTUM MECHANICS: Principles and Formalism, Roy McWeeny. Graduate student–oriented volume develops subject as fundamental discipline, opening with review of origins of Schrödinger's equations and vector spaces. Focusing on main principles of quantum mechanics and their immediate consequences, it concludes with final generalizations covering alternative "languages" or representations. 1972 ed. 15 figures. xi+155pp. 5⅜ x 8½. 42829-X

INTRODUCTION TO QUANTUM MECHANICS WITH APPLICATIONS TO CHEMISTRY, Linus Pauling & E. Bright Wilson, Jr. Classic undergraduate text by Nobel Prize winner applies quantum mechanics to chemical and physical problems. Numerous tables and figures enhance the text. Chapter bibliographies. Appendices. Index. 468pp. 5⅜ x 8½. 64871-0

METHODS OF THERMODYNAMICS, Howard Reiss. Outstanding text focuses on physical technique of thermodynamics, typical problem areas of understanding, and significance and use of thermodynamic potential. 1965 edition. 238pp. 5⅜ x 8½. 69445-3

TENSOR ANALYSIS FOR PHYSICISTS, J. A. Schouten. Concise exposition of the mathematical basis of tensor analysis, integrated with well-chosen physical examples of the theory. Exercises. Index. Bibliography. 289pp. 5⅜ x 8½. 65582-2

THE ELECTROMAGNETIC FIELD, Albert Shadowitz. Comprehensive undergraduate text covers basics of electric and magnetic fields, builds up to electromagnetic theory. Also related topics, including relativity. Over 900 problems. 768pp. 5⅝ x 8¼. 65660-8

GREAT EXPERIMENTS IN PHYSICS: Firsthand Accounts from Galileo to Einstein, Morris H. Shamos (ed.). 25 crucial discoveries: Newton's laws of motion, Chadwick's study of the neutron, Hertz on electromagnetic waves, more. Original accounts clearly annotated. 370pp. 5⅜ x 8½. 25346-5

RELATIVITY, THERMODYNAMICS AND COSMOLOGY, Richard C. Tolman. Landmark study extends thermodynamics to special, general relativity; also applications of relativistic mechanics, thermodynamics to cosmological models. 501pp. 5⅜ x 8½. 65383-8

STATISTICAL PHYSICS, Gregory H. Wannier. Classic text combines thermodynamics, statistical mechanics, and kinetic theory in one unified presentation of thermal physics. Problems with solutions. Bibliography. 532pp. 5⅜ x 8½. 65401-X

Paperbound unless otherwise indicated. Available at your book dealer, online at **www.doverpublications.com**, or by writing to Dept. GI, Dover Publications, Inc., 31 East 2nd Street, Mineola, NY 11501. For current price information or for free catalogs (please indicate field of interest), write to Dover Publications or log on to **www.doverpublications.com** and see every Dover book in print. Dover publishes more than 500 books each year on science, elementary and advanced mathematics, biology, music, art, literary history, social sciences, and other areas.